IFIP Advances in Information and Communication Technology 385

T0189715

IFIP – The International Federation for Information Processing

IFIP was founded in 1960 under the auspices of UNESCO, following the First World Computer Congress held in Paris the previous year. An umbrella organization for societies working in information processing, IFIP's aim is two-fold: to support information processing within ist member countries and to encourage technology transfer to developing nations. As ist mission statement clearly states,

> *IFIP's mission is to be the leading, truly international, apolitical organization which encourages and assists in the development, exploitation and application of information technology for the benefit of all people.*

IFIP is a non-profitmaking organization, run almost solely by 2500 volunteers. It operates through a number of technical committees, which organize events and publications. IFIP's events range from an international congress to local seminars, but the most important are:

- The IFIP World Computer Congress, held every second year;
- Open conferences;
- Working conferences.

The flagship event is the IFIP World Computer Congress, at which both invited and contributed papers are presented. Contributed papers are rigorously refereed and the rejection rate is high.

As with the Congress, participation in the open conferences is open to all and papers may be invited or submitted. Again, submitted papers are stringently refereed.

The working conferences are structured differently. They are usually run by a working group and attendance is small and by invitation only. Their purpose is to create an atmosphere conducive to innovation and development. Refereeing is less rigorous and papers are subjected to extensive group discussion.

Publications arising from IFIP events vary. The papers presented at the IFIP World Computer Congress and at open conferences are published as conference proceedings, while the results of the working conferences are often published as collections of selected and edited papers.

Any national society whose primary activity is in information may apply to become a full member of IFIP, although full membership is restricted to one society per country. Full members are entitled to vote at the annual General Assembly, National societies preferring a less committed involvement may apply for associate or corresponding membership. Associate members enjoy the same benefits as full members, but without voting rights. Corresponding members are not represented in IFIP bodies. Affiliated membership is open to non-national societies, and individual and honorary membership schemes are also offered.

Zhongzhi Shi   David Leake   Sunil Vadera (Eds.)

Intelligent Information Processing VI

7th IFIP TC 12 International Conference, IIP 2012
Guilin, China, October 12-15, 2012
Proceedings

Volume Editors

Zhongzhi Shi
Chinese Academy of Sciences
Institute of Computing Technology
Beijing 100190, China
E-mail: shizz@ics.ict.ac.cn

David Leake
Indiana University
Computer Science Department
Bloomington, IN 47405, USA
E-mail: leake@cs.indiana.edu

Sunil Vadera
University of Salford
School of Computing Science and Engineering
Salford M5 4WT, UK
E-mail: s.vadera@salford.ac.uk

ISSN 1868-4238   e-ISSN 1868-422X
ISBN 978-3-642-44729-7   ISBN 978-3-642-32891-6 (eBook)
DOI 10.1007/978-3-642-32891-6
Springer Heidelberg Dordrecht London New York

CR Subject Classification (1998): I.2.3-4, I.2.6, F.4.1, H.2.8, H.3, I.5, F.3.1, C.2, D.2

Typesetting: Camera-ready by author, data conversion by Scientific Publishing Services, Chennai, India

Printed on acid-free paper

Springer is part of Springer Science+Business Media (www.springer.com)

Preface

This volume comprises the 7th IFIP International Conference on Intelligent Information Processing. As the world proceeds quickly into the Information Age, it encounters both successes and challenges, and it is well recognized today that intelligent information processing provides the key to the Information Age and to mastering many of these challenges. Intelligent information processing supports the most advanced productive tools that are said to be able to change human life and the world itself. However, the path is never a straight one and every new technology brings with it a spate of new research problems to be tackled by researchers; as a result we are not running out of topics, rather the demand is ever increasing. This conference provides a forum for engineers and scientists in academia, university and industry to present their latest research findings in all aspects of intelligent information processing.

This is the 7th IFIP International Conference on Intelligent Information Processing. We received more than 70 papers, of which 39 are included in this program as regular papers and five as short papers. We are grateful for the dedicated work of both the authors and the referees, and we hope these proceedings will continue to bear fruit over the years to come. All papers submitted were reviewed by two referees.

A conference such as this cannot succeed without help from many individuals who contributed their valuable time and expertise. We want to express our sincere gratitude to the Program Committee members and referees, who invested many hours for reviews and deliberations. They provided detailed and constructive review reports that significantly improved the papers included in the program.

We are very grateful to have had the sponsorship of the following organizations: IFIP TC12, Guilin University of Electronic Technology and Institute of Computing Technology, Chinese Academy of Sciences.

Finally, we hope you find this volume inspiring and informative.

August 2012

Zhongzhi Shi
David Leake
Sunil Vadera

Organization

General Chairs

T. Dillon (Australia)
T. Gu (China)
A. Aamodt (Norway)

Program Chairs

Z. Shi (China)
D. Leake (USA)
S. Vadera (UK)

Program Committee

A. Aamodt (Norway)
A. Bernardi (Germany)
N. Bredeche (France)
C. Bryant (UK)
L. Cao (Australia)
E. Chang (Australia)
L. Chang (China)
E. Chen (China)
H. Chen (UK)
F. Coenen (UK)
Z. Cui (China)
S. Dustdar (Austria)
S. Ding (China)
Y. Ding (USA)
Q. Duo (China)
J. Ermine (France)
P. Estraillier (France)
W. Fan (UK)
Y. Gao (China)
L. Hansen (Denmark)
T. Hong (Taiwan)
Q. He (China)
T. Honkela (Finland)

Z. Huang (The
  Netherlands)
P. Ibarguengoyatia
  (Mexico)
G. Kayakutlu (Turkey)
D. Leake (USA)
J. Liang (China)
Y. Liang (China)
H. Leung (HK)
S. Matwin (CA)
E. Mercier-Laurent
  (France)
F. Meziane (UK)
Z. Meng (China)
S. Nefti-Meziani (UK)
T. Nishida (Japan)
G. Osipov (Russia)
M. Owoc (Poland)
A. Rafea (Egypt)
K. Rajkumar (India)
M. Saraee (UK)
F. Segond (France)
Q. Shen (UK)

ZP. Shi (China)
K. Shimohara (Japan)
A. Skowron (Poland)
M. Stumptner
  (Australia)
E. Succar (Mexico)
H. Tianfield (UK)
IJ. Timm (Germany)
S. Tsumoto (Japan)
G. Wang (China)
S. Vadera (UK)
Y. Xu (Australia)
H. Xiong (USA)
J. Yang (Korea)
Y. Yao (Canada)
J. Yu (China)
J. Zhang (China)
X. Zhao (China)
J. Zhou (China)
Z.-H. Zhou (China)
J. Zucker (France)

Table of Contents

Automatic Reasoning

Semantic Web

Information Retrieval

Knowledge Representation

Social Networks

Trust Software

Internet of Things

Image Processing

Pattern Recognition

The AI Journey: The Road Traveled and the (Long) Road Ahead

Ramon Lopez de Mantaras

Artificial Intelligence Research Institute,
Spanish National Research Council (CSIC), Spain
mantaras@iiia.csic.es

Abstract. In this talk I will first briefly summarize the many impressive results we have achieved along the road so far traveled in the field of AI including some concrete results obtained at the IIIA-CSIC. Next I will describe some of the future challenges to be faced along the (long) road we still have ahead of us with an emphasis on integrated systems, a necessary step towards human-level AI. Finally I will comment on the importance of interdisciplinary research to build such integrated systems (for instance, sophisticated robots having artificial cartilages, artificial muscles, artificial skin, etc) using some examples related to materials science.

Z. Shi, D. Leake, and S. Vadera (Eds.): IIP 2012, IFIP AICT 385, p. 1, 2012.
© IFIP International Federation for Information Processing 2012

Transfer Learning and Applications

Qiang Yang

Department of Computer Science and Engineering,
Hong Kong University of Science and Technology, Hong Kong
qyang@cse.ust.hk

Abstract. In machine learning and data mining, we often encounter situations where we have an insufficient amount of high-quality data in a target domain, but we may have plenty of auxiliary data in related domains. Transfer learning aims to exploit these additional data to improve the learning performance in the target domain. In this talk, I will give an overview on some recent advances in transfer learning for challenging data mining problems. I will present some theoretical challenges to transfer learning, survey the solutions to them, and discuss several innovative applications of transfer learning, including learning in heterogeneous cross-media domains and in online recommendation, social media and social network mining.

Z. Shi, D. Leake, and S. Vadera (Eds.): IIP 2012, IFIP AICT 385, p. 2, 2012.
© IFIP International Federation for Information Processing 2012

Semantics of Cyber-Physical Systems

Tharam Dillon[1], Elizabeth Chang[2], Jaipal Singh[3], and Omar Hussain[2]

[1] La Trobe University, Australia
tharam.dillon7@gmail.com
[2] School of Information Systems, Curtin University, Australia
{Elizabeth.Chang,O.Hussain}@curtin.edu.au
[3] Dept. of Electrical and Computer Engineering, Curtin University, Australia
J.Singh@curtin.edu.au

Abstract. The very recent development of Cyber-Physical Systems (CPS) provides a smart infrastructure connecting abstract computational artifacts with the physical world. The solution to CPS must transcend the boundary between the cyber world and the physical world by providing integrated models addressing issues from both worlds simultaneously. This needs new theories, conceptual frameworks and engineering practice. In this paper, we set out the key requirements that must be met by CPS systems, review and evaluate the progress that has been made in the development of theory, conceptual frameworks and practical applications. We then discuss the need for semantics and a proposed approach for addressing this. Grand challenges to informatics posed by CPS are raised in the paper.

1   Introduction

The very recent development of Cyber-Physical Systems (CPS) provides a unified framework connecting the cyber world with the physical world. CPS allows for robust and flexible systems with multi-scale dynamics and integrated wired and wireless networking for managing the flows of mass, energy, and information in a coherent way through integration of computing and communication capabilities with the monitoring and/or control of entities in the physical world in a dependable, safe, secure, efficient and real-time fashion.

CPS has recently been listed as the No.1 research priority by the U.S. President's Council of Advisors on Science and Technology [2]. This led the US National Science Foundation to organize a series of workshops on CPS [3]. The CPS framework has the capability to tackle numerous scientific, social and economic issues. The three applications for CPS are in future distributed energy systems, future transportation systems and future health care systems [1,4,14]. These applications will require seamless and synergetic integration between sensing, computation, and control with physical devices and processes.

Building CPS is not a trivial task. It requires a new ground-breaking theory that models cyber and physical resources in a unified framework. This is a huge challenge that none of the current state-of-the-art methods are able to overcome due to the fact that computer science and control theory are independently developed based on

Z. Shi, D. Leake, and S. Vadera (Eds.): IIP 2012, IFIP AICT 385, pp. 3–12, 2012.

overly-simplified assumptions of each other. For example, many key requirements (e.g. uncertainty, inaccuracy, etc.) crucial to physical systems are not captured and fully dealt with in the computer science research agenda. In a similar vein, computational complexity, system evolution and software failure are often ignored from the physical control theory viewpoint, which treats computation as a precise, error-free, static 'black-box'. The solution to CPS must transcend the boundary between the cyber world and the physical world by providing a unified infrastructure that permits integrated models addressing issues from both worlds simultaneously. This paper begins by setting out the requirements and then evaluates the progress that has been made in addressing these [2]. This paper will show that CPS calls for new theories and some fundamental changes to the existing computing paradigm [4-6].

In particular it will look at the need for semantics and propose a framework for such semantics.

2   Requirements of CPS

The National Science Foundation (NSF) CPS Summit held in April 2008 [4] defines CPS as "physical and engineered systems whose operations are monitored, coordinated, controlled and integrated by a computing and communication core". Researchers from multiple disciplines such as embedded systems and sensor networks have been actively involved in this emerging area.

Our vision of CPS is as follows: networked information systems that are tightly coupled with the physical process and environment through a massive number of geographically distributed devices [1]. As networked information systems, CPS involves computation, human activities, and automated decision making enabled by information and communication technology. More importantly, these computation, human activities and intelligent decisions are aimed at monitoring, controlling and integrating physical processes and environment to support operations and management in the physical world. The scale of such information systems range from micro-level, embedded systems to ultra-large systems of systems. Devices provide the basic interface between the cyber world and the physical one.

The discussions in the NSF Summit [4] can be summarized into eleven scientific and technological challenges for CPS solutions. These challenges constitute the top requirements for building cyber-physical systems and are listed below: (1) Compositionality (2) Distributed Sensing, Computation and Control; (3) Physical Interfaces and Integration; (4) Human Interfaces and Integration; (5) Information: From Data to Knowledge; (6) Modeling and Analysis: Heterogeneity, Scales, Views; (7) Privacy, Trust, Security; (8) Robustness, Adaptation, Reconfiguration; (9) Software; (10) Verification, Testing and Certification; (11) Societal Impact.

Based on the challenges listed above, a new unified cyber-physical systems foundation that goes beyond current computer mediated systems needs to be developed. We explain how this can be achieved, in-line with the challenges to CPS identified by the NSF summit report.

CPS need to support resource composition that integrates heterogeneous components from both the physical and cyber world. This requires: (1) a "plug-n-play" framework that provides a high degree of adaptivity and (re-) programmability supported by flexible interface definitions and integration styles (e.g. synchronous, asynchronous, continuous, and discrete) and (2) the capability to model and predict run-time performance and behavior of the evolving composite.

CPS need to make distributed control, sensing, and communication. This requires: (1) information collection and decision making in a distributed network environment with high latency and high uncertainty, (2) new theories are needed on the use of network communication at the micro-level to prevent adverse effects at the macro-scale in a distributed environment, (3) new theories are needed on control without centralized administration , measurement and perhaps incomplete information.

CPS need to stay in constant touch with physical objects. This requires: (1) models that abstract physical objects with varying levels of resolutions, dimensions, and measurement scales, (2) mathematical representation of these models and understanding of algorithmic, asymptotic behavior of these mathematical models, and (3) abstractions that captures the relationships between physical objects and CPS.

Humans have to play an essential role (e.g. influence, perception, monitoring, etc.) in CPS. This requires: (1) seamless integration and adaptation between human scales and physical system scales. (2) support for local contextual actions pertinent to specific users, who are part of the system rather than just being the "users" of the system, (3) new theories on the boundary (e.g. hand-over or switch) between human control and (semi-) automatic control.

Many CPS are aimed at developing useful knowledge from raw data. This requires (1) algorithms for sensor data fusion that also deal with data cleansing, filtering, validation, etc. (2) data stream mining in real-time (3) storage and maintenance of different representations of the same data for efficient and effective (e.g. visualization) information retrieval and knowledge extraction.

CPS needs to deal with massive heterogeneity when integrating components of different natures from different sources. This requires (1) integration of temporal, event-driven, and spatial data defined in significantly different models (asynchronous vs. synchronous) and scales (e.g. discrete vs. continuous), (2) new computation models that characterize dimensions of physical objects such as time (e.g. to meet real-time deadline), location, energy, memory footprint, cost, uncertainty from sensor data, etc., (3) new abstractions and models for cyber-physical control that can deal with - through compensation, feedback processing, verification, etc. - uncertainty that is explicitly represented in the model as a "first-class citizen" in CPS, (4) new theories on "design for imperfection" exhibited by both physical and cyber objects in order to ensure stability, reliability, and predictability of CPS, (5) system evolution in which requirements and constraints are constantly changing and need to be integrated into different views of CPS, and (6) new models for dealing with issues in large-scaled systems such as efficiency trade-offs between local and global, emergent behavior of complex systems, etc.

CPS in general reveal a lot of physical information, create a lot of data concerning security (e.g. new types of attacks), privacy (e.g. location), and trust (e.g. heterogeneous resources). This requires: (1) new theories and methods on design principles for resilient CPS, threat/hazard analysis, cyber-physical inter-dependence anatomy, investigation/prediction of gaming plots at different layers of CPS, (2) formal models for privacy specification that allow reasoning about and proof of privacy properties, (3) new mathematical theories on information hiding for real-time streams, (4) light-weight security solutions that work well under extremely limited computational resources (e.g. devices), (5) new theories on confidence and trust maps, context-dependent trust models, and truth/falseness detection capabilities.

Due to the unpredictability in the physical world, CPS will not be operating in a controlled environment, and must be robust to unexpected conditions and adaptable to subsystem failures. This requires: (1) new concepts of robust system design that deals with and lives on unexpected uncertainties (of network topology, data, system, etc.) occurring in both cyber and physical worlds, (2) the ability to adapt to faults through (self-) reconfiguration at both physical and cyber levels, (3) fault recovery techniques using the most appropriate strategies that have been identified, categorized, and selected, (4) system evolvement through learning faults and dealing with uncertainties in the past scenarios, (5) system evolvement through run-time reconfiguration and hot deployment.

One important omission from the above requirements is the need for semantics. In particular semantics that are capable of bridging the real physical world and the virtual world.

3   Brief Overview of Architectural Framework for CPS Systems

We have previously proposed a Web-of-Things (WoT) framework for CPS systems [1] that augments the Internet-of-Things in order to deal with issues such as information-centric protocol, deterministic QoS, context-awareness, etc. We argue that substantial extra work such as our proposed WoT framework is required before IoT can be utilized to address technical challenges in CPS Systems.

The building block of WoT is Representational State Transfer (REST), which is a specific architectural style [4]. It is, in effect, a refinement and constrained version of the architecture of the Web and the HTTP 1.1 protocol [5], which has become the most successful large-scale distributed application that the world has known to date. Proponents of REST style argue that existing RPC (Remote Procedure Call)-based Web services architecture is indeed not "Web-oriented". Rather, it is merely the "Web" version of RPC, which is more suited to a closed local network, and has serious potential weakness when deployed across the Internet, particularly with regards to scalability, performance, flexibility, and implementability [6]. Structured on the original layered client-server style [4], REST specifically introduces numerous architectural constraints to the existing Web services architecture elements in order to: a) simplify interactions and compositions between service requesters and providers; b) leverage the existing WWW architecture wherever possible.

The WoT framework for CPS is shown in Fig. 1, which consists of five layers – WoT Device, WoT Kernel, WoT Overlay, WoT Context and WoT API. Underneath the WoT framework is the cyber-physical interface (e.g. sensors, actuators, cameras) that interacts with the surrounding physical environment. The cyber-physical interface is an integral part of the CPS that produces a large amount of data. The proposed WoT framework allows the cyber world to observe, analyze, understand, and control the physical world using these data to perform mission / time-critical tasks.

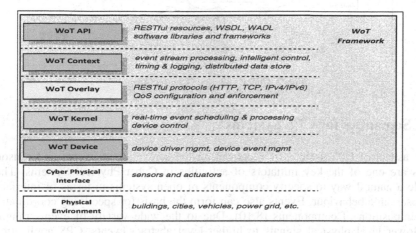

Fig. 1. WoT Framework for CPS

As shown in Figure 1, the proposed WoT based CPS framework consists of five layers:

(1) WoT Device: This layer constitute the cyber-physical interface of the system. It is a resource-oriented abstraction that unifies the management of various devices. It states the device semantics in terms of RESTful protocol. (2) WoT Kernel: This layer provides low level run-time capability for communication, scheduling, and WoT resources management. It identifies events and allocates the required resources, i.e. network bandwidth, processing power and storage capacity for dealing with a large amount of data from the WoT Device layer. (3) WoT Overlay: This layer is an application-driven, network-aware logical abstraction atop the current Internet infrastructure. It will manage volatile network behavior such as latency, data loss, jitter and bandwidth by allowing nodes to select paths with better and more predictable performance. (4) WoT Context: This layer provides semantics for events captured by the lower layers of WoT framework. This layer is also responsible for decision making and controlling the behaviour of the CPS applications. (5) WoT API: This layer provides abstraction in the form of interfaces that allow developers to interact with the WoT framework.

Based on the WoT framework in Fig. 1, the CPS reference architecture is shown in Fig. 2, which aims to capture both domain requirements and infrastructure requirements at a high level of abstraction. It is expected that CPS applications can be built atop the CPS reference architecture.

More details about the CPS Fabric structure and the CPS node structure are given in Dillon et. al. [1].

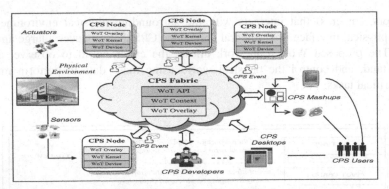

Fig. 2. CPS Reference Architecture

4   Semantics for CPS Systems

A key aspect of CPS systems is representing the semantics of events and sensors. Events are one of the key initiators of activities in Cyber-Physical Systems. They provide a natural way to specify components of open systems in terms of interfaces and observable behaviour. Events also can form the basis for specifying coordination and composition of components [8,10]. Due to the wide variety of events ranging from lower level physical signals to higher-level abstract events, CPS applications and users may be interested in certain conditions of interest in the physical world, according to which certain predefined operations are executed by the CPS. Detection of these conditions of interest (events) will lead to the desired predefined operations. As a result, any CPS task can be represented as an "Event-Action" relation [11]. Some of the challenges involved in the representation events in dynamic real-time CPS are [7]:

A. Abstractions for Sensors and Event Representations
Sensor data requires processing to generate information which can then be used for event identification and representation. Abstraction of sensor instances and event instances are required for representing sensors and events respectively. However, this is a challenging task that requires a framework that can deal with this problem in its entirety. Sensors can be represented as a resource with a unique identifier that allows them to be located through the Internet. In addition, it is important to develop the notion of (1) Sensors classes which represent collections of sensors instances with particular properties. e.g. a temperature sensors monitor and alarm class. This class could consist of several instances. To allow for mobility, this class would have both location and time as two amongst other properties. (2) Event classes which represent collections of event instances with particular properties. e.g. an intrusion event detection class which could consist of several event instances. Event classes too would have location and time (or at least position in a sequence in time) as two amongst other properties.

There will be relationships between these classes, which will allow for representation of generalization/specialization, composition and associations. As mentioned earlier, interoperability is an important challenge to be considered for the seamless integration of different high level applications. Successful implementation of this allows us to meet the following challenges: (1) Management of raw sensor data that is kept, maintained and exported by disparate sources. (2) Interpreting events associated with particular sensor configurations and outputs. (3) Transforming system level knowledge from distinct sensors into higher-level management applications. (4) Upgrade of existing sensors with new advanced sensors using standardised interfaces by using the same abstract level representation.

B. Compositions of Sensors and Events to Address Complex Requirements
Depending upon specific application requirements, composition of event data from multiple sensors may be required. This is a challenging task that requires a dedicated framework on how information from multiple sensors is composed and correlated for meeting the QoS requirements of the specific application scenario.

A decomposition technique is needed for decomposing complex functionality into lower level resources which can then be met by specific sensors or events. This decomposition requires the specification of items in the aggregation and their dynamic sequence of execution or arrangement. The dynamics in the case of services can be modelled by using workflows specified in a language such as BPEL and in the case of resources by using Mashups which allow easier configuration of these workflows.

The composition of events remains a challenging issue with a focus on sequencing them to produce a composite event.

C. Semantics for Compositions of Sensors and Events
Semantics need to be provided for automatic sensor discovery, selection and composition. Semantics here is the study of the meaning of Things (resources that represent sensors) and events. It represents the definitions of the meaning of elements, as opposed to the rules for encoding or representation. Semantics describe the relationship between the syntactical elements and the model of computation. It is used to define what entities mean with respect to their roles in a system. This includes capabilities, or features that are available within a system.

Two approaches can be used to represent semantics. They are: (1) ontologies and (2) lightweight semantic annotations.

Ontologies are a formal, explicit specifications of a shared semantic conceptualization that are machine understandable and abstract models of consensual knowledge, can be used. Using an ontology, it is possible to define concepts through uniquely identifying their specifications and their dynamic and static properties. Concepts, their details and their interconnections are defined as an ontology specification. Ontology compositions are typically formed from many interconnected ontology artifacts that are constructed in an iteratively layered and hierarchical manner. It is necessary to use a graphical representation of ontologies at a higher level of abstraction to make it easier for the domain experts to capture the semantic richness of the defined ontology and understand and critique each model.

Ontologies can be used to define the following system properties: (1) Information and Communication, refers to the basic ability to gather and exchange information between the parties involved. (2) Integrability, relates to the ability of sensors and devices from different sources to mesh relatively seamlessly and the ability to integrate these sensors to build the final solution in a straightforward way. (3) Coordination, focuses on scheduling and ordering tasks performed by these parties. (4) Awareness and Cooperation, refer to the implicit knowledge of the operation process that is being performed and the state of the system.

For representing the knowledge in a given domain ontologies and for adding semantics to individual resources annotation may suffice. A special challenge here is developing ontologies for events. Lightweight semantic RDF Metadata Annotations are an attractive alternative for providing semantics for CPS systems. The Resource Description Framework (RDF) provides an enhanced representation over XML including: (1) The triples (Resources, Property, Value) or (Subject, Predicate, Object). (2) Defining relationships such as the concept of class and subclass

RDF is extensible, which means that descriptions can be enriched with additional descriptive information. RDF is metadata that can be inserted into XML code or device or vice versa . An example of an Extensible representation of a sensor instance is given in Figure 3 below and associated URL Hierarchy is given in Figure 4..

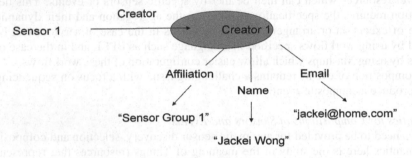

Fig. 3. Extensible RDF representation of a Sensor instance

This RDF statement may also be described in terms of triple (Subject, Predicate, Object). Example:

- Subject: http://www.webofthings.com/~test
- Predicate: The element <creator>
- Value: The string "Jackei Wong"

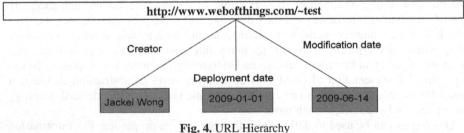

Fig. 4. URL Hierarchy

Another important aspect here would be the use of RDF profiles. A sample RDF Profile is given in Figure 5.

Semantics for CPS systems must be able to represent: Timeliness requirements; Resources descriptions (e.g. hardware or software applications); Quality of service; Schedule of tasks; Other functional issues that must be addressed. Thus one can distinguish four main semantic components namely:Time- provides a detailed definition of time; Resources- provides a wide range of primitives for resource description including physical resources such as device, CPU or communication devices; as well as resource service such as application or services description; QoS -At this stage we have provided a profile for illustration of QoS concepts for Resource Instances description; Causality-The description of action execution and resource service instance.

Title	Definition
Identifier	An unambiguous reference to the resource within a given context
Description	Purpose of resource and what it does eg. measures xx quantity
Coverage	The extent or scope of the content of the resource
Creator	An entity primarily responsible for creating & deploying the resource
Format	The physical or digital manifestation of the resource
Date	A date of an event in the lifecycle of the resource
Type of resource	The nature or genre of the resource
Relation	A reference to a related resource
Access rights	Information about rights to modify & access the resource

Fig. 5. Example of a RDF Profile

D. Event Models for CPS

Approaches have been proposed in the literature that model events in CPS. They classify events as temporal or spatial [11]. They further define different dimensions such as (a) punctual or interval, (b) single or stream, (c) action or observation, (d) point or field, and (e) causal, discrete, or continuous [10,11].

In a CPS, the events can be further categorised as follows [11]: (1) Physical-events: Physical event models the occurrence of the end-user interest in the physical world and can be any change in attribute, temporal or spatial status of one or more physical objects or physical phenomena. These events are captured through physical observation, which is a snapshot of attribute, temporal, or spatial status of the target physical event. (2) Cyber-Physical events: The physical event captured using sensors collect the sensor event instances from other sensor motes as input observations and generate cyber-physical event instances based on the cyberphysical event conditions. (3) Cyber-events: The top level CPS control unit serves as the highest level of observer in CPS event model. It may combine cyber-physical event instances from other CPS components (sensors) and other control units as input observations to generate the cyber event. Some of these approaches utilise a Spatial-Temporal event models for CPS in 2-dimensions [11] and in another uses 3-dimensions [12]. Another approach provides semantics to the events detected by using first order logic, such as an adaptive discrete event calculus [13].

5   Conclusion

In this paper we discussed a semantic framework for Cyber Physical Systems based on the reference architecture previously presented. A key element here is the semantics of Events and Sensors. Both Ontologies and lightweight semantics based on RDF could be used depending on the specific application and domain.

References

1. Dillon, T.S., Zhuge, H., Wu, C., Singh, J., Chang, E.: Web-of-things framework for cyber–physical systems. Concurrency and Computation: Practice and Experience 23(9), 905–923 (2011)
2. President's Council of Advisors on Science and Technology (PCAST), Leadership under challenge: Information technology r&d in a competitive world (August 2007), http://www.nitrd.gov/pcast/reports/PCAST-NIT-FINAL.pdf
3. National Science Foundation, Cyber-physical systems (CPS) workshop series, http://varma.ece.cmu.edu/Summit/Workshops.html
4. National Science Foundation, Cyber-physical systems summit report, Missouri, USA (April 24-25, 2008), http://precise.seas.upenn.edu/events/iccps11/_doc/CPS_Summit_Report.pdf
5. Lee, E.: Cyber physical systems: Design challenges. In: IEEE Object Oriented Real-Time Distributed Computing, pp. 363–369 (2008)
6. Lee, E.: Computing needs time. Communications of the ACM 52(5), 70–79 (2009)
7. Singh, J., Hussain, O., Chang, E., Dillon, T.S.: Event Handling for Distributed Real-Time Cyber-Physical Systems. In: IEEE 15th International Symposium on Object/Component/Service-Oriented Real-Time Distributed Computing (ISORC), China, pp. 23–30 (2012)
8. Dillon, T.S., Talevski, A., Potdar, V., Chang, E.: Web of Things as a Framework for Ubiquitous Intelligence and Computing. In: Zhang, D., Portmann, M., Tan, A.-H., Indulska, J. (eds.) UIC 2009. LNCS, vol. 5585, pp. 2–13. Springer, Heidelberg (2009)
9. Dillon, T.: Web-of-things framework for cyber-physical systems. In: The 6th International Conference on Semantics, Knowledge & Grids (SKG), Ningbo, China (2010) (Keynote)
10. Talcott, C.: Cyber-Physical Systems and Events. In: Wirsing, M., Banâtre, J.-P., Hölzl, M., Rauschmayer, A. (eds.) Soft-Ware Intensive Systems. LNCS, vol. 5380, pp. 101–115. Springer, Heidelberg (2008)
11. Tan, Y., Vuran, M.C., Goddard, S.: Spatio-temporal event model for cyber-physical systems. In: 29th IEEE International Conference on Distributed Computing Systems Workshops, pp. 44–50 (2009)
12. Tan, Y., Vuran, M.C., Goddard, S., Yu, Y., Song, M., Ren, S.: A concept lattice-based event model for Cyber-Physical Systems. Presented at the Proceedings of the 1st ACM/IEEE International Conference on Cyber-Physical Systems, Stockholm, Sweden (2010)
13. Yue, K., Wang, L., Ren, S., Mao, X., Li, X.: An Adaptive Discrete Event Model for Cyber-Physical System. In: Analytic Virtual Integration of Cyber-Physical Systems Workshop, USA, pp. 9–15 (2010)
14. Yu, X., Cecati, C., Dillon, T., Godoy Simões, M.: Smart Grids: An Industrial Electronics Perspective. In: IEEE Industrial Electronics Magazine IEM-02-2011 (2011)

Big Data Mining in the Cloud

Zhongzhi Shi

Key Laboratory of Intelligent Information Processing, Institute of Computing Technology,
Chinese Academy of Sciences, Beijing, China
shizz@ics.ict.ac.cn

Abstract. Big Data is the growing challenge that organizations face as they deal
with large and fast-growing sources of data or information that also present a
complex range of analysis and use problems. Digital data production in many
fields of human activity from science to enterprise is characterized by an exponential growth. Big data technologies will become a new generation of technologies and architectures which is beyond the ability of commonly used software
tools to capture, manage, and process the data within a tolerable elapsed time.

Massive data sets are hard to understand, and models and patterns hidden
within them cannot be identified by humans directly, but must be analyzed by
computers using data mining techniques. The world of big data present rich
cross-media contents, such as text, image, video, audio, graphics and so on. For
cross-media applications and services over the Internet and mobile wireless
networks, there are strong demands for cross-media mining because of the
significant amount of computation required for serving millions of Internet or
mobile users at the same time. On the other hand, with cloud computing boom-
ing, new cloud-based cross-media computing paradigm emerged, in which users
store and process their cross-media application data in the cloud in a distributed
manner. Cross-media is the outstanding characteristics of the age of big data
with large scale and complicated processing task. Cloud-based Big Data platforms will make it practical to access massive compute resources for short time
periods without having to build their own big data farms. We propose a framework for cross-media semantic understanding which contains discriminative
modeling, generative modeling and cognitive modeling. In cognitive modeling,
a new model entitled CAM is proposed which is suitable for cross-media
semantic understanding. A Cross-Media Intelligent Retrieval System (CMIRS),
which is managed by ontology-based knowledge system KMSphere, will be
illustrated.

This talk also concerns Cloud systems which can be effectively employed to
handle parallel mining since they provide scalable storage and processing
services, as well as software platforms for developing and running data analysis
environments. We exploit Cloud computing platforms for running big data
mining processes designed as a combination of several data analysis steps to be
run in parallel on Cloud computing elements. Finally, the directions for further
researches on big data mining technology will be pointed out and discussed.

Z. Shi, D. Leake, and S. Vadera (Eds.): IIP 2012, IFIP AICT 385, pp. 13–14, 2012.

Acknowledgement. This work is supported by Key projects of National Natural Science Foundation of China (No. 61035003, 60933004), National Natural Science Foundation of China (No. 61072085, 60970088, 60903141), National Basic Research Program (2007CB311004).

Research on Semantic Programming Language

Shi Ying

State Key Laboratory of Software Engineering,
Wuhan University, Wuhan, China
yingshi@whu.edu.cn

Abstract. As technologies of Semantic Web Service are gradually matured, developing intelligent web applications with Semantic Web Services becomes an important research topic in Software Engineering. This speech introduces our efforts on Semantic Web Service oriented programming. Employing the concept of semantic computing into service-oriented programming, we proposed a programming language SPL, Semantic Programming Language, which supports the expression and process of semantic information. Based on collaboration of semantic space and information space, the running mechanism of SPL program is presented, which provides SPL program with higher flexibility and stronger adaptability to changes. Furthermore, with the introduction of semantic operators, a kind of searching conditional expression is offered to facilitate the search of Semantic Web Services with greater preciseness and higher flexibility. Besides, semantic based policy and exception mechanism are also brought in to improve the intelligence of policy inference and exception handing in SPL program. At the same time, a platform that supports design and running of SPL program is developed.

Z. Shi, D. Leake, and S. Vadera (Eds.): IIP 2012, IFIP AICT 385, p. 15, 2012.
© IFIP International Federation for Information Processing 2012

Effectively Constructing Reliable Data for Cross-Domain Text Classification

Fuzhen Zhuang, Qing He, and Zhongzhi Shi

The Key Laboratory of Intelligent Information Processing, Institute of Computing Technology,
Chinese Academy of Sciences, Beijing, China
{zhuangfz,heq,shizz}@ics.ict.ac.cn

Abstract. Traditional classification algorithms often fail when the *independent and identical distributed* (i.i.d.) assumption does not hold, and the cross-domain learning emerges recently is to deal with this problem. Actually, we observe that though the trained model from training data may not perform well over all test data, it can give much better prediction results on a subset of the test data with high prediction confidence. Also this subset of data from test data set may have more similar distribution with the test data. In this study, we propose to construct the reliable data set with high prediction confidence, and use this reliable data as training data. Furthermore, we develop an EM algorithm to refine the model trained from the reliable data. The extensive experiments on text classification verify the effectiveness and efficiency of our methods. It is worth to mention that the model trained from the reliable data achieves a significant performance improvement compared with the one trained from the original training data, and our methods outperform all the baseline algorithms.

Keywords: Cross-domain Learning, Reliable Data, EM Algorithm.

1 Introduction

Classification techniques play an important role in intelligent information process, such as the analysis of World-Wide-Web pages, images and so on. This requires the trained models from the training data (also referred as *source domain*) to give correct prediction or classification on unlabeled test data (also referred as *target domain*). Traditional classification techniques, e.g., Support Vector Machine (SVM) [2,3] and Naïve Bayesian (NBC) [15,19] and Logistical Regression (LR) [9], are proved to perform very well when the training and test data are drawn from the *independent* and *identical distribution* (i.i.d.). However, this assumption always actually does not hold in real-world applications, since the test data usually come from the information sources with different distribution caused by the sample selection bias, concept drift and so on.

In recent years, cross-domain learning[1] [23] has attracted a great attention, which focuses on the model adaptation from source domain to target domain with distribution mismatch. These works include feature selection learning [4,29,21,25], transfer learning via dimensionality reduction [20,7], model combination [6,30] and sample selection bias [26,28] et al.

---

[1] Also referred as transfer learning or domain adaptation in the previous research.

Z. Shi, D. Leake, and S. Vadera (Eds.): IIP 2012, IFIP AICT 385, pp. 16–27, 2012.

Table 1. The Prediction Probabilities for Two-class Classification on 20 Documents (The label in bracket is the true label of document)

		$d_1(c_1)$	$d_2(c_1)$	$d_3(c_1)$	$d_4(c_1)$	$d_5(c_2)$	$d_6(c_2)$	$d_7(c_2)$	$d_8(c_2)$	$d_9(c_2)$	$d_{10}(c_2)$
Classes	c_1	1.00	0.99	0.98	0.985	0.97	0.015	0.01	0.02	0.005	0.025
	c_2	0.00	0.01	0.02	0.015	0.03	0.985	0.99	0.98	0.995	0.975
		$d_{11}(c_2)$	$d_{12}(c_2)$	$d_{13}(c_2)$	$d_{14}(c_2)$	$d_{15}(c_2)$	$d_{16}(c_1)$	$d_{17}(c_1)$	$d_{18}(c_1)$	$d_{19}(c_1)$	$d_{20}(c_1)$
Classes	c_1	0.91	0.12	0.10	0.85	0.89	0.92	0.09	0.20	0.86	0.13
	c_2	0.09	0.88	0.90	0.15	0.11	0.08	0.91	0.80	0.14	0.87

Unlike previous research, our work is motivated by the observation that although the trained model from source domain may give poor performance over all the target domain data, it might perform well on an elaborately selected subset. Let's look at an intuitive example in Table 1, there are prediction probabilities (shown in columns 2, 3, 5 and 6) for classes c_1 and c_2 of 20 documents ($d_1 \sim d_{20}$) given by the model. The documents $d_1 \sim d_{10}$ are with probabilities higher than 0.96 when their labels are predicted as c_1 or c_2, while the rest documents $d_{11} \sim d_{20}$ are with much lower probabilities (lower than 0.93). If we compute the prediction accuracy over all the test documents, we can only obtain a low accuracy 65%. However, if the samples with higher prediction probabilities are selected (e.g., higher than 0.96), we can get a much better performance 90% on this sub data set (e.g., $d_1 \sim d_{10}$). Indeed, this coincides with the human sentiment that people always give higher confidence when they make sure of something, and vice versa they give lower confidence when not sure. Thus, we can trust the prediction results when the test samples with high prediction probabilities. Under these observations, in this paper we propose a two-step method for cross-domain text classification. First, we construct the reliable data set with high prediction confidence from target domain, and then use them as "labeled" data[2] to train a model. Second, we further propose an EM algorithm to refine the model trained from the reliable data set. This is an intuitively appealing fact, the model trained from reliable data set may perform much better than the one from source domain, since the reliable data are selected from target domain and with more similar distribution. The experimental results in Section 4 validate that.

A word for the outline of this paper. Section 2 survey some related works, and followed by the detailed description of the proposed method in Section 3. In Section 4, we evaluate our method on a large amount of experiments on text classification. Finally, we conclude the paper in Section 5.

2   Related Works

Here we summarize some previous works which are mostly related to this paper, including cross-domain learning and learning positive and unlabeled samples.

---

[2] The reliable data are not really labeled data, since their labels are predicted by the trained model from source domain.

2.1   Cross-Domain Learning

Cross-domain Learning studies how to deal with the classification problem when the source and target domain data obey different distributions. There are many papers appear in recent years, and they can be grouped into three types of techniques used for knowledge transfer, namely feature selection based [11,4,29], feature space mapping [20,22,7], weight based [5,6,10].

For the feature selection based methods, Jiang et al. [11] developed a two-step feature selection framework for domain adaptation. They first selected the general features to build a general classifier, and then considered the unlabeled target domain to select specific features for training target classifier. Dai et al. [4] proposed a Co-clustering based approach for this problem. In this method, they identified the word clusters among the source and target domains, via which the class information and knowledge propagated from source domain to target domain. Feature space mapping based methods are to map the original high-dimensional features into a low-dimensional feature space, under which the source and target domains comply with the same data distribution. Pan et al. [20] proposed a dimensionality reduction approach to find out this latent feature space, in which supervised learning algorithms can be applied to train classification models. Gu et al. [7] learnt the shared subspace among multiple domains for clustering and transductive transfer classification. In their problem formulation, all the domains have the same cluster centroid in the shared subspace. The label information can also be injected for classification tasks in this method. For the weight based methods, Jiang et al. [10] proposed a general instance weighting framework, which has been validated to work well on NLP tasks. Gao et al. [6] proposed a dynamic model weighting method for each test example according to the similarity between the model and the local structure of the test example in the target domain.

The most related work is [28]. Instead of selecting reliable data set from target domain, they used the labeled data from target domain to select useful data points in source domain, and then combined them with a few labeled data from target domain to build a good classifier. The main difference from our work is that, they needed some labeled data from target domain, while the target domain data in our problem are totally unlabeled.

2.2   Learning from Positive and Unlabeled Samples

The research of learning from positive and unlabeled (*LPU*) samples focuses on the application scenarios that there are only labeled positive samples but not any negative ones. Several techniques were proposed in [18,14,17,16,27,8].

Most of these methods take a two-step strategy. First, they adopted the techniques, e.g., Rocchio algorithm [24], Support Vector Machine (SVM) [3] and Naïve Baysian method [19] et al., to extract some reliable negative examples from the unlabeled data set, and then used the positive and likely negative samples to train a model. Second, the EM algorithm or iterative SVM were used to update the model. For example, Liu et al. [18] proposed a S-EM method for *LPU* learning, in which they first selected some documents from positive samples as "spy" documents to select more reliable negative data, and then used EM algorithm to build the final classifier. Although we also develop a two-step method and is similar with the techniques in *LPU* learning, there are two main differences from *LPU* learning. 1) Our method is to find reliable data set from target domain data for cross-domain learning, in which there are not any labeled data from target domain and the distributions of labeled source domain and unlabeled target domain are different, also *LPU* learning aims to build a binary classifier; 2) In the second step of our approach, the prediction probabilities of all target domain data (including the reliable data set) are updated during the EM iteration, while in *LPU* learning the probabilities of labeled positive samples retain unchanged. We adapt S-EM method to select the reliable data for our problem setting, but the performance is poor. To the best of our knowledge this is the first time to construct reliable data set for cross-domain learning, and the experimental results verify its effectiveness.

Another related work is Co-training [1]. Co-training assumed there were two views of feature set for data instances, and either view of the examples would be sufficient for learning if given enough labeled data. Also in Co-training algorithm, the unlabeled samples were selected as labeled data according to the consensus prediction of the two classifiers learnt from two views of labeled data. Instead, in our method we select the unlabeled data with high confident prediction as reliable data set.

3 The Proposed Algorithms

In this paper we propose two-step approaches for cross-domain classification: the first step is to select the reliable data set from target domain data which are predicted with high confidence by the trained model from source domain; then secondly, an EM algorithm is used to refine the model trained on the reliable data.

It is worth to note that selecting reliable data set is very important for our method, so we adopt the state-of-the-art supervised classifiers Logistical Regression [9] and Naïve Baysian method [19], which not only can produce probabilistic predictions, but also are approved to perform very well when the distributions between training and test data are the same.

Given the source and target domains, denoted as $\mathcal{D} = \{\mathcal{D}_s, \mathcal{D}_t\}$. The source domain with label information $\mathcal{D}_s = (x_i^s, y_i^s)|_{i=1}^{n_s}$ and the target domain without any label information $\mathcal{D}_t = (x_i^t)|_{i=1}^{n_t}$, where y_i^s is the true label of instance x_i^s in source domain, n_s and n_t are respectively the number of data instances in source and target domains. Our goal is to build a model that can predict target domain correctly.

3.1 The Techniques Used in Step I

We apply Logistical Regression [9] and Naïve Baysian method [19] to select reliable data set from the target domain according to the produced probabllistic predictions.

Logistical Regression. Logistic regression is an approach to learn functions of $P(Y|\mathbf{X})$ in the case where Y is discrete-valued, and $\mathbf{X}$ is any vector containing discrete or continuous random variables. Logistic regression assumes a parametric form for the distribution $P(Y|\mathbf{X})$, then directly estimates its parameters from the training data. The parametric model assumed by logistic regression in the case where Y is Boolean is

$$P(y = \pm 1|\mathbf{x}; \mathbf{w}) = \sigma(y\mathbf{w}^T\mathbf{x}) = \frac{1}{1 + \exp(-y\mathbf{w}^T\mathbf{x})}, \tag{1}$$

where $\mathbf{w}$ is the parameter of the model. Under the principle of *Maximum A-Posteriori* (MAP), $\mathbf{w}$ is estimated under the Laplacian prior. Given a data set $\mathcal{D} = \{(\mathbf{x}_i, y_i)\}_{i=1}^{N}$, we want to find the parameter $\mathbf{w}$ which maximizes:

$$\sum_{i=1}^{N} \log \frac{1}{1 + \exp(-y_i \mathbf{w}^T \mathbf{x}_i)} - \frac{\lambda}{2} \mathbf{w}^T \mathbf{w}. \tag{2}$$

After $\mathbf{w}$ is estimated, Equation (1) can be used to compute the probabilities of an instance belonging to the positive and negative classes. Though the Logistical Regression introduced here can only deal with two-class classification problem, it can be naturally extended to tackle multi-class case.

Naïve Baysian. Naïve Baysian is one of popular methods for text classification. Given a set of training documents $\mathcal{D} = (x_i, y_i)|_{i=1}^{n}$ (n is the number of documents) with m distinct words, each document d_i is considered an ordered list of words, and $x_{i,k}$ denotes the word in position k of x_i. We also have a set of pre-defined classes $C = \{c_1, \cdots, c_l\}$ (l is the number of classes), and need to compute the posterior probability $P(c_j|d_i)$. Based on the Baysian probability and multinomial model, we have

$$P(c_j) = \sum_{i=1}^{n} P(c_j|d_i)/n, \tag{3}$$

and with Laplacian smoothing,

$$P(w_t|c_j) = \frac{1 + \sum_{i=1}^{n} O(w_t, d_i) P(c_j|d_i)}{m + \sum_{s=1}^{m} \sum_{i=1}^{n} O(w_s, d_i) P(c_j|d_i)}, \tag{4}$$

where $O(w_t, d_i)$ is the co-occurrence of word w_t and document d_i, m is the total number of words. Finally, we can compute the posterior probability $P(c_j|d_i)$ under the independent assumption of the probabilities of the words as follows,

$$P(c_j|d_i) = \frac{P(c_j) \prod_{k=1}^{|d_i|} P(x_{i,k}|c_j)}{\sum_{r=1}^{l} P(c_r) \prod_{k=1}^{|d_i|} P(x_{i,k}|c_r)}. \tag{5}$$

The document d_i is predicted as

$$\max_{j} P(c_j|d_i). \tag{6}$$

When applying the trained model from source domain $\mathcal{D}_s$ to the target domain $\mathcal{D}_t$, we can obtain the resultant prediction probability matrix $A \in R_+^{n_t \times l}$, where n_t is the number of documents in target domain and l is the number of classes, R_+ denotes the set of nonnegative real numbers. Note that we normalize the elements in each row of A, $\sum_{j=1}^{l} A_{i,j} = 1$. Then according to the probability matrix A, the class labels of documents can be predicted by Equation (6). For the example in Table 1, documents $\{d_1, d_2, d_3, d_4, d_5, d_{11}, d_{14}, d_{15}, d_{16}, d_{19}\}$ are predicted as c_1, while documents $\{d_6, d_7, d_8, d_9, d_{10}, d_{12}, d_{13}, d_{17}, d_{18}, d_{20}\}$ are predicted as c_2. Finally, we sort the prediction probabilities of the documents in each class, and select a portion of documents with highest prediction probabilities as reliable data given a selecting rate r.

E.g., $r = 0.5$, the selected documents in c_1 are $\{d_1, d_2, d_3, d_4, d_5\}$, and the ones in c_2 are $\{d_6, d_7, d_8, d_9, d_{10}\}$. Note that we select a portion of documents with highest prediction confidence from each class to construct reliable data is trying to ensure that we have "label" data for each class when training a classifier.

3.2 The EM Algorithm in Step II

In this step we develop an EM Algorithm to build a final model which can predict the target domain more correctly. Specifically, we iteratively retrain a model during the EM iterating process according to the prediction results produced by the model in last iteration.

The EM algorithm also contains two steps, the *Expectation* step (E step) and the *Maximization* step (M step). In our algorithm, the E step is to estimate the model, while M step is to maximize the posterior probability of target domain data. Our EM algorithm is based on the Naïve Baysian method, so the E step corresponds to Equations (3) and (4), and Equation (5) is for M step. The description of the proposed two-step method is detailed in Algorithm 1. Note that when the EM algorithm converges, the predicted labels of reliable data may be changed. The reason we update them is that they are not really labeled data, though there is high prediction accuracy on reliable data.

Algorithm 1. Effectively Constructing Reliable Data for Cross-domain Text Classification

Input: Given labeled source domain $\mathcal{D}_s$ and unlabeled target domain $\mathcal{D}_t$. K, the number of iterations. r, the selecting rate.
Output: the model can correctly predict target domain data.

1. Apply the supervised learning algorithm (e.g., NBC or LR) to train a model on source domain, and then predict the target domain $\mathcal{D}_t$ to obtain prediction probability matrix A.
2. Select the reliable data RD based on the prediction probability matrix A and the selecting rate r.
3. Use RD as "labeled" data set to train a model (e.g., NBC), and make a prediction on target domain $\mathcal{D}_t$, we obtain the prediction results $P^{(0)}(c_j|d_i)$.
4. $k := 1$
5. Update $P^{(k)}(c_j)$ and $P^{(k)}(w_t|c_j)$ $(1 \leq j \leq l, 1 \leq t \leq m)$ according to Equations (3) and (4) in E step;
6. Update $P^{(k)}(c_j|d_i)$ $(1 \leq i \leq n_t)$ according to Equation (5) in M step;
7. $k := k + 1$, if $k < K$, turn to Step 5.
8. Output the final model.

---

4 Experiments

We conduct experiments on a large amount of classification problems, and focus on binary classification.

4.1 Data Preparation

20Newsgroup[3] is one of the benchmark data sets for text categorization. Since the data set is not originally designed for cross-domain learning, we need to do some data preprocessing. The data set is partitioned evenly cross 20 different newsgroups, and some very related newsgroups are grouped into certain top category. For example, the top category *sci* contains four subcategories *sci.crypt*, *sci.electronics*, *sci.med* and *sci.space*.

We select three top-categories *sci*, *talk* and *rec* (they all have four sub-categories) to perform two-class classification experiments. Any two top categories can be selected to construct two-class classification problems, and we can construct three data sets *sci vs. talk*, *rec vs. sci* and *rec vs. talk* in the experimental setting. For the data set *sci vs. talk*, we randomly select one subcategory from *sci* and one subcategory from *talk*, which denote the positive and negative data, respectively. The test data set is similarly constructed as the training data set, except that they are from different subcategories. Thus, the constructed classification task is suitable for cross-domain learning due to the facts that 1) the training and test data are from different distributions since they are from different subcategories; 2) they are also related to each other since the positive (negative) instances in the training and test set are from the same top categories. For the data set *sci vs. talk*, we totally construct 144 ($P_4^2 \cdot P_4^2$) classification tasks. The data sets *rec vs. sci* and *rec vs. talk* are constructed similarly with *sci vs. talk*.

4.2 Compared Approaches

In this paper, the supervised learning algorithms Naïve Bayesian (NBC) [19] and Logistical Regression (LR) [9] are used to select the reliable data set, thus we have two methods, called NBC_R_EM and LR_R_EM, respectively. The baseline methods include: 1) Using Naïve Bayesian (NBC) [19] and Logistic Regression (LR) [9] to learn classifiers; 2) Training the classifiers only on the selected reliable data set produced by NBC and LR, respectively denoted as NBC_R and LR_R. 3) The Transductive Support Vector Machine (TSVM) [12] and the state-of-the-art cross-domain classification method (CoCC) [4].

Actually, we can combine and make full use of all the reliable data selected by NBC and LR, and train classifiers on this combination of reliable data[4], denoted as Com_R and Com_R_EM.

The baseline methods LG is implemented by the package[5], and TSVM are given by SVM^{light}[6]. The parameter settings of CoCC is the same as their original paper. The selecting rates $r_1 = 0.2$ for NBC_R_EM and $r_2 = 0.2$ for LR_R_EM, the maximal number of EM iterations is 50. We use the classification accuracy on target domain data to evaluate all compared algorithms.

---

[3] http://people.csail.mit.edu/jrennie/20Newsgroups/
[4] In the combination process, we get rid of the data instances that these two algorithms NBC and LR do not give the same prediction label.
[5] http://research.microsoft.com/~minka/papers/logreg/
[6] http://svmlight.joachims.org/

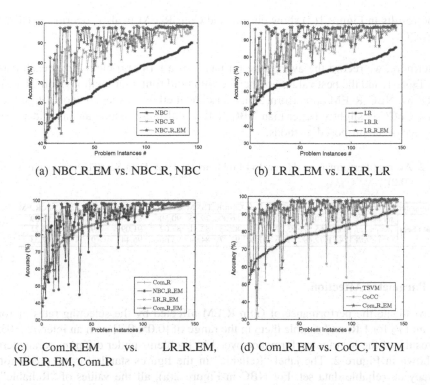

(a) NBC_R_EM vs. NBC_R, NBC

(b) LR_R_EM vs. LR_R, LR

(c)   Com_R_EM   vs.   LR_R_EM, NBC_R_EM, Com_R

(d) Com_R_EM vs. CoCC, TSVM

Fig. 1. The Performance Comparison of All Classification Algorithms on Data Set *sci vs. talk* (The parameters $r_1 = 0.2, r_2 = 0.2$)

4.3   Experimental Results

We evaluate all the compared approaches on three date sets *sci vs. talk*, *rec vs. sci* and *rec vs. talk*, and only list the detailed results of *sci vs. talk* due to the space limit. The results are shown in Figure 1, and we have following findings:

1) In Figure 1(a) and 1(b), the methods NBC_R and LR_R are better than NBC and LR, respectively, which indicates that the reliable data selected from target domain are very effective. Also NBC_R_EM outperforms NBC_R and LR_R_EM outperforms LR_R show the accuracy gains of EM algorithm in the second step. Moreover, NBC_R_EM (LR_R_EM) performs significantly better than NBC (LR), which notes that our proposed methods are more successful to handle cross-domain classification problems.

2) From Figure 1(c), we can find that all the two-step methods outperform Com_R. Also LR_R_EM is better than NBC_R_EM, which may indicate that the reliable data selected by LR is more informative. Though the results of combination Com_R_EM is very similar with LR_R_EM, it seems much more stable. So we can use Com_R_EM as the final classifier when we do not know which one of the classifiers NBC_R_EM and LR_R_EM is better.

3) The results in Figure 1(d) show our method Com_R_EM is also superior to TSVM and CoCC.

Furthermore, we record the average performance of all 144 problems from each data set in Table 1, and the best values are marked with bold font. Our methods Com_R_EM, LR_R_EM, NBC_R_EM outperform all the baseline methods, except that on data *set sci vs. talk* CoCC is slightly better than NBC_R_EM. All these results again validate the advantage of the proposed methods.

Table 2. Average Performances (%) on 144 Problems for Each Data Set (The parameters $r_1 = 0.2, r_2 = 0.2$)

Compared Algorithms		NBC	LR	NBC_R	LR_R	Com_R	TSVM	CoCC	NBC_R_EM	LR_R_EM	Com_R_EM
	sci vs. talk	70.93	70.64	85.20	84.23	86.00	79.35	90.50	89.65	91.19	**91.39**
Data Sets	*rec vs. sci*	68.75	65.57	83.07	80.31	82.43	82.81	87.02	91.09	91.41	**93.21**
	rec vs. talk	74.71	72.49	91.36	91.20	92.87	84.94	93.66	96.44	97.31	**97.33**

4.4 Parameter Affection

We investigate the performance of Com_R_EM affected by the selecting rates r_1 for NBC and r_2 for LR, and sample them in the range of $[0.05, 0.7]$ with an interval 0.05. The results of the average performance over 144 problems under different parameters are shown in Figure 2. The label "Reliable" in the figures stands for the prediction accuracy on reliable data set. For NBC in Figure 2(a), all the values of "Reliable", NBC_R and NBC_R_EM first increases then decreases with the increasing of selected rate r_1. While for LR in Figure 2(b), the value "Reliable" decreases along with the increasing of r_2, and NBC_R, NBC_R_EM keep stable when r_2 is large enough. From these results, to ensure the good performance of our methods we recommend that the selecting rate should not be set too large or too small, since too small value of selecting rate may not include sufficient information and too large value of selecting rate will introduce more false prediction results when constructing the reliable data set. Also, it is shown again that LR is more stable and much safer to select the reliable data than NBC. In this paper, we set $r_1 = 0.2$ and $r_2 = 0.2$, and Com_R_EM reaches its peak of performance in Figure 2(c).

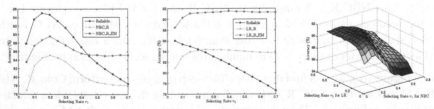

(a) The Performance Affection of Rate r_1 on Selecting Reliable Data by NBC

(b) The Performance Affection of Rate r_2 on Selecting Reliable Data by LR

(c) The Performance Affection of Rate r_1 and r_2 on Com_R_EM

Fig. 2. The Performance Affection of Rate r_1 and r_2 on Data Set *sci vs. talk*

Running Time. Since in each EM iteration, we essentially train a naïve baysian classifier. As you know, the training of naïve baysian classifier is very fast, so our method also can run very fast. Moreover, the EM algorithm can almost converge within 30 iterations on all the classification problems, which shows its efficiency.

4.5 Distribution Mismatch Investigation

In this subsection we study the distribution mismatch between the source domain and target domain data, and the reliable data and target domain data. K-L divergence [13] is one of the popular evaluation criteria to measure data distribution differences from different domains.

The K-L divergence [13] is computed as follows,

$$K\text{-}L(\mathcal{D}_1 \| \mathcal{D}_2) = \sum_w P_1(w) \log \frac{P_1(w)}{P_2(w)}, \tag{7}$$

where $P_1(w)$ $(P_2(w))$ is the estimation of word w on $\mathcal{D}_1$ $(\mathcal{D}_2)$. If we randomly split the data set from the same domain into training data and test data, then the K-L divergence has a value of nearly zero. The K-L divergence values on three data sets are shown in Figure 3, in which RD_{NBC} and RD_{LR} denote the reliable data set selected by NBC and LR models, respectively. From these results, we can find that the K-L divergence values between the source and target domain data are much larger than the ones between the reliable data and target domain data, which indicates that the selected reliable data are more similar with the target domain data. This is why the performance of models trained from reliable data are much better than the ones trained from source domain (i.e., see the results in Table 2). Also it can be seen that the K-L divergence values between RD_{LR} and $\mathcal{D}_t$ is smaller and more stable than the one between RD_{NBC} and $\mathcal{D}_t$, which validates the findings in Section 4.3 and 4.4 that the reliable data selected by LR are much more informative and safer.

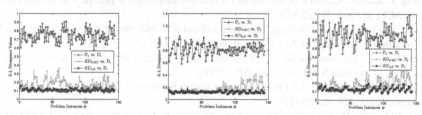

(a) The Distribution Mismatch (b) The Distribution Mismatch on Data Set *scivs.talk* match on Data Set *recvs.sci* (c) The Distribution Mismatch on Data Set *recvs.talk*

Fig. 3. The Distribution Mismatch Investigation on Three Data Sets

5 Conclusions

In this paper we investigate the prediction results on target domain predicted by the model learnt from source domain, and find the model can give very good predictions on a certain subset of the target domain data with high prediction confidence. Along this line, we propose a new two-step method for cross-domain learning, in which, we first construct reliable data set with high prediction confidence, and then develop an EM algorithm to build the final classifier. Experimental results show that our methods can handle well the cross-domain classification problems.

Acknowledgments. This work is supported by the National Natural Science Foundation of China (No. 60933004, 60975039, 61175052, 61035003, 61072085), National High-tech R&D Program of China (863 Program) (No.2012AA011003).

References

1. Blum, A., Mitchell, T.: Combining labeled and unlabeled data with co-training. In: Proc. of the 11th Annual Conference on Computational Learning Theory, pp. 92–100 (1998)
2. Boser, B.E., Guyou, I., Vapnik, V.: A training algorithm for optimal margin classifiers. In: Proc. of the 5th AWCLT (1992)
3. Chang, C.C., Lin, C.J.: LIBSVM: a library for support vector machines (2001) Software available at, http://www.csie.ntu.edu.tw/~cjlin/libsvm
4. Dai, W., Xue, G., Yang, Q., Yu, Y.: Co-clustering based classification for out-of-domain documents. In: Proc. of the 13th ACM SIGKDD, pp. 210–219 (2007)
5. Dai, W., Yang, Q., Xue, G., Yu, Y.: Boosting for transfer learning. In: Proc. of the 24th International Conference on Machine Learning (ICML), Corvallis, OR, pp. 193–200 (2007)
6. Gao, J., Fan, W., Jiang, J., Han, J.W.: Knowledge transfer via multiple model local structure mapping. In: Proc. of the 14th ACM SIGKDD, pp. 283–291 (2008)
7. Gu, Q.Q., Zhou, J.: Learning the shared subspace for multi-task clustering and transductive transfer classification. In: Proc. of the ICDM (2009)
8. He, J.Z., Zhang, Y., Li, X., Wang, Y.: Naive bayes classifier for positive unlabeled learning with uncertainty. In: Proceedings of the 10th SIAM SDM, pp. 361–372 (2010)
9. Hosmer, D., Lemeshow, S.: Applied Logistic Regression. Wiley, New York (2000)
10. Jiang, J., Zhai, C.X.: Instance weighting for domain adaptation in nlp. In: Proceedings of the 45th ACL, pp. 264–271 (2007)
11. Jiang, J., Zhai, C.X.: A two-stage approach to domain adaptation for statistical classifiers. In: Proceedings of the 16th ACM CIKM, pp. 401–410 (2007)
12. Joachims, T.: Transductive inference for text classification using support vector machines. In: Proc. of the 16th ICML, pp. 200–209 (1999)
13. Kullback, S., Leibler, R.A.: On information and sufficiency. The Annals of Mathematical Statistics 22(1), 79–86 (1951)
14. Lee, W.S., Liu, B.: Learning with positive and unlabeled examples using weighted logistic regression. In: Proceedings of the 20th ICML (2003)
15. Lewis, D., Riguette, M.: A comparison of two learning algorithms for text categorization. In: Proc. 3rd Annual Symposium on Document Analysis and Information Retrieval, pp. 81–93 (1994)
16. Li, X.L., Liu, B., Ng, S.K.: Negative training data can be harmful to text classification. In: Proceedings of the 2010 Conference on EMNLP, pp. 218–228 (2010)
17. Liu, B., Dai, Y., Li, X.L., Lee, W.S., Yu, P.S.: Building text classifiers using positive and unlabeled examples. In: Proceedings of the 3rd IEEE ICDM, pp. 179–186 (2002)
18. Liu, B., Lee, W.S., Yu, P.S., Li, X.L.: Partially supervised classification of text documents. In: Proceedings of the 19th ICML, pp. 387–394 (2002)

19. McCallum, A., Nigam, K.: A comparison of event models for naive bayes text classification. In: Proceedings of the AAAI Workshop on Learning for Text Categorization (1998)
20. Pan, S.J., Kwok, J.T., Yang, Q.: Transfer learning via dimensionality reduction. In: Proceedings of the 23rd AAAI, pp. 677–682 (2008)
21. Pan, S.J., Ni, X.C., Su, J.T., Yang, Q., Chen, Z.: Cross-domain sentiment classification via spectral feature alignment. In: Proceedings of the 19th WWW, pp. 751–760 (2010)
22. Pan, S.J., Tsang, I.W., Kwok, J.T., Yang, Q.: Domain adaptation via transfer component analysis. In: Proceedings of the 21st IJCAI, pp. 1187–1192 (2009)
23. Pan, S.J., Yang, Q.: A survey on transfer learning. IEEE TKDE 22(10), 1345–1359 (2010)
24. Rocchio, J.: Relevance feedback in information retrieval. The SMART Retrieval System, 313–323 (1971)
25. Uguroglu, S., Carbonell, J.: Feature Selection for Transfer Learning. In: Gunopulos, D., Hofmann, T., Malerba, D., Vazirgiannis, M. (eds.) ECML PKDD 2011. LNCS, vol. 6913, pp. 430–442. Springer, Heidelberg (2011)
26. Zadrozy, B.: Learning and evaluating classifiers under sample selection bias. In: Proceedings of the 21th ICML, pp. 114–121 (2004)
27. Zhang, B.Z., Zuo, W.L.: Learning from positive and unlabeled examples: A survey. In: Proceedings of ISIP, pp. 650–654 (2008)
28. Zhen, Y., Li, C.Q.: Cross-domain knowledge transfer using semi-supervised classification. In: Proceedings of the 21st AJCAI, pp. 362–371 (2008)
29. Zhuang, F.Z., Luo, P., Xiong, H., He, Q., Xiong, Y.H., Shi, Z.Z.: Exploiting associations between word clusters and document classes for cross-domain text categorization. In: Proc. of the SIAM SDM, pp. 13–24 (2010)
30. Zhuang, F.Z., Luo, P., Xiong, H., Xiong, Y.H., He, Q., Shi, Z.Z.: Cross-domain learning from multiple sources: A consensus regularization perspective. IEEE TKDE, 1664–1678 (2010)

Improving Transfer Learning by Introspective Reasoner

Zhongzhi Shi[1], Bo Zhang[1,2], and Fuzhen Zhuang[1]

[1] Key Laboratory of Intelligent Information Processing, Institute of Computing Technology,
Chinese Academy of Sciences, 100190, Beijing, China
[2] Graduate School of the Chinese Academy of Sciences, 100039, Beijing, China
{shizz,zhangb,zhuangfz}@ics.ict.ac.cn

Abstract. Traditional learning techniques have the assumption that training and test data are drawn from the same data distribution, and thus they are not suitable for dealing with the situation where new unlabeled data are obtained from fast evolving, related but different information sources. This leads to the cross-domain learning problem which targets on adapting the knowledge learned from one or more source domains to target domains. Transfer learning has made a great progress, and a lot of approaches and algorithms are presented. But negative transfer learning will cause trouble in the problem solving, which is difficult to avoid. In this paper we have proposed an introspective reasoner to overcome the negative transfer learning.

Introspective learning exploits explicit representations of its own organization and desired behavior to determine when, what, and how to learn in order to improve its own reasoning. According to the transfer learning process we will present the architecture of introspective reasoner for transductive transfer learning.

Keywords: Introspective reasoned, Transfer learning, Negative transfer.

1   Introduction

Introspection method is early a psychological research approach. It investigates the psychological phenomena and process according to the report of the tested person or the experience described by himself. The introspection learning is to introduce introspection concept into machine learning. That is to say, by checking and caring about knowledge processing and reasoning method of intelligence system itself and finding out problems from failure or poor efficiency, the introspection learning forms its own learning goal and then improves the method to solving problems [1].

Cox's paper mentioned that research on introspective reasoning has a long history in artificial intelligence, psychology, and cognitive science [2]. Leake et al. pointed that in introspective learning approaches, a system exploits explicit representations of its own organization and desired behavior to determine when, what, and how to learn in order to improve its own reasoning [3]

Introspective learning has become more and more important in recent years as AI systems have begun to address real-world problem domains, characterized by a high

Z. Shi, D. Leake, and S. Vadera (Eds.): IIP 2012, IFIP AICT 385, pp. 28–39, 2012.

degree of complexity and uncertainty. Early of 1980s, introspective reasoning is implemented as planning within the meta-knowledge layer. SOAR [4] employs a form of introspective reasoning by learning meta-rules which describe how to apply rules about domain tasks and acquire knowledge. Birnbaum et al. proposed the use of self-models within case-based reasoning [5]. Cox and Ram proposed a set of general approaches to introspective reasoning and learning, automatically selecting the appropriate learning algorithms when reasoning failures arise [6]. They defined a taxonomy of causes of reasoning failures and proposed a taxonomy of learning goals, used for analyzing the traces of reasoning failures and responding to them. The main step in the introspective learning process is to looking for the reason of failure according to the features of failure. The introspective learning system should evaluate reasoning besides finding errors. The blame assignment is like troubleshooting with a mapping function from failure symptom to failure cause. A series of processes based on case-based reasoning, such as search, adjustment, evaluation and reservation can attain blame assignment and explanation failure, and improve the efficiency of evaluation and explanation, so case-based reasoning is an effective approach.

Introspective reasoning can be a useful tool for autonomously improving the performance of a CBR system by reasoning about system problem solving failures. Fox and Leake described a case-based system called ROBBIE which uses introspective reasoning to model, explain, and recover from reasoning failures [7]. Fox and Leake also took a model-based approach to recognize and repair reasoning failures. Their particular form of introspective reasoning focuses on retrieval failures and case index refinement. Work by Oehlmann, Edwards and Sleeman addressed on the related topic of re-indexing cases, through introspective questioning, to facilitate multiple viewpoints during reasoning [8]. Leake, Kinley, and Wilson described how introspective reasoning can also be used to learn adaptation knowledge in the form of adaptation cases [9]. Zhang and Yang adopted introspective learning to maintain feature weight in case-based reasoning [10]. Craw applied introspective learning to build case-based reasoning knowledge containers [11]. Introspective reasoning to repair problems may also be seen as related to the use of confidence measures for assessing the quality of the solutions proposed by a CBR system [12]. Arcos, Mulayim and Leake proposed an introspective model for autonomously improving the performance of a CBR system by reasoning about system problem solving failures [13]. The introspective reasoner monitors the reasoning process, determines the causes of the failures, and performs actions that will affect future reasoning processes. Soh et al. developed CBRMETAL system which integrates case-based reasoning and meta-learning to introspectively refine the reasoning of intelligent tutoring system [14]. CBRMETAL system specifically improves the casebase, the similarity heuristics and adaptation heuristics through reinforcement and adheres to a set of six principles to minimize interferences during meta-learning. Leake and Powell proposed WebAdapt approach which can introspectively correct failures in its adaptation process and improve its selection of Web sources to mine [15].

Transfer learning has already achieved significant success. One of the major challenges in developing transfer methods is to produce positive transfer between appropriately related tasks while avoiding negative transfer between tasks that are less related. So far there are some approaches for avoiding negative transfer.

Negative transfer happens when the source domain data and task contribute to the reduced performance of learning in the target domain. Despite the fact that how to avoid negative transfer is a very important issue, little research work has been published on this topic. Torrey and Shavlik summarized into three methods, that is, rejecting bad information, choosing a source task, modeling task similarity [16]. Croonenborghs et al. proposed an option-based transfer in reinforcement learning is an example of the approach that naturally incorporates the ability to reject bad information [17]. Kuhlmann and Stone looked at finding similar tasks when each task is specified in a formal language [18]. They constructed a graph to represent the elements and rules of a task. This allows them to find identical tasks by checking for graph isomorphism, and by creating minor variants of a target-task graph, they can also search for similar tasks. Eaton and DesJardins proposed to choose from among candidate solutions to a source task rather than from among candidate source tasks [19]. Carroll and Seppi developed several similarity measures for reinforcement learning tasks, comparing policies, value functions, and rewards[20].

Besides above mentioned methods, Bakker and Heskes adopted a Bayesian approach in which some of the model parameters are shared for all tasks and others more loosely connected through a joint prior distribution that can be learned from the data [21]. The data are clustered based on the task parameters, where tasks in the same cluster are supposed to be related to each other. Argyriou et al. divided the learning tasks into groups [22]. Tasks within each group are related by sharing a low-dimensional representation, which differs among different groups. As a result, tasks within a group can find it easier to transfer useful knowledge. Though there have been lots of transfer learning algorithms, few of them can guarantee to avoid negative transfer. In this paper we propose a method which adopts introspective reasoner to avoid negative transfer.

The rest of this paper is organized as follows. The following section describes transfer learning. Section 3 introduces the transductive transfer learning. Section 4 discusses introspective reasoner. Failure taxonomy in introspective reasoner is described in Section 5. Finally, the conclusions and further works are given.

2   Transfer Learning

Transfer learning is a very hot research topic in machine learning and data mining area. The study of transfer learning is motivated by the fact that people can intelligently apply knowledge learned previously to solve new problems faster or with better solutions. The fundamental motivation for transfer learning in the field of machine learning was discussed in a NIPS-95 workshop on Learning to Learn. Since 1995 Research on transfer learning has attracted more and more attention. In 2005, the Broad Agency Announcement (BAA) 05-29 of Defense Advanced Research Projects Agency (DARPA)'s Information Processing Technology Office (IPTO) gave a new mission of transfer learning: the ability of a system to recognize and apply knowledge and skills learned in previous tasks to novel tasks. In this definition, transfer learning aims to extract the knowledge from one or more source tasks and apply the knowledge to a target task[23]. Fig. 1 shows you the transfer learning process.

The formal definitions of transfer learning are given follows. For simplicity, we only consider the case where there is one source domain D_S, and one target domain, D_T.

Definition 1. Source domain data DS

$$D_S = \{(x_{S_1}, y_{S_1}), \cdots (x_{S_n}, y_{S_n})\} \tag{1}$$

where $x_{S_i} \in X_S$ is the data instance and $y_{S_i} \in Y_S$ is the corresponding class label.

Definition 2. Target domain data DT

$$D_T = \{(x_{T_1}, y_{T_1}), \cdots (x_{T_m}, y_{T_m})\} \tag{2}$$

where the input $x_{T_i} \in X_T$ and $y_{T_i} \in Y_T$ is the corresponding output.

Definition 3. Transfer learning: Given a source domain DS and learning task TS, a target domain DT and learning task TT , transfer learning aims to help improve the learning of the target predictive function in DT using the knowledge in DS and TS, where $DS \neq DT$, and $TS \neq TT$.

Definition 4. Negative transfer: In transfer learning if the source task is not sufficiently related or if the relationship is not well leveraged by the transfer method, the performance may not only fail to improve but also result in actually decreasing.

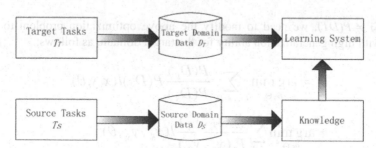

Fig. 1. Transfer learning process

The goal of our introspective reasoning system is to detect reasoning failures, refine the function of reasoning mechanisms, and improve the system performance for future problems. To achieve this goal, the introspective reasoner monitors the reasoning process, determines the possible causes of its failures, and performs actions that will affect future reasoning processes.

In this paper, we only focus on the negative transfer in the transductive transfer learning, that means, the source and target tasks are the same, while the source and target domains are different. In this situation, no labeled data in the target domain are available while a lot of labeled data in the source domain are available.

3   Transductive Transfer Learning

In the transductive transfer learning, the source and target tasks are the same, while the source and target domains are different. In this situation, no labeled data in the target domain are available while a lot of labeled data in the source domain are available. Here we give the definition of transductive transfer learning as follows:

Definition 5. Transductive transfer learning: Given a source domain DS and a corresponding learning task TS, a target domain DT and a corresponding learning task TT , transductive transfer learning aims to improve the learning of the target predictive function f_T (•) in DT using the knowledge in DS and TS, where $DS \neq DT$ and $TS = TT$.

Definition 6. is clear that transductive transfer learning has different data sets and same learning tasks. In the transductive transfer learning, we want to learn an optimal model for the target domain by minimizing the expected risk [23],

$$\theta^* = \arg\min_{\theta \in \Theta} \sum_{(x,y) \in D_T} P(D_T) l(x, y, \theta) \tag{3}$$

where $l(x, y, \theta)$ is a loss function that depends on the parameter θ. Since no labeled data in the target domain are observed in training data, we have to learn a model from the source domain data instead. If $P(DS) = P(DT)$, then we may simply learn the model by solving the following optimization problem for use in the target domain,

$$\theta^* = \arg\min_{\theta \in \Theta} \sum_{(x,y) \in D_S} P(D_S) l(x, y, \theta) \tag{4}$$

If $P(DS) \neq P(DT)$, we need to modify the above optimization problem to learn a model with high generalization ability for the target domain, as follows:

$$\theta^* = \arg\min_{\theta \in \Theta} \sum_{(x,y) \in D_S} \frac{P(D_T)}{P(D_S)} P(D_S) l(x, y, \theta)$$

$$\approx \arg\min_{\theta \in \Theta} \sum_{i=1}^{n_S} \frac{P_T(x_{T_i}, y_{T_i})}{P_S(x_{S_i}, y_{S_i})} l(x_{S_i}, y_{S_i}, \theta) \tag{5}$$

By adding different penalty values to each instance (x_{Si}, y_{Si}) with the corresponding weight $\frac{P_T(x_{T_i}, y_{T_i})}{P_S(x_{S_i}, y_{S_i})}$, we can learn a precise model for the target domain. Since $P(YT|XT) = P(YS|XS)$, the differences between $P(DS)$ and $P(DT)$ are caused by $P(XS)$ and $P(XT)$ and

$$\frac{P_T(x_{T_i}, y_{T_i})}{P_S(x_{S_i}, y_{S_i})} = \frac{P(x_{T_i})}{P(x_{S_i})}$$

If we can estimate $\frac{P(x_{S_i})}{P(x_{T_i})}$ for each instance, we can solve the transductive transfer learning problems.

In transductive transfer learning the distributions on source domain and target domain do not match, we are facing sample selection bias or covariate shift. Specifically, given a domain of patterns X and labels Y, we obtain source training samples $Z_S=\{(x_{S_1}, y_{S_1}), ..., (x_{S_n}, y_{S_n})\} \subseteq X \times Y$ from a Borel probability distribution PS(x, y), and target samples $Z_T=\{(x_{T_1}, y_{T_1}), ..., (x_{T_m}, y_{T_m})\}$ drawn from another such distribution $P_T(x, y)$. Huang et al. proposed a kernel-mean matching (KMM) algorithm to learn $\frac{P(x_{S_i})}{P(x_{T_i})}$ directly by matching the means between the source domain data and the target domain data in a reproducing-kernel Hilbert space (RKHS) [24]. KMM can be rewritten as the following optimization problem.

$$\min_{\beta} \frac{1}{2} \beta^T K \beta - k^T \beta$$

(6)

$$\text{s.t. } \beta_i \in [0, B] \text{ and } |\sum_{i=1}^{n_S} \beta_i - n_S| \le n_S \varepsilon$$

where

$$K = \begin{bmatrix} K_{S,S} & K_{S,T} \\ K_{T,S} & K_{T,T} \end{bmatrix}$$

and $K_{i,j}=k(x_i, x_j)$, $K_{S,S}$ and $K_{T,T}$ are kernel matrices for the source domain data and the target domain data, respectively. Note that (6) is a quadratic program which can be solved efficiently using interior point methods or any other successive optimization procedure. Huang et al. proved that

$$P(x_{T_i}) = \beta_i P(x_{S_i})$$

(7)

The advantage of using KMM is that it can avoid performing density estimation of either $P(x_{Si})$ or $P(x_{Ti})$, which is difficult when the size of the data set is small.

A common assumption in supervised learning is that training and test samples follow the same distribution. However, this basic assumption is often violated in practice and then standard machine learning methods do not work as desired. The situation where the input distribution $P(x)$ is different in the training and test phases but the conditional distribution of output values $P(y|x)$ remains unchanged is called covariate shift. Sugiyama et al. proposed an algorithm known as Kullback-Leibler Importance Estimation Procedure(KLIEP) to estimate $\frac{P(x_{S_i})}{P(x_{T_i})}$ directly[25], based on the minimization of the Kullback-Leibler divergence. The optimization criterion is as follows:

$$\max_{\{\theta_l\}_{l=1}^b} \left[\sum_{j=1}^m \log \sum_{l=1}^b \theta_l \phi_l(\mathbf{x}_j^m) \right]$$

$$\text{subject to } \sum_{i=1}^n \sum_{l=1}^b \theta_l \phi_l(x_i^n) \tag{8}$$

It can be integrated with cross-validation to perform model selection automatically in two steps: a) estimating the weights of the source domain data and b) training models on the reweighted data.

The Naïve Bayes classifier is effective for text categorization [26]. Regarding the text categorization problem, a document $d \in D$ corresponds to a data instance, where D denotes the training document set. The document d can be represented as a bag of words. Each word $w \in d$ comes from a set W of all feature words. Each document d is associated with a class label $c \in C$, where C denotes the class label set. The Naïve Bayes classifiers estimate the conditional probability as follows:

$$P(c \mid d) = P(c) \prod_{w \in d} P(w \mid c) \tag{9}$$

Dai et al. proposed a transfer-learning algorithm for text classification based on an EM-based Naïve Bayesian classifiers NBTC [27]. Dai et al. also proposed co-clustering based classification for out-of-domain documents algorithm CoCC which takes co-clustering as a bridge to propagate the knowledge from the in-domain to out-of-domain [28].

Ling et al. proposed a spectral classification based method CDSC, in which the labeled data from the source domains are available for training and the unlabeled data from target domains are to be classified [29]. Based on the normalized cut cost function, supervisory knowledge is transferred through a constraint matrix, and the regularized objective function finds the consistency between the source domain supervision and the target domain intrinsic structure. Zhuang et al. proposed CCR_3 approach which is a consensus regularization framework to exploit the distribution differences and learn the knowledge among training data from multiple source domains to boost the learning performance in a target domain [30].

4   Introspective Reasoner

The goal of our introspective reasoning system is to detect negative transfer , refine the function of reasoning mechanisms, and improve the system performance for future problems. In order to reach this goal, the introspective reasoner should solve following problems [31]:

1. There are standards that determine when the reasoning process should be checked, i.e. monitoring the reasoning process;
2. Determine whether failure reasoning takes place according to the standards;

3. Confirm the final reason that leads to the failure;
4. Change the reasoning process in order to avoid the similar failure in the future.

Introspective reasoner for transfer learning will monitor the transfer learning process, determine the possible causes of its failures, and perform actions that will affect future transfer processes. The architecture of introspective reasoner is shown in Fig. 2.

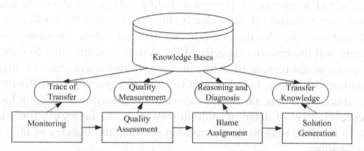

Fig. 2. Architecture of Introspective Reasoner

Introspective reasoner consists of monitoring, quality assessment, blame assignment, solution generation and ontology-based knowledge bases. The monitoring task tracks the transductive transfer learning process. For each learning task solved by the transductive transfer learning system, the monitor generates a trace containing: a) the source data retrieved; b) the ranking criteria applied to the data, together with the values that each criterion produced and the final ranking; and c) the transfer operators which were applied, with the sources to which they were applied and the target changes produced.

When the user's final solution is provided to the system, quality assessment is triggered to determine the real quality of the system-generated solution, by analyzing the differences between the system's proposed solution and the final solution. Quality assessment provides a result in qualitative terms: positive quality or negative quality.

Blame assignment starts by negative quality assessment. It takes as input the differences between the source data and target data, and tries to relate the solution differences to the source data. The system searches the ontology-based knowledge bases and selects those that apply to the observed solution differences.

Solution generation identifies the learning goals related to the negative transfer failure selected in the blame assignment stage. Each failure may be associated with more than one learning goal. For each learning goal, a set of plausible source data changes in the active policies are generated by using a predefined ontology-based knowledge base.

Failure taxonomy is an important problem in introspective learning system. It can provide a clue for the failure explanation and the revise learning target. In the introspective reasoner system, the failures are put into knowledge bases which use ontology as knowledge representation. In this way, the knowledge is organized as conceptualization, formal, semantics explicit and shared [32].

5   Failure Taxonomy in Introspective Reasoner

Failure taxonomy is an important problem in introspective learning system. Hierarchical failure taxonomy is a method that can solve the problem classification is too fine or too coarse in failure taxonomy. Failure can be represented into large class according to the different reasoning step. This can improve the introspective revision of the system and accelerate the process of failure comparing. If failure taxonomy is fine, failures can be described more clearly, in order to provide valuable clue. Giving a rational treatment for the relationship between the failure taxonomy and failure interpretation will increase the capacity of the system introspection. System not only need to reason out the reasons of failure and give introspective learning target accord to the symptoms of failure, but also need to have all capacities to deal with different problems. Similarly, failure explanation also can be classed into different layers, such as abstract level, level of detail or multidimensional. The layers of failure taxonomy contribute to form the reasonable relationship between the features of failure and the failure explanation.

In introspective reasoner knowledge bases contain failure taxonomy and special knowledge coming from previous experiences or expertise. Solving the failure taxonomy in introspective learning through ontology-based knowledge bases can make the failure taxonomy in introspective learning more clear and retrieval process more efficient.

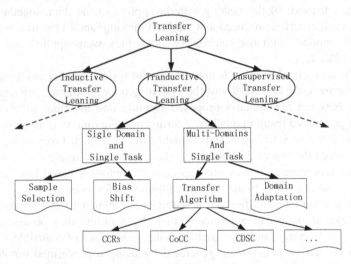

Fig. 3. Failure Taxonomy In Introspective Reasoner

Based on different situations between the source and target domains and tasks, transfer learning can be categorized three types, such as inductive transfer learning, transductive transfer learning and unsupervised transfer learning. According to the basic situation, the failure taxonomy in introspective reasoner is constructed as Fig. 3. In this paper we only focus on the transductive transfer learning which can be divided two categories, that is, single source domain single task and multi-domain single task. At the present time, researchers make a lot approaches and algorithms in multi-domain multi task. And also domain adaptation is the key problem in the transductive transfer learning. When transfer learning failure happens we can do blame assignment.

Here we make an example from domain adaptation. Domain adaptation arises in a variety of modern applications where limited or no labeled data is available for a target application. Obtaining a good adaptation model requires the careful modeling of the relationship between $P(D_S)$ and $P(D_T)$. If these two distributions are independent, then the target domain data DT is useless for building a model. On the other hand, if $P(D_S)$ and $P(D_T)$ are identical, then there is no adaptation necessary and we can simply use a standard learning algorithm. In practical problems $P(D_S)$ and $P(D_T)$ are neither identical nor independent.

In text classification, Daume III et al. proposed the Maximum Entropy Genre Adaptation Model, which is called MEGA Model [33]. The failure taxonomy can be illustrated in Table 1. When the transductive transfer learning is failed, the diagnosis is provided as input for the blame assessment with failure categories used to suggest points to avoid negative transfer or low performance transfer.

Table 1. Sample possible sources of failure

Failed knowledge goal	Failure point	Explanation
Only target domain	No knowledge	Source domain choice
Only source domain	No transfer need	Target domain choice
Extra source domain data need	Parameter is on held out	Source domain data not enough
Adapted source domain	Negative transfer	Source domain selection

6   Conclusions

Traditional learning techniques have the assumption that training and test data are drawn from the same data distribution, and thus they are not suitable for dealing with the situation where new unlabeled data are obtained from fast evolving, related but different information sources. This leads to the cross-domain learning problem which targets on adapting the knowledge learned from one or more source domains to target domains. Transfer learning has made a great progress and a lot of approaches and algorithms are presented. But negative transfer learning will cause trouble in the problem solving and it is difficult to avoid. In this paper we have proposed an introspective reasoner to overcome the negative transfer learning.

Introspective learning exploits explicit representations of its own organization and desired behavior to determine when, what, and how to learn in order to improve its own reasoning. According to the transfer learning process, we presented the architecture of introspective reasoner with four components, which is monitoring, quality assessment, blame assignment and solution generation. All of these components are supported by knowledge bases. The knowledge bases are organized in terms of ontology. This paper only focuses on transductive transfer learning. In the future we are working on introspective reasoner for multi-source and multi-tasks transfer learning.

Acknowledgements. This work is supported by Key projects of National Natural Science Foundation of China (No. 61035003, 60933004), National Natural Science Foundation of China (No. 61072085, 60970088, 60903141), National Basic Research Program (2007CB311004).

References

1. Shi, Z.: Intelligence Science. World Scientific, Singapore (2011)
2. Cox, M.: Metacognition in computation: A selected research review. Artificial Intelligence 169(2), 104–141 (2005)
3. Leake, D.B., Wilson, M.: Extending introspective learning from self- models. In: Cox, M.T., Raja, A. (eds.) Metareasoning: Thinking About Thinking (2008)
4. Laird, J.E., Newell, A., Rosenbloom, Soar, P.S.: A Architecture for General Intelligence. Artificial Intelligence (1987)
5. Birnbaum, L., Collins, G., Brand, M., Freed, M., Krulwich, B., Pryor, L.: A model-based approach to the construction of adaptive case-based planning systems. In: Bareiss, R. (ed.) Proceedings of the DARPA Case-Based Reasoning Workshop, pp. 215–224. Morgan Kaufmann, San Mateo (1991)
6. Cox, M.T., Ram, A.: Introspective multistrategy learning on the construction of learning strategies. Artificial Intelligence 112, 1–55 (1999)
7. Fox, S., Leake, D.B.: Using Introspective Reasoning to Refine Indexing. In: Proceedings of the 14th International Joint Conference on Artificial Intelligence, pp. 391–397 (1995)
8. Oehlmann, R., Edwards, P., Sleeman, D.: Changing the Viewpoint: Re-Indexing by Introspective Question. In: Proceedings of the 16th Annual Conference of the Cognitive Science Society, pp. 381–386 (1995)
9. Leake, D.B., Kinley, A., Wilson, D.: Learning to Improve Case Adaptation by Introspective Reasoning and CBR. Case-Based Reasoning Research and Development. In: Aamodt, A., Veloso, M.M. (eds.) ICCBR 1995. LNCS, vol. 1010. Springer, Heidelberg (1995)
10. Zhang, Z., Yang, Q.: Feature Weight Maintenance in Case Bases Using Introspective Learning. Journal of Intelligent Information Systems 16(2), 95–116 (2001)
11. Craw, S.: Introspective Learning to Build Case-Based Reasoning (CBR) Knowledge Containers. In: Perner, P., Rosenfeld, A. (eds.) MLDM 2003. LNCS, vol. 2734, pp. 1–6. Springer, Heidelberg (2003)
12. Cheetham, W., Price, J.: Measures of Solution Accuracy in Case-Based Reasoning Systems. In: Funk, P., González Calero, P.A. (eds.) ECCBR 2004. LNCS (LNAI), vol. 3155, pp. 106–118. Springer, Heidelberg (2004)
13. Arcos, J.L., Mulayim, O., Leake, D.: Using Introspective Reasoning to Improve CBR System Performance. In: AAAI Metareasoning Workshop, pp. 21–28. AAAI Press (2008)
14. Soh, L.-K., Blank, T.: Integrated Introspective Case-Based Reasoning for Intelligent Tutoring Systems. In: Association for the Advancement of Artificial Intelligence. AAAI 2007 (2007)
15. Leake, D.B., Powell, J.H.: Enhancing Case Adaptation with Introspective Reasoning and Web Mining. In: IJCAI 2011, pp. 2680–2685 (2011)

16. Torrey, L., Shavlik, J.: Transfer Learning. In: Handbook of Research on Machine Learning Applications and Trends: Algorithms, Methods, and Techniques, pp. 242–264 (2010)
17. Croonenborghs, T., Driessens, K., Bruynooghe, M.: Learning Relational Options for Inductive Transfer in Relational Reinforcement Learning. In: Blockeel, H., Ramon, J., Shavlik, J., Tadepalli, P. (eds.) ILP 2007. LNCS (LNAI), vol. 4894, pp. 88–97. Springer, Heidelberg (2008)
18. Kuhlmann, G., Stone, P.: Graph-Based Domain Mapping for Transfer Learning in General Games. In: Kok, J.N., Koronacki, J., Lopez de Mantaras, R., Matwin, S., Mladenič, D., Skowron, A. (eds.) ECML 2007. LNCS (LNAI), vol. 4701, pp. 188–200. Springer, Heidelberg (2007)
19. Eaton, E., DesJardins, M.: Knowledge transfer with a multiresolution ensemble learning of classifiers. In: ICML Workshop on Structural Knowledge Transfer for Machine (2006)
20. Carroll, C., Seppi, K.: Task similarity measures for transfer in reinforcement works learning task libraries. In: IEEE International Joint Conference on Neural Net (2005)
21. Bakker, B., Heskes, T.: Task Clustering and Gating for Bayesian Multitask Learning. J. Machine Learning Research 4, 83–99 (2003)
22. Argyriou, A., Maurer, A., Pontil, M.: An Algorithm for Transfer Learning in a Heterogeneous Environment. In: Daelemans, W., Goethals, B., Morik, K. (eds.) ECML PKDD 2008, Part I. LNCS (LNAI), vol. 5211, pp. 71–85. Springer, Heidelberg (2008)
23. Pan, S.J., Yang, Q.: A Survey on Transfer Learning. IEEE Trans. Knowl. Data Eng. 22(10), 1345–1359 (2010)
24. Huang, J., Smola, A., Gretton, A., Borgwardt, K.M., Scholkopf, B.: Correcting Sample Selection Bias by Unlabeled Data. In: Proc. 19th Ann. Conf. Neural Information Processing Systems (2006)
25. Sugiyama, M., Nakajima, S., Kashima, H., Buenau, P.V., Kawanabe, M.: Direct Importance Estimation with Model Selection and its Application to Covariate Shift Adaptation. In: Proc. 20th Ann. Conf. Neural Information Processing Systems (December 2008)
26. Lewis, D.D.: Representation and learning in information retrieval. Ph.D. Dissertation, Amherst, MA, USA (1992)
27. Dai, W., Xue, G., Yang, Q., Yu, Y.: Transferring Naive Bayes Classifiers for Text Classification. In: Proc. 22nd Assoc. for the Advancement of Artificial Intelligence (AAAI) Conf. Artificial Intelligence, pp. 540–545 (July 2007)
28. Dai, W.Y., Xue, G.R., Yang, Q., et al.: Co-clustering based Classification for Out-of-domain Documents. In: Proceedings of 13th ACM International Conference on Knowledge Discovery and Data Mining, pp. 210–219. ACM Press, New York (2007)
29. Ling, X., Dai, W.Y., Xue, G.R., et al.: Spectral Domain-Transfer Learning. In: Proceedings of 14th ACM International Conference on Knowledge Discovery and Data Mining, pp. 488–496. ACM Press, New York (2008)
30. Fuzhen, Z., Ping, L., Hui, X., Yuhong, X., Qing, H., Zhongzhi, S.: Cross-domain Learning from Multiple Sources: A Consensus Regularization Perspective. IEEE Transactions on Knowledge and Data Engineering (TKDE) 22(12), 1664–1678 (2010)
31. Shi, Z., Zhang, S.: Case-Based Introspective Learning. In: Proceedings of the Fourth IEEE International Conference on Cognitive Informatics. IEEE (2005)
32. Dong, Q., Shi, Z.: A Research on Introspective Learning Based on CBR. International Journal of Advanced Intelligence 3(1), 147–157 (2011)
33. Daume III, H., Marcu, D.: Domain Adaptation for Statistical Classifiers. J. Artificial Intelligence Research 26, 101–126 (2006)

PPLSA: Parallel Probabilistic Latent Semantic Analysis Based on MapReduce

Ning Li[1,2,3], Fuzhen Zhuang[1], Qing He[1], and Zhongzhi Shi[1]

[1] The Key Laboratory of Intelligent Information Processing,
Institute of Computing Technology, Chinese Academy of Sciences, Beijing, China
[2] Graduate University of Chinese Academy of Sciences, Beijing, China
[3] Key Lab. of Machine Learning and Computational Intelligence, College of Mathematics and
Computer Science, Hebei University, Baoding, China
{lin,heq}@ics.ict.ac.cn

Abstract. PLSA(Probabilistic Latent Semantic Analysis) is a popular topic
modeling technique for exploring document collections. Due to the increasing
prevalence of large datasets, there is a need to improve the scalability of computation in PLSA. In this paper, we propose a parallel PLSA algorithm called
PPLSA to accommodate large corpus collections in the MapReduce framework.
Our solution efficiently distributes computation and is relatively simple to
implement.

Keywords: Probabilistic Latent Semantic Analysis, MapReduce, EM, Parallel.

1   Introduction

In many text collections, we encounter the scenario that a document contains multiple
topics. Extracting such topics subtopics/themes from the text collection is important for
many text mining tasks[1]. The traditional modeling method is "bag of words" model
and the VSM(Vector Space Model) is always used as the representation. However, this
kind of representation ignores the relationship between the words. For example, "actor"
and "player" are different word in the "bag of words" model but have the similar
meaning. Maybe they should be put into one word which means the topic. To deal with
this problem, a variety of probabilistic topic models have been used to analyze the
content of documents and the meaning of words[2]. PLSA is a typical one, which is
also known as Probabilistic Latent Semantic Indexing (PLSI) when used in information
retrieval. The main idea is to describe documents in terms of their topic compositions.
Complex computation need to be done in the PLSA solving process. There is a need to
improve the scalability of computation in PLSA due to the increasing prevalence of
large datasets. Parallel PLSA is a good way to do this.

MapReduce is a patented software framework introduced by Google in 2004. It is a
programming model and an associated implementation for processing and generating
large data sets in a massively parallel manner [5,9]. Users specify a map function that
processes a key/value pair to generate a set of intermediate key/value pairs, and a

Z. Shi, D. Leake, and S. Vadera (Eds.): IIP 2012, IFIP AICT 385, pp. 40–49, 2012.

reduce function that merges all intermediate values associated with the same intermediate key [5]. MapReduce is used for the generation of data for Google's production web search service, sorting, data mining, machine learning, and many other systems [5].

Two kinds of parallel PLSA with MapReduce have been propose in [10], which are P^2LSA and P^2LSA+ respectively. In P^2LSA, the Map function is adopted to perform the E-step and Reduce function is adopted to perform the M-step. Transferring a large amount of data between the E-step and the M-step increases the burden on the network and the overall running time. Differently, the Map function in P^2LSA+ performs the E-step and M-step simultaneously. However, the parallel degree is still not well. Different from these two algorithms, we have different parallel strategies. We design two kinds of jobs, one is for counting all the occurrences of the words and the other is for updating probabilities.

In this paper, we first present PLSA in Section 2. In Section 3 we introduce the MapReduce framework. We then present parallel PLSA (PPLSA). Section 5 uses large-scale application to demonstrate the scalability of PPLSA. Finally, we draw a conclusion and discuss future research plans in Section 6.

2   Probabilistic Latent Semantic Analysis

2.1   The Main Idea of PLSA

For extracting topics from the text collection, a well accepted practice is to explain the generation of each document with a probabilistic topic model. In such a model, every topic is represented by a multinomial distribution on the vocabulary. Correspondingly, such a probabilistic topic model is usually chosen to be a mixture model of k components, each of which is a topic[1]. One of the standard probabilistic topic models is the Probabilistic Latent Semantic Analysis (PLSA).

The basic idea of PLSA is to treat the words in each document as observations from a mixture model where the component models are the topic word distributions. The selection of different components is controlled by a set of mixing weights. Words in the same document share the same mixing weights.

For a text collection $D = \{d_1,...,d_N\}$, each occurrence of a word w belongs to $W = \{w_1,...,w_M\}$. Suppose there are totally K topics, the topic of document d is the sum of the K topics, i.e. $p(z_1 \mid d), p(z_2 \mid d),..., p(z_K \mid d)$ and $\sum_{k=1}^{K} p(z_K \mid d) = 1$. In other words, each document may belong to different topics. Every topic z is represented by a multinomial distribution on the vocabulary. For example, if the words such as "basketball" and "football" occur with a high probability, it should be considered that it is a topic about " physical education". Each w in document d can be generated as follows. First, pick a latent topic z_k with probability $p(z \mid d)$. Second, generate a word w with probability $p(w \mid z_k)$. Fig.1(a) is the graphic model and Fig.1(b) is the symmetric version with the help of Bayes' rule.

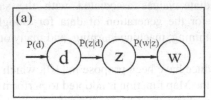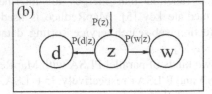

Fig. 1. The graphic model of PLSA

2.2   Solving PLSA with EM Algorithm

The standard procedure for maximum likelihood estimation in PLSA in the Expectation Maximization(EM) algotithm[3]. Acoording to EM algorithm and the PLSA model, the E-step is the following equation.

$$P(z \mid d, w) = \frac{P(z)P(d \mid z)P(w \mid z)}{\sum_{z'} P(z')P(d \mid z')P(w \mid z')} \tag{1}$$

It is the probability that a word w in a particular document d is explained by the factor corresponding to z.

The M-step re-estimation equations are as follows.

$$P(w \mid z) = \frac{\sum_d n(d, w)P(z \mid d, w)}{\sum_{d,w'} n(d, w')P(z \mid d, w')} \tag{2}$$

$$P(d \mid z) = \frac{\sum_w n(d, w)P(z \mid d, w)}{\sum_{d',w} n(d', w)P(z \mid d', w)} \tag{3}$$

$$P(z) = \frac{1}{R}\sum_{d,w} n(d, w)P(z \mid d, w), \quad R \equiv \sum_{d,w} n(d, w) \tag{4}$$

3   MapReduce Overview

MapReduce is a programming model and an associated implementation for processing and generating large data sets. As the framework showed in Fig.2, MapReduce speci-fies the computation in terms of a map and a reduce function, and the underlying runtime system automatically parallelizes the computation across large-scale clusters of machines, handles machine failures, and schedules inter-machine communication to make efficient use of the network and disks.

Essentially, the MapReduce model allows users to write Map/Reduce components with functional-style code.

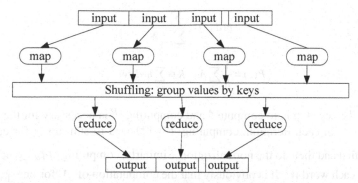

Fig. 2. Illustration of the MapReduce framework: the "map" is applied to all input records, which generates intermediate results that are aggregated by the "reduce"

Map takes an input pair and produces a set of intermediate key/value pairs. The MapReduce library groups together all intermediate values associated with the same intermediate key and passes them to the reduce function [6]. That is, a map function is used to take a single key/value pair and outputs a list of new key/value pairs. It could be formalized as:

$$map :: (key_1, value_1) \Rightarrow list(key_2, value_2)$$

The reduce function, also written by the user, accepts an intermediate key and a set of values for that key. It merges together these values to form a possibly smaller set of values. Typically just zero or one output value is produced per reduce invocation. The intermediate values are supplied to the users reduce function via an iterator. This allows us to handle lists of values that are too large to fit in memory. The reduce function is given all associated values for the key and outputs a new list of values. Mathematically, this could be represented as:

$$reduce :: (key_2, list(value_2)) \Rightarrow (key_3, value_3)$$

The MapReduce model provides sufficient high-level parallelization. Since the map function only takes a single record, all map operations are independent of each other and fully parallelizable. Reduce function can be executed in parallel on each set of intermediate pairs with the same key.

4   Parallel PLSA Based on MapReduce

As described in section2, EM algorithm is used to estimate parameters of the PLSA model. Our purpose is to compute $P(w|z)$, $P(d|z)$ and $P(z)$. The whole procedure is an iteration process. We set A to represent $n(d,w)P(z|d,w)$, and so the equation (2) to equation (4) can be rewritten as follows.

$$P(w|z) = \frac{\sum_d A}{\sum_{d,w'} A} \tag{5}$$

$$P(d \mid z) = \frac{\sum_w A}{\sum_{d',w} A} \qquad (6)$$

$$P(z) = \frac{1}{R} \sum_{d,w} A, \quad R \equiv \sum_{d,w} n(d,w) \qquad (7)$$

Note that the key step is to compute A and computing R is necessary for the computation of $P(z)$. In each iteration, computing $P(w \mid z)$ need to sum up A for each document d first and then do the normalization. Similarly, computing $P(d \mid z)$ need to sum up A for each word w. It is obviously that the computation of A for one document is irrelevant with the result of another one in the same iteration. For the same reason, computation of $\sum_w A$ and R is in the same situation. So the computation of A, $\sum_w A$ and R could be parallel executed. Therefore, we design two kinds of MapReduce job, one is to compute $\sum_d A$ and $\sum_w A$, the other is for computing R.

For the job computing $\sum_d A$ and $\sum_w A$, the map function, shown as **Map1**, performs the procedure of computing A and $\sum_w A$ for each document and thus the map stage realizes the computation of A and $\sum_w A$ for all the documents in a parallel way. The reduce function, shown as **Reduce1**, performs the procedure of summing up A to get $\sum_d A$.

For the job computing R, the map stage realizes the computation of $\sum_w n(d,w)$ for each document, and $R \equiv \sum_{d,w} n(d,w)$ is gotten in the reduce stage. The map function and reduce function are shown as **Map2** and **Recuce2** respectively.

As the analysis above, the procedure of PPLSA is shown in the following.

Procedure PPLSA

---

1. Input: Global variable numCircle, the number of latent topics
2. Initialize p(z), p(w|z), p(d|z);
3. The job computing R is carried out;
4. **for** (circleID = 1; circleID <= numCircle; circleID++)
5. /*numCircle is the max number of interations.*/
6. The job computing $\sum_d A$ and $\sum_w A$ is carried out.
7. Compute $\sum_{d,w} A$;
8. Update p(z), p(w|z), p(d|z);
9. **end**
10. Output p(z), p(w|z), p(d|z).

Map1   Map(*key, value*)

Input: pz, pdz, pwz, the offset *key*, the sample *value*
Output: <*key', value'*> pair

1. Initialize arrayzd[] which is used for storing $\sum_w A$
2. **for** (int i = 1;i<strarry1.length;i++) /*strarry1.length= the number of words in *value**/
3.   compute p(z)*p(d|z)*p(w|z) for each topic and proceed the normalization
4.   **for**(int j = 0;j<number of topics;j++)
5.   tmp= n(d,w)P(z|d,w)
6.   Output(**key'**=j+"-w-"+(WordID-1),**value'**=tmp);/*for computing P(w|z) */
7.   arrayzd[j] += tmp;   /*compute $\sum_w n(d,w)P(z|d,w)$ */
8.   **end**
9. **end**
10. **for** (i=0;i<number of topics;i++)
11.   Ouput(**key'**=j+"-d-"+(DocID-1),**value'**=arrayzd[j]); /*for computing P(d|z) */
12. **End**

Reduce 1   Reduce (*key, value*)

1. sum=0;
2. **for**(Text value:values)
3.   sum+=*value* /* sum up the values with the same key */
4. **end**
5. output(key,sum);

Map2   Map(*key, value*)

Input: the offset *key*, the sample *value*
Output: <*key', value'*> pair

1. nCount=0;
2. **for** (int i = 1;i< the number of words in *value* ;i++)
3.   get each word frequence freq[i];
4.   nCount+=freq[i];
5. **end**
6. Output(**key'**=a random number belong (0,100), **value'**=nCount)

Reduce 2   Reduce (*key, value*)

```
1. sum=0;
2. for(Text value:values)
3.     sum+=value;
4. end
5. output(key,sum);
```

---

5   Experimental Analysis

In this section, we evaluate the performance of PPLSA. Performance experiments were run on a cluster of 4 computers, each of which has four 2.8GHz cores and 4GB memory. Hadoop version 0.20.2 and Java 1.5.0_14 are used as the MapReduce system for all the experiments. Experiments were carried on 10 times to obtain stable values for each data point.

5.1   The Datasets

We performed experiments on two datasets: a subset of the TREC AP corpus containing 2246 documents with 10,473 unique terms and a dataset extracted from internet about stock, containing 316 html documents with 27,925 terms.

5.2   The Evaluation Measure

We use scaleup, sizeup and speedup to evaluate the performance of PPLSA algorithm.

Scaleup: Scaleup is defined as the ability of an m-times larger system to perform an m-times larger job in the same run-time as the original system. The definition is as follows.

$$Scaleup(data, m) = \frac{T_1}{T_{mm}} \tag{8}$$

Where, T_1 is the execution time for processing data on 1 core, T_{mm} is the execution time for processing m*data on m cores.

Sizeup: Sizeup measures how much longer it takes on a given system, when the dataset size is m-times larger than the original dataset. It is defined by the following formula:

$$Sizeup(data, m) = \frac{T_m}{T_1} \tag{9}$$

Where, T_m is the execution time for processing m*data, T_1 is the execution time for processing data.

Speedup: Speedup refers to how much a parallel algorithm is faster than a corresponding sequential algorithm. It is defined by the following formula:

$$Speedup = \frac{T_1}{T_p} \tag{10}$$

Where, p is the number of processors, T_1 is the execution time of the algorithm with one processor, T_p is the execution time of the parallel algorithm with p processors.

5.3   The Performance and Analysis

To demonstrate how well the PPLSA algorithm handles larger datasets when more cores of computers are available, we have performed scaleup experiments where the increase of the datasets size is in direct proportion to the number of computer cores in the system. We ran the datasets which are 60-times, 120-times and 240-times of the original ones on 4, 8, 16 distributed machines respectively. The scaleup performance of PPLSA is shown in Fig.3.

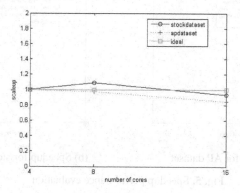

Fig. 3. Scaleup performance evaluation

We have plotted scaleup which is the execution time normalized with respect to the execution time for 4 machines. Clearly the PPLSA algorithm scales very well, being able to keep the execution time almost constant as the dataset and machine sizes increase.

To measure the performance of sizeup, we fix the number of cores to 4, 8 and 16 respectively. Fig.4 shows the sizeup results on different cores. The results show sublinear performance for the PPLSA algorithm, the program is actually more efficient as the dataset size is increased. Increasing the size of the dataset simply makes the noncommunication portion of the code take more time due to more I/O and more documents processing. This has the result of reducing the percentage of the overall time spent in communication. Since I/O and CPU processing scale well with sizeup, we get sublinear performance.

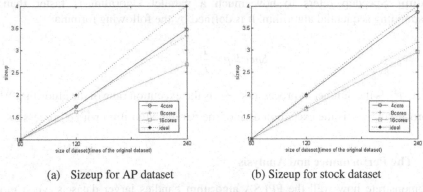

(a)   Sizeup for AP dataset   (b) Sizeup for stock dataset

Fig. 4. Sizeup performance evaluation

We used 4 cores as the baseline to measure the speedup of using more than 4 cores. The speedup performances are shown in Fig.5.

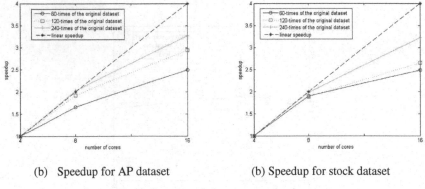

(b)   Speedup for AP dataset   (b) Speedup for stock dataset

Fig. 5. Speedup performance evaluation

From Fig.5 we can see that PPLSA can achieve linear speedup when the core is small. However, the improvement becomes gradually undramatic as the number of processors grows. This is expected due to both the increase in the absolute time spending in communication between machines, and the increase in the fraction of the communication time in the entire execution time. When the fraction of the computation part dwindles, adding more machines (CPUs) cannot improve much speedup.

Moreover, the speedup performance shows better on the large datasets. This is an artifact of the large amount of data each node processing. In this case, computation cost becomes a significant percentage of the overall response time. Therefore, PPLSA algorithm can deal with large datasets efficiently.

6   Conclusion

In this paper, we presented a parallel implementation of PLSA based on MapReduce. We use scaleup, sizeup and speedup to evaluate the performance. The experimental results show that it scales well through the machine cluster and has a nearly linear speedup. However, due to the limit memory, the algorithm do not work well when the dataset are too large. In the future, we will look into strategies to solve this problem.

Ackonwledgement. This work is supported by the National Natural Science Foundation of China (No. 60933004, 60975039, 61175052, 61035003, 61072085, 60903088), National High-tech R&D Program of China (863 Program) (No.2012AA011003).

References

1. Mei, Q., Zhai, C.: A note on EM algorithm for probabilistic latent semantic analysis. In: Proceedings of the International Conference on Information and Knowledge Management, CIKM (2001)
2. Steyvers, M.: Probabilistic Topic Models. In: Landauer, T., McNamara, D., Dennis, S., Kintsch, W. (eds.) Latent Semantic Analysis: A Road to Meaning. Laurence Erlbaum
3. Hofmann, T.: Probabilistic Latent Semantic Analysis. In: Proc. of 15th Conference on Uncertainty in Artificial Intelligence, pp. 289–296. Morgan Kaufmann, San Francisco (1999)
4. Hofmann, T.: Unsupervised learning by probabilistic latent semantic analysis. Machine Learning 42(1), 177–196 (2001)
5. Dean, J., Ghemawat, S.: MapReduce: Simplified Data Processing on Large Clusters. In: Proc. of Operating Systems Design and Implementation, San Francisco, CA, pp. 137–150 (2004)
6. Ghemawat, S., Gobioff, H., Leung, S.: The Google File System. In: Symposium on Operating Systems Principles, pp. 29–43 (2003)
7. Hadoop: Open source implementation of MapReduce (June 24, 2010), http://hadoop.apache.org
8. Han, J., Kamber, M.: Data Mining, Concepts and Techniques. Morgan Kaufmann (2001)
9. Lammel, R.: Google's MapReduce Programming Model - Revisited. Science of Computer Programming 70, 1–30 (2008)
10. Jin, Y., Gao, Y., Shi, Y., Shang, L., Wang, R., Yang, Y.: P²LSA and P²LSA+: Two Paralleled Probabilistic Latent Semantic Analysis Algorithms Based on the MapReduce Model. In: Yin, H., Wang, W., Rayward-Smith, V. (eds.) IDEAL 2011. LNCS, vol. 6936, pp. 385–393. Springer, Heidelberg (2011)

Analysis on Limitation Origins of Information Theory

Yong Wang[1], Huadeng Wang[1], and Qiong Cao[2]

[1] School of Computer Science and Engineering, Guilin University of Electronic Technology,
Guilin 541004, Guangxi Province, China
[2] Computer Science and Engineering, Chongqing University of Technology,
Chongqing 400050, China
hellowy@126.com

Abstract. The limitations of Shannon information theory are pointed out from new perspectives. The limitations mainly exist in the neglects of the information reliability and completeness. The significances of the information reliability to the information measurements are further illustrated through example analysis. It is pointed out that such limitations originate from neglects of multi-level information uncertainties, uncertainty of the model and other objects of information system, and insufficient knowledge on uncertainties of probability values.

Keywords: information theory, communication, reliability, model, uncertainty, probability.

1   Introduction

Shannon information theory is aimed at the communication issues, and is not quite necessarily applicable to the information issues in the reality [1], and his infor-mation theory is called the special information theory by the later researchers. Aimed at the general information, some researchers have developed comprehensive information theory, generalized information theory, unified information theory etc [2-5], in which some limitations of the special information theory are identified, but no one of these information theories can eliminate all the limitations of information theories. In this paper, we attempt to analyze the limitations of the special information theory from some new perspectives.

2   Limitations of Shannon Information Theory in the Reality

The currently recognized limitations of Shannon information theory are mainly as follows: Firstly, only the random uncertainty is considered in his information theory, while the uncertainties such as the limitations of sets in information expressions and the information fuzziness etc are not considered. Aimed at this issue, some research-ers have developed the theories of fuzzy sets and rough sets. Secondly, neither the semantic nor the pragmatic aspects are considered in Shannon information theory,

Z. Shi, D. Leake, and S. Vadera (Eds.): IIP 2012, IFIP AICT 385, pp. 50–57, 2012.

which were considered as the main origin of the limitation of information theory by some researchers [2, 3]. ZHONG Yixin developed his comprehensive information theory which includes syntactic information, semantic information and pragmatic information [2].

We find out the following limitations of information theory:

Firstly, the information reliability is not considered in information theory, while most information in the reality is unreliable. The reliability of information is the pre-condi-tion to the information values, for example, the reliability of the intelligence informa-tion is very important. The reliability is a key indicator of information. however, it is not enough considered in information theory, which mainly emphasizes the uncertainty of information and the reliability of information transmission.

Secondly, the information completeness is not fully considered in information theory, while in the reality information is always incomplete (insufficient) and needs to be fused. When the more perfect information is unavailable, people cannot help but expediently take partial information temporarily as complete information. In this case, the information that may be deemed that the incomplete information is unreliable compared with the complete information [6].

Thirdly, in information theory, some simple channels may be combined into a channel through parallel or serial connections, for example, the channel matrix of two simple serial channels may be directly multiplied into one channel. However, the complex and multi-level information transmissions are not considered in information theory. for example, the information may be transmitted from one source to an intermediate sink, and forwarded from the intermediate sink to a final sink, the information may be fused from many sources to many sinks, the transition probability matrix may be uncertain, furthermore, during such transmissions, the information may be converted, during multi-level transmissions or complex transmission, multiple uncertainty may occur. In that case, not only the message or code is uncertain, but the probabilities that express the uncertainty are uncertain. The information in the reality often needs to be conveyed through this kind of multiplex transmissions, during which multiple uncertainties may occur. If the above-mentioned fuzzy sets etc are also considered, such multiple uncertainties will become more complicated. The complexity of the parameters such as the channel matrix transmission probability etc is not considered in information theory, while in the reality, such transmission property may not be necessarily maintained certain and unchanged, it may be a random variable or more complex variable. Therefore the uncertainty of information system model should be considered.

Fourthly, the objects studied in information theory are the communications, whose transmission signals are certain. While in the reality, there are many uncertainties. During communications, it does not matter to define the information as something which reduces uncertainty. However, when confronted with the intrinsically uncertain information, the information is distorted if its uncertainty is eliminated. And the trifles are attended but the essentials are neglected during such eliminations. The quantum information theory serves as one kind of popularizations of the classical information theory. while the quanta bytes may include 0, 1 and superposed numbers between 0 and 1. In the reference [4], the information reliability is considered, and it is pointed out that the information reliability is more important than its certainty. In the reality, people tend to select more uncertain but more reliable information, rather than select-ing more unreliable but more certain information.

Fifthly, the conditions in information theory are comparatively simple, and mostly expressed in the probability of conditions. However, int the reality the conditions of information are often very complicated. For example, the given conditions may be knowledge, laws etc. Based on a known and prior probability, another law may be acquired, but it is impossible to easily calculate any corresponding probability of conditions through this law.

Sixthly, the known information is expressed with the prior probability in information theory, however in the reality, a lot of known information may not be expressed with the prior probability, for example, an unknown variable may be included, which may be a restrictive condition, or may be a law.

Seventhly, as the semantic aspects are not considered in information theory, it is not considered that the information may be mutually incompatible and contradictive. In the reality, a large quantity of information may be inconsistent and contradictive each other.

3   Analyses on Information Examples

Considering the above limitations, and in order to make them look more obvious, we analyzed the following examples.

Example 1: If a sender sends a receiver an important intelligence saying: "Some two nations are going to start a war". When speaking from an information theory perspective, such information itself is the information that involves a large quantity of information. But if this information is not true, it may cause great disasters. People tend to emphasize more on the information reliability, if the information is unreliable, further communications become insignificant, and the information becomes valueless. In information theory, efforts are extended only to ensure that a receiver receives the original information. while it is not studied whether the information from an informa-tion source is truly reliable. If the above-exemplified information is unreliable, then the probability of a war between such two nations may be not at 1, but a number be-tween 0 and 1. in this case, the probability value itself is randomly uncertain. In order to avoid confusing the probability uncertainty and the information uncertainty, it may be assumed that the sent intelligence is rewritten as: "the probability of a war to be started between some two nations is at 0.7". If this intelligence is not absolutely relia-ble, then this probability value is most likely not at 0.7, but close to 0.7. in this case, the probability itself is not certain, and the value at 0.7 may be just an average. This kind of reliability issues may be more complex. for example, in addition to that the war possibility is unreliable, it is possible that the subjects are unreliable. for example, maybe the war is possible between three nations instead of two nations, or between two persons etc.

Example 2: The probability of an event m is determined by some conditions, supposing that these conditions are $c_1, c_2, \ldots, c_n$, and supposing that the probability of the event m may be expressed as

$$P(m) = f(c_1, c_2, \ldots, c_n)$$

If the only condition c_1 is unknown to us, while all of the others are known, maybe the average value of P(m) can be calculated according to probability distributions of the condition c_1. Based on specific cases, the condition c_1 is known and certain. But when this condition is unknown, we can only expediently replace the true probability with the average probability value calculated when this condition is unknown. In this case, the two probability values are close, but they are not equal. When the average value is used to replace a specific value, the information becomes obviously unreliable. Therefore, when the information in this example is incomplete, it may become an information unreliability issue. Generally speaking, the more conditions become known, the more complete the conditions are. and the probability well become more closer to the probability when the conditions are fully complete, therefore the information becomes more reliable. In this example, the case may become more complex. For example, the probability of m still may be a random variable even when all of the conditions are certain. It is like the inaccurate measurement principle in quanta dynamics, which is not caused by existences of hidden parameters or the incompleteness of quanta dynamics, but is one kind of natural uncertainties.

Example 3: The occurrence probability of an event m may be tested through experiments. When the known condition is t: "the experiment result is P(m|t) =0.7", we can't affirm P(m|t), which is the occurrence probability of the event m, we can only be sure that P(m|t) is a random variable close to 0.7. If this random variable has to be replaced with the fixed value 0.7, this value will cause reliability issues. The joint probability distributions and condition probabilities frequently appear in the probability theory and information theory. but it is not considered that under many conditions, the probability value itself in the condition probabilities or joint probability distributions may be a random variable, or partially unknown, or even completely unknown, and then the uncertain probability value will be improperly taken as a certain probability value.

Example 4: A student always gets very high scores with an excellence probability at 0.875, but he gets sick right before an exam. As his learning is delayed due to his sickness, the probability of his excellence may reduce to 0.75. Before we know that the student gets sick influencing his learning, the uncertainty of the prior probability that we known is less than the uncertainty of the posterior probability after we know that his learning is influenced. From an information theory perspective, the latter is reversely reduced in the information contents. In the reality, people will not select an prior probability on the ground that its uncertainty is low, but will select a posterior probability determined when the conditions become more complete as it is closer to the reality. This example shows that measurements on information per uncertainties are limited, and such measurements are not applicable to the uncertain issues. in addition, it also shows that the information needs to be specified in measurements on its reliability and completeness, and that this specification is more important than the information entropy used to measure the information uncertainties.

Example 5: A receives from B an intelligence saying: "it is 99% sure that the enemy will attack us on tomorrow morning". later, A receives the same intelligence from C. If seen from information theory perspective, the uncertainty of the question "will the enemy attack us on tomorrow morning?" is the same in such two cases. therefore the information contents equal to each other, while it seems that C does not provide any new information. However, people will still feel that they receive the information from C, and this kind of information makes A feel more certain "it is 99% sure that the enemy will attack us on tomorrow morning". this example further shows that the information reliability shall be taken as one specification to measure the information.

Example 6: When the information that "all events occur at equivalent probabilities" is received, and if speaking from what contents are included in this sentence, or the question "how are the occurrence probabilities of all events distributed?" the uncertainties hereof are eliminated. However, if speaking from what events will occur, it is impossible that such information becomes more certain. the information contents cannot be increased, but only be decreased. This point shows that the information contents are only measured on the information uncertainties, while the uncertainties of the questions derived from this information are not certainly related to the information contents. therefore, the measurements through an information entropy are applicable only to limited scopes, but not to the daily information issues.

Example 7: Brillouin once developed one paradox: If one information paragraph is sent in a text whose last part tells the receiver otherwise that all of the previous information is not true. In this case, is any information transmitted? Brillouin recommends that attentions be made on the negative information. In this paradox, there are two information parts, the second of which negates the first. I believe that such two information parts contradict each other and their reliabilities are both relative. But in general cases, if the sender is honest and not making purposeful jokes, then according to the context and reasoning analysis, the possibility of the latter information part being true becomes much higher. In this case, when the former and the latter information parts are summed up, it may be deemed that no useful information has been sent. Of course, the possibility that the sender made a mistake in the latter part or sent the mistaken information cannot be absolutely eliminated. In information theory, the associations between various information symbols are expressed with redundancies. but in general cases, this information may be integrally looked as one kind of redundancy codes, namely the codes that are impossible to be sent (the codes with a sending probability at 0). As time delays exist during communications (if no time delays exist. in general cases, the information will be self appropriate), such delays enable sending the information that is originally incompatible or impossible to be sent through the complete information.

Through the above analyses, some limitations of information theory are revealed to provide foundations for identifying the origins of information theory limitations. Of course, there are many other limitations, which will not be described here further.

4   Limitations Origins of Information Theory

As analyzed through the above examples, it may be concluded that in information theory, the information reliability is not considered, while the information reliability is a very important indicator. During communications, as the information is certain, a certain relationship exists between eliminations of uncertainties and increases of reliabilities. In fact, it is easy for us to eliminate the uncertainty. while in Shannon information theory, an uncertainty is eliminated on the foundation that the information reliability and completeness are ensured. for example, error correction codes are used to correct errors, and the posterior probability is used to strengthen the information completeness. If the information certainty is taken as the sole specification, and the information reliability is disregarded. then the probability of an event may be randomly determined as 1, while the probabilities of the other events may be randomly determined as 0. Furthermore, if we take the information certainty as the primary target to be considered, and the reliability as the secondary one. then we may also designate the probability of the most probable event as 1, and of the other events as 0. In this case, the certainty is firstly ensured. while the reliability is also satisfied to certain extents. If so, information theory and the information processing become quite simple. obviously in the reality, people do not treat them in this way. According to many of the above analysis, the reliability is one of the primary information specifications.

The above information reliability, information completeness, and the incompliance of classical sets with the reality may be all summed up into the neglects of the multi-level information uncertainties. For example, during example analysis, we discovered the unreliable information, whose expression itself was not fixed, and whose probability value might be a random variable. while the incomplete information is also similar. The non-classical sets in the categories of fuzzy sets and rough sets then may be deemed to have been caused by one set that includes uncertain objects. For example, in rough sets, an object a may or may not belong to the set X. then a random uncertainty exists as to whether the object a belongs to the set X. When some uncertainties are accumulated on the original uncertainties in information theory, the multi-level uncertainties will occur. The uncertainties in this case may include the other uncertainties in more forms, in addition to the random uncertainty and the fuzzy uncertainty. they also include some uncertainties caused by incomplete restrictive conditions. It can be seen that the neglects of the multi-level information uncertainties are important origins of information theory limitations. while the neglect of the information reliability is also an important reason why it is impossible for information theory to be widely applied. Whereas all of the information is rarely reliable and complete, so we may attribute the information reliability and completeness to the information relativity. In fact, it is very rare that people in the reality receive the completely reliable information, they can only expediently apply the comparatively reliable information. when the more reliable information becomes available, people will apply the more reliable information to replace the previously experienced information. As the reliability is also related to the uncertainty of the probability values, the information reliability may be measured with reference to the measurements by Shannon on the information uncertainties. However, the calculations of the probability uncertainties will be more complex than the calculations of the information entropies, as the probabilities have to satisfy more restrictive conditions.

From Brillouin paradox, example 4 and example 6, we can know that the formula of Shannon information contents is used only to measure the minimum transmission when the special information is being transmitted. during communications, it also reflects the reliability to some extents. However, when the meanings correspondingly derived from the information are considered, neither the information contents nor the information meaning certainties are certainly related to the sizes of the information contents. while the certain information may end up increasing the uncertainties of the other information. Some information may even increase the uncertainties of some information and meanwhile decrease the uncertainties of the other information. for example, the information, "it rains today" may decrease the uncertainty of "are the roads wet today?", but meanwhile increase the uncertainty of "will the students come late today?". Obviously, it deserves our deliberations on whether the negative infor-mation needs to be considered.

Of course, information theory is highly similar to the information issues in the reality. the methods applied in information theory provide very valuable references for re-searches on the information issues in the reality (including researches on the informa-tion reliabilities). For example, the mutual contradictions in Brillouin para-dox are very similar to the inconsistence caused by the mistaken error correction codes. simi-larly, the most probable events may be taken as the events that really occur. while when all probabilities are close to each other, the information may be "deleted" in the similar way how the information channels are deleted.

Information theory limitations originate from the nature that information theory is oriented to communications issues, and are the intrinsic limitations of information theory models. Of course, such limitations are also related to the limitations of the probability theory. Due to insufficient researches on the randomness and multi-level random uncertainties of the probability values, people are likely to fall into the thinking tendencies that "the probabilities (including distributions of the joint probabilities) are the certain values which never happen to be random variables in any case", that "the condition probabilities may be determined whenever the conditions are given" etc. while these thinking tendencies are only applicable to some parts of the probability theory issues in the reality. As these restrictive conditions in information theory can better satisfy the communication demands, consequently information theory has been successfully applied in the communication fields. If we wish to popularize it in the general information fields, then corresponding restrictive conditions need to be removed according to its limitations.

Most of the limitations is rooted in the overlooking of the ubiquitous uncertainty of the objects about the information system, such as uncertainty of model, parameters, algorithms, functions, participator and process. We can simplify the information problem by using certain object to replace uncertain objects, this is very beneficial for research and can transform the reality problems to mathematical model so as to be processed by mathematical method, but this poses limitations and restricts the freedom of problems.

5   Conclusion

This paper analyzes some limitations of Shannon information theory. Through the researches on its limitations, on one hand, its applicable scopes can be determined to avoid abusing information theory and to apply information theory in suitable fields. On the other hand, directions may be provided for popularizing and referring to information theory. For example, the information expressions may be improved according to the limitations that the probabilities in the information expressions are fixed values, but not random variable or more complex variable, so as to accommodate the needs to research the information reliabilities, and to further popularize information theory. In addition, new directions are also proposed for the developments of the probability theory. Information theory is also similar, in many aspects, to the information issues in the reality, such as the afore-mentioned considerations of the information reliabilities and completeness. And some of these issues can be solved by referencing information theory.

References

1. Shannon, C.E.: A mathematical theory of communication. Bell System Technical Journal 27, 379–429, 623–656 (1948)
2. Zhong, Y.: Principles of information science. Fujian People's Publishing House, Fuzhou (1988)
3. Lu, C.: Generalized information theory. Publishing House of, University of Science and Technology of China (1993)
4. Klir, G.J.: Uncertainty and Information: Foundations of Generalized Information Theory. Wiley (2006)
5. Federico, D.: Contributions towards a unified concept of information. Doctorial Thesis of University Bern (1995)
6. Wang, Y.: Analysis and Betterment of Shannon's Information Definition. Journal of Information 27(8), 57–60 (2008)

Intelligent Inventory Control: Is Bootstrapping Worth Implementing?

Tatpong Katanyukul[1], Edwin K.P. Chong[2], and William S. Duff[3]

[1] Embedded System Research Group and Department of Computer Engineering,
Faculty of Engineering, Khon Kaen University, Khon Kaen, Thailand
tatpong@kku.ac.th
[2] Department of Electrical and Computer Engineering, College of Engineering,
Colorado State University, Fort Collins, CO, USA
edwin.chong@colostate.edu
[3] Department of Mechanical Engineering, College of Engineering, Colorado State University,
Fort Collins, CO, USA
bill@engr.colostate.edu

Abstract. The common belief is that using Reinforcement Learning methods (RL) with bootstrapping gives better results than without. However, inclusion of bootstrapping increases the complexity of the RL implementation and requires significant effort. This study investigates whether inclusion of bootstrapping is worth the effort when applying RL to inventory problems. Specifically, we investigate bootstrapping of the temporal difference learning method by using eligibility trace. In addition, we develop a new bootstrapping extension to the Residual Gradient method to supplement our investigation. The results show questionable benefit of bootstrapping when applied to inventory problems. Significance tests could not confirm that bootstrapping had statistically significantly reduced costs of inventory controlled by a RL agent. Our empirical results are based on a variety of problem settings, including demand correlations, demand variances, and cost structures.

Keywords: approximate dynamic programming, inventory control, reinforcement learning, bootstrapping, eligibility trace, intelligent agent.

1 Introduction

Inventory management is one of the major business activities. A well managed inventory can help a business stay competitive by keeping its cash flow at a controllable level. A stochastic multiperiod inventory problem—one of the most common inventory problems—can be modeled as a Markov Decision Problem (MDP). Approximate Dynamic Programming (ADP) is a method to solve practical Markov decision problems. Most previous studies on ADP applied to inventory management focused on learning scheme, which is also known as Reinforcement Learning (RL). Due to its effectiveness and its link to mammal learning processes, temporal difference learning, or sometimes called "one-step temporal-difference learning", TD(0), is one of the most widely studied RL methods.

Z. Shi, D. Leake, and S. Vadera (Eds.): IIP 2012, IFIP AICT 385, pp. 58–67, 2012.
© IFIP International Federation for Information Processing 2012

Eligibility trace has been used to bootstrap the learning process of TD(0). The integration of the eligibility trace into the temporal difference learning leads to the TD(λ) method. The success of TD(λ) in many applications has lead to a general belief (see, e.g., Prestwich et al. (2008)) that using TD(λ) with λ between 0 and 1 will give better results than TD(0). However, the effort required to implement TD(λ) is considerably higher than to implement TD(0). The value and benefit of using TD(λ) compared to TD(0) have never been investigated especially for inventory management.

Our study investigates the application of Sarsa(λ), a widely studied implementation of the TD(λ) method. To supplement the investigation, our study develops the Direct Credit Back method (DCB)—the bootstrapped version of the Residual Gradient method (RG). Finally, we evaluate both bootstrapping methods—Sarsa(λ) and DCB—and confer to their non-bootstrapped counterparts, Sarsa and RG. The underlying methods are evaluated with various structures of inventory problems. Two other inventory control methods—Look-Ahead and Rollout—are also included in our study.

The findings here provide a practical approach to apply an ADP method for an inventory problem. The results reveal questionable benefits of bootstrapping. The understanding exposed here will help promote efficient inventory management and aid in the transfer of intelligent system research into practice.

2 Literature Review

ADP has been introduced recently into inventory management research. Kim et al. (2005), Kim et al. (2008), Kwon et al. (2008), and Jiang and Sheng (2009) implemented TD(0) with a look-up table as a cost-to-go approximation—a crucial component of most RL methods. A look-up table is simple to implement and works well with TD(0). However, a look-up table suffers from a scalability issue: its memory requirement grows exponentially as the dimension of the problem space grows.

Some studies, e.g., Van Roy et al. (1997) and Shervais et al. (2003), apply TD(0) with nonlinear function approximation, such as artificial neural network, to mitigate a scalability issue. However, applying TD(0) with nonlinear function approximation requires a high level of expertise in both application and techniques and it might result in divergence leading to instability of the control. Baird (1995) proposed Residual Gradient method (RG) to be used with nonlinear function approximation. However, RG is reported to underperform TD(0) in overall.

In addition to extend RL method with scalable function approximation, bootstrapping is used to speed up the learning process of TD(0) and leads to a more general method, denoted TD(λ). Successes of applying TD(λ)—the most famous work is Tesauro (1994)—lead to common belief of the benefit of bootstrapping. However, due to complexity of implementation and extra computational costs, there is only work of Prestwich et al. (2008) related to bootstrapping and inventory management. Prestwich et al. (2008) studied the viability of combining Sarsa(λ) and Noisy Genetic Algorithm by using the Cultural Algorithm, introduced by Reynolds (1994). They claimed the viability of their approach for partially observable Markov decision problems.

Among previous authors applying ADP to inventory problems, many authors have used TD(0) or methods related to TD(0), but none[1] has investigated the effectiveness of bootstrapping. Leng et al. (2009) also mentioned that the mechanism of Eligibility Trace in ADP with function approximation has not been sufficiently investigated.

Inspired by the development of eligibility trace to bootstrap TD(0), we develop an extension to RG based on bootstrapping, called "Direct Credit Back" (DCB). The new method DCB provides a contribution in its own right as well as supplements the investigation of an effect of bootstrapping in applying RL to inventory management.

3 Background

Sarsa. Sarsa, as discussed by Sutton and Barto (1998), uses an approximate state-action cost $Q(s_t, a_t)$, often called the Q-value, to determine an action and updates $Q(s_t, a_t)$ based on TD(0). The equations for Sarsa are $Q(s,a) \leftarrow Q(s,a) + \beta\, \psi(s,a)$ where $Q(s, a)$ approximates the state-action cost of state s and action a; β is a Sarsa parameter, called the learning rate; the temporal difference $\psi(s,a) = c(s, a) + \alpha\, Q(s', a') - Q(s, a)$ when $c(s, a)$ is a single period cost and action a_t is determined by a chosen policy based on state s_t and the current Q-value.

Eligibility Trace. TD(0) uses newly observed information, $c(s_t, a_t)$, to update a Q-value of the most recent state-action. To utilize trajectory information, Eligibility Trace (ET) is developed. ET[2] updates its values, such that when (s, a) is visited, eligibility variable $e(s, a) \leftarrow 1$, $e(s, \hat{a}) \leftarrow 0$ for each action $\hat{a} \neq a$, and $e(\hat{s}, b) \leftarrow \alpha\, \lambda\, e(\hat{s}, b)$ for each state $\hat{s} \neq s$ and each action $b \in A(\hat{s})$ where α is a discount factor; λ is an eligibility factor; and $A(\hat{s})$ is a feasible action set for the given state $\hat{s}$. Sarsa(λ) is an implementation of TD(λ), whose the update equation is $Q(s, a) \leftarrow Q(s, a) + \beta\, \psi(s, a)\, e(s, a)$ for all s and a.

Residual Gradient Method. Sarsa was originally designed to be implemented using a look-up table and later was extended for use with an approximation function. Although there are some success, associated risk of instability is commonly known, see Baird (1995), Barreto and Anderson (2008), and Maei et al. (2009). The Residual Gradient method (RG), introduced by Baird (1995), is designed to be used with various approximation functions, including ones belonging to a non-linear family. RG is developed to minimize the approximation error[3] $\xi(\theta) = \sum_{t=1,...,T} \xi_t(\theta)$ where $\xi(\theta)$ is the total approximation error and $\xi_t(\theta) = \{C(s_t) - Q(s_t\,|\theta)\}^2/2$ is the approximation error of period t. Therefore, the parameter values can be determined by gradient descend method: $\theta \leftarrow \theta - \beta \cdot \{ C(s_t) - Q(s_t|\theta) \}\, \nabla_\theta \{ C(s_t) - Q(s_t\,|\theta) \}$. In the development of Sarsa, partial differentiation is applied, then the real state cost $C(s_t)$ is approximated by $c(s_t) + \alpha\, Q(s_{t+1})$. In development of RG, the approximation is

---

[1] Prestwich et al. (2008) used Sarsa(λ) only as a competing method with hill-climbing method to determine the parameter values of Sarsa(λ), but have not investigated the effectiveness of bootstrapping.

[2] ET presented here is based on Singh and Sutton (1996)'s replacing trace.

[3] To be concise, the content here is presented based on state costs. The development based on state-action costs can be conducted in a similar manner.

applied before differentiation. The RG update equation is $\theta \leftarrow \theta + \beta \cdot \psi(s_t) \cdot \{\; \alpha \nabla_\theta Q(s_{t+1}|\theta) - \nabla_\theta Q(s_t|\theta) \}$ where $\psi(s_t) = c(s_t) + \alpha Q(s_{t+1}|\theta) - Q(s_t|\theta)$.

4 Direct Credit Back

Baird (1995) claims that RG always converges. However, it has been criticized for delivering an inferior solution compared to TD(0) by Maei et al. (2009). To improve the solution obtained from RG, we develop a Direct Credit Back (DCB) method based on bootstrapping. The DCB method uses the newly observed datum $c(s_t)$ to update the approximate costs of the most recent state as well as other prior states.

The temporal difference error of the approximate cost of state s_t is

$$\psi(s_t) = c(s_t) + \alpha Q(s_{t+1}|\theta) - Q(s_t|\theta). \tag{1}$$

After the period cost $c(s_t)$ is observed, $Q(s_t|\theta)$ can be approximated by $c(s_t) + \alpha Q(s_{t+1}|\theta)$. Therefore, the temporal difference error of the approximate cost of state s_{t-1} after $c(s_t)$ is observed is defined as

$$\psi(s_{t-1}|c(s_t)) = c(s_{t-1}) + \alpha \{\; c(s_t) + \alpha Q(s_{t+1}|\theta) \;\} - Q(s_{t-1}|\theta). \tag{2}$$

Equation (2) can be rearranged as shown in Equation (3),

$$\psi(s_{t-1}|c(s_t)) = c(s_{t-1}) + \alpha \{\; \psi(s_t|c(s_t)) + Q(s_t|\theta) \;\} - Q(s_{t-1}|\theta) \tag{3}$$

where $\psi(s_t|c(s_t)) = \psi(s_t)$.

The update equations for other prior states can be obtained in a similar manner. After $c(s_t)$ is observed, the temporal difference error of approximate cost of state s_{t-i}, for $i = 1, 2, ..., t-1$, is shown in Equation (4):

$$\psi(s_{t-i}|c(s_t)) = c(s_{t-i}) + \alpha \{\; \psi(s_{t-i+1}|c(s_t)) + Q(s_{t-i+1}|\theta) \;\} - Q(s_{t-i}|\theta). \tag{4}$$

Minimizing the squared temporal difference errors of all prior states is equivalent to minimizing $\Sigma_{i=0, ..., t-1} \psi^2(\; s_{t-i} \mid c(s_t) \;)$. The DCB update equation can be obtained by the gradient descent method. The update can be truncated to only a specific number of prior approximate state costs,

$$\theta \leftarrow \theta - \beta \sum_{i=0,..., \min\{t-1, N\}} \psi(s_{t-i}|c(s_t)) \cdot \{\; \alpha^{i+1} \cdot \nabla_\theta Q(s_{t+1}|\theta) - \nabla_\theta Q(s_{t-i}|\theta) \} \tag{5}$$

where $N \in \{0, 1, 2, ...\}$ is the number of periods crediting back. The update equation for approximate state-action costs can be obtained in a similar manner. When the parameter $N = 0$, the DCB method reduces to RG.

5 Experiments

Our study uses computer simulations to conduct numerical experiments. The inventory problem investigated here is a periodic review single-echelon problem with nonzero leadtime and a setup cost.

The demand is modeled as AR1/GARCH(1,1). Therefore, a state is composed of the previous demand D_{t-1}, the previous demand error ε_{t-1}, the variance of the previous demand error σ_{t-1}^2, the on-site inventory x_t, and the in-transit inventory $B^{(t)}$ whose length depends on a leadtime L. The state space of this problem is $\{0, I^+\} \times R \times R \times I \times \{0, I^+\}^L$ for D_{t-1}, ε_{t-1}, σ_{t-1}^2, x_t, and $B^{(t)}$, respectively. An action $u_t \in \{0, I^+\}$ is a replenishment order. State transitions are specified by (1) $D_t = a_0 + a_1 D_{t-1} + \varepsilon_t$, (2) $\varepsilon_t = e_t \sigma_t$, (3) $\sigma_t^2 = v_0 + v_1 \varepsilon_{t-1}^2 + v_2 \sigma_{t-1}^2$, (4) $x_{t+1} = x_t + B^{(t)}_1 - D_t$, and (5) $B^{(t+1)} = [B^{(t+1)}_1 \cdots B^{(t+1)}_{L-1} B^{(t+1)}_L]^T = [B^{(t)}_2 \cdots B^{(t)}_L u_t]^T$ where a_0 and a_1 are AR1 model parameters; v_0, v_1, and v_2 are GARCH(1,1) parameters; e_t is white noise distributed according to $N(0, 1)$; and u_t is the replenishment order.

The period cost consists of the replenishment cost and the inventory handling cost, $c_{t+1} = K_t\,\delta(u_t) + g_t\,u_t + h_t\,(x_{t+1})^+ + b_t\,(-x_{t+1})^+$ where c_{t+1} is the period cost, whose value will be known at time $t + 1$ (the end of period t); K_t is the setup cost; g_t is the unit replenishment cost; h_t is the unit holding cost; b_t is the unit backlogging cost; $\delta(\cdot)$ is the step function; $(\cdot)^+$ is the positive function defined by $(a)^+ = a\,\delta(a)$; and other variables are as mentioned earlier.

Our study investigated 13 different inventory problem structures (as shown in Table 1) with other parameters held fixed: AR1's $a_0 = 2$, GARCH(1,1)'s $v_1 = 0.1$ and $v_2 = 0.8$, a discount factor $\alpha = 1$, a leadtime $L = 1$, and the unit replenishment cost $g = \$100/\text{unit}$. Each experiment is initialized at $D_0 = 50$, $\varepsilon_0 = 10$, $\sigma_0^2 = 400$, $x_1 = 10$, and $B^{(1)} = 0$. Each method's performance is evaluated based on its aggregate cost of Periods 13–60. Using an aggregate cost allows ADP methods to have initial learning

Table 1. A summary of demand and cost variables investigated

	demand variables		cost variables		
	correlation (a₁)	noise variance offset (v₀)	set up cost (K: $/transaction)	penalty cost (b: $/item)	holding cost (h: $/item)
P1	0.8	70	100	200	1
P2	0	70	100	200	1
P3	-0.8	70	100	200	1
P4	0.8	40	100	200	1
P5	0.8	10	100	200	1
P6	0.8	70	150	100	1
P7	0.8	70	100	100	1
P8	0.8	70	50	100	1
P9	0.8	70	0	100	1
P10	0.8	70	0	200	1
P11	0.8	70	0	50	1
P12	0.8	70	0	50	10
P13	0.8	70	0	50	25

periods and thus provides better performance evaluation of ADP methods. The experiments run each inventory controller for 50 replications of each of 60 time-unit-indeterminate periods.

Implementation. Sarsa, Sarsa(λ), RG, and DCB methods are used with the Radial Basis Function (RBF) as the approximation function. RBF has three sets of parameters: centers, scales, and weights. RBF centers and scales are set up as Katanyukul et al. (2011). RBF weights are adjustable parameters whose values are determined by the method under investigated. Variables ε_{t-1} and σ_{t-1}^2 are excluded from state parameters for reasons discussed in Katanyukul et al. (2011).

6   Experimental Results

We measure performance of each inventory control by aggregate costs. Figures 1 and 2 show averages and confidence intervals of aggregate costs obtained from different methods under problem scenarios P1–P13 (Table 1). A '*' marks an average aggregate cost and line beside it represents a 90% confidence interval, based on t-test statistics. On the y-axis, labels 'H', 'S', 'SL', 'RG', and 'DCB', and 'Roll' indicate results obtained from the 12-period Look-Ahead, Sarsa, Sarsa(λ), RG, and the DCB method, respectively[4]. The Roll-out method was used with perfect system information: Rollout simulation parameter values match those of the actual problem.

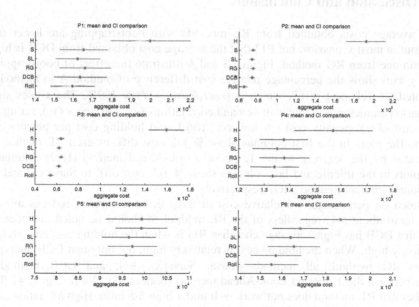

Fig. 1. Means and Confidence Intervals of aggregate costs

---

[4]   Results of each method are presented with the best performing set of parameter values.

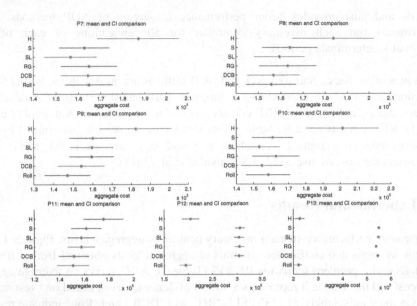

Fig. 2. Means and Confidence Intervals of aggregate costs

7   Discussion and Conclusions

The average costs obtained from RL methods with bootstrapping are lower than without in most scenarios, but P13 that the average cost obtained from DCB is higher than one from RG method. Figures 3 and 4 illustrate the effect of bootstrapping. The y-axes show the percentage relative cost difference of method A to method B, denoted "% rel. cost diff.", which is $(cost_A/cost_B - 1) \times 100\%$. The x-axes show scenario parameters as indicated: demand correlations (a_1), variances (v_0), set up per unit cost (K/g), penalty cost per unit cost (b/g), and holding cost per penalty cost (h/b). The plots in the first columns show % rel. cost diff. of each RL method, as indicated by the legends, to the 12-period Look-Ahead method (H12). Similarly, the plots in the middle and last columns show % rel. cost diff. to Sarsa (S) and the Residual Gradient method (RG), respectively.

Based on percentages of relative cost difference, bootstrapping reduces average cost up to about 5%, regardless of the RL method. It should be noted that scenario P13 that DCB has higher average cost than RG is when the holding cost ratio (h/b) is relatively high. When the holding cost is relatively high, not only that DCB underperforms RG method, all methods—Sarsa, Sarsa(λ), RG, and DCB methods—underperform the 12-period Look-Ahead method (bottom left plot of Figure 4). This implies that RL method does not work well under high h/b ratio. High h/b ratios make inventory decisions more critical—stocking more inventory has more negative effect. This may cause slower convergence of all RL methods. The study of suitable ADP methods for critical decision problems deserves further investigation.

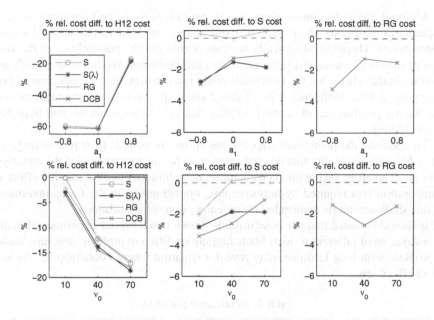

Fig. 3. Relative cost differences of ADP methods on different demand correlations (upper row) and variances (lower row)

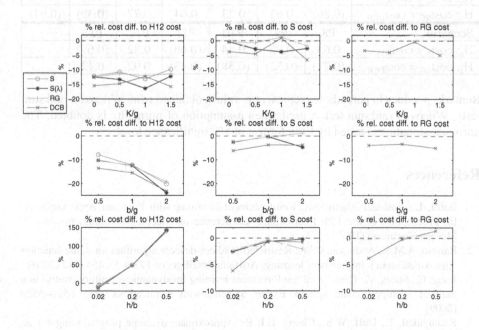

Fig. 4. Relative cost differences of ADP methods on different set up costs (top row), penalty costs (middle row) and holding costs (bottom row)

Although the reduction of average aggregate costs by using bootstrapping seems apparent in most cases, the reduction is still within ranges of variation (at 95% confidence level). Therefore, statistically we cannot rule out the possibilities of the variation of results due to stochastic variations in the problems. We have run significance tests to double-check, but the test results could not confirm the cost reduction of bootstrapping at 95% confidence level. Table 2 shows p-values of the significance tests. The higher p-value, i.e., closer to 1, implies that evidence against the null hypothesis is insufficient.

To conclude the point, bootstrapping has shown to be able to reduce average cost up to 5% with possibility that this reduction may be due to variation of the underlying process. Therefore, taking into consideration the additional implementation effort and computation cost required by bootstrapping, bootstrapping in the forms investigated in this study are not recommended for inventory control applications.

It should be noted that our conclusion is drawn based on our experimental results. Investigation of different form of bootstrapping or different problem structure, such as a problem with long leadtime, may reveal a situation where bootstrapping can work more effectively.

Table 2. Significance test results

p-values obtained from t-tests							
Scenario	P1	P2	P3	P4	P5	P6	P7
H_a: $cost_S \neq cost_{S(\lambda)}$	0.76	0.60	0.39	0.75	0.57	0.66	(0.45)
H_a: $cost_{RG} \neq cost_{DCB}$	(0.86)	0.65	0.23	0.64	0.77	(0.39)	(0.97)
Scenario	P8	P9	P10	P11	P12	P13	
H_a: $cost_S \neq cost_{S(\lambda)}$	0.63	0.96	0.44	(0.66)	0.82	(0.69)	
H_a: $cost_{RG} \neq cost_{DCB}$	(0.71)	(0.52)	(0.58)	(0.43)	0.92	0.12	

Remark: p-values in parentheses represent p-values obtained from Wilcoxon ranksum tests. Wilcoxon ranksum test is used when assumption of normality is doubted. The normal assumption is tested by Lilliefor test at 5% significance level.

References

1. Baird, L.: Residual Algorithms: Reinforcement Learning with Function Approximation. In: Proceedings of the 12th International Conference on Machine Learning, pp. 30–37. Morgan Kaufmann (1995)
2. Barreto, A.M.S., Anderson, C.W.: Restricted gradient-descent algorithm for value-function approximation in reinforcement learning. Artificial Intelligence 172(4-5), 454–482 (2008)
3. Jiang, C., Sheng, Z.: Case-based reinforcement learning for dynamic inventory control in a multi-agent supply chain system. Expert Systems with Applications 36(3), 6520–6526 (2009)
4. Katanyukul, T., Duff, W.S., Chong, E.K.P.: Approximate dynamic programming for an inventory problem: Empirical comparison. Computers & Industrial Engineering 60(4), 719–743 (2011)

5. Kim, C.O., Jun, J., Baek, J.K., Smith, R.L., Kim, Y.D.: Adaptive inventory control models for supply chain management. International Journal of Advanced Manufacturing Technology 26(9-10), 1184–1192 (2005)
6. Kim, C.O., Kwon, I.H., Baek, J.G.: Asynchronous action-reward learning for nonstationary serial supply chain inventory control. Applied Intelligence 28(1), 1–16 (2008)
7. Kwon, I.H., Kim, C.O., Jun, J., Lee, J.H.: Case-based myopic reinforcement learning for satisfying target service level in supply chain. Expert Systems with Applications 35(1-2), 389–397 (2008)
8. Leng, J., Jain, L., Fyfe, C.: Experimental analysis of eligibility traces strategies in temporal difference learning. International Journal of Knowledge Engineering and Soft Data Paradigms 1(1), 26–39 (2009)
9. Maei, H.R., Szepesvari, C., Bhatnagar, S., Precup, D., Silver, D., Sutton, R.S.: Convergent Temporal-Difference Learning with Arbitrary Smooth Function Approximation. In: Advances in Neural Information Processing Systems. MIT Press, Vancouver (2009)
10. Prestwich, S.D., Tarim, S.A., Rossi, R., Hnich, B.: A Cultural Algorithm for POMDPs from Stochastic Inventory Control. In: Blesa, M.J., Blum, C., Cotta, C., Fernández, A.J., Gallardo, J.E., Roli, A., Sampels, M. (eds.) HM 2008. LNCS, vol. 5296, pp. 16–28. Springer, Heidelberg (2008)
11. Reynolds, R.G.: An Introduction to Cultural Algorithms. In: Proceedings of the 3rd Annual Conference on Evolutionary Programming, pp. 131–139. World Scientific Publishing (1994)
12. Shervais, S., Shannon, T.T., Lendaris, G.G.: Intelligent Supply Chain Management Using Adaptive Critic Learning. IEEE Transactions on Systems, Man, and Cybernetics-Part A: Systems and Humans 33(2), 235–244 (2003)
13. Singh, S.P., Sutton, R.S.: Reinforcement Learning with Replacing Eligibility Traces. Machine Learning 22(1-3), 123–158 (1996)
14. Sutton, R.S., Barto, A.G.: Reinforcement Learning. MIT Press (1998)
15. Tesauro, G.J.: TD-Gammon, a self-teaching backgammon program, achieves master level play. Neural Computation 6(2), 215–219 (1994)
16. Van Roy, B., Bertsekas, D.P., Lee, Y., Tsitsiklis, J.N.: A Neuro-Dynamic Programming Approach to Retailer Inventory Management. In: Proceedings of the IEEE Conference on Decision and Control (1997)

Support Vector Machine with Mixture
of Kernels for Image Classification

Dongping Tian[1,2], Xiaofei Zhao[1], and Zhongzhi Shi[1]

[1] Key Laboratory of Intelligent Information Processing, Institute of Computing Technology,
Chinese Academy of Sciences, Beijing, 100190, China
[2] Graduate University of the Chinese Academy of Sciences, Beijing, 100049, China
{tiandp,zhaoxf,shizz}@ics.ict.ac.cn

Abstract. Image classification is a challenging problem in computer vision. Its performance heavily depends on image features extracted and classifiers to be constructed. In this paper, we present a new support vector machine with mixture of kernels (SVM-MK) for image classification. On the one hand, the combined global and local block-based image features are extracted in order to reflect the intrinsic content of images as complete as possible. SVM-MK, on the other hand, is constructed to shoot for better classification performance. Experimental results on the Berg dataset show that the proposed image feature representation method together with the constructed image classifier, SVM-MK, can achieve higher classification accuracy than conventional SVM with any single kernels as well as compare favorably with several state-of-the-art approaches.

Keywords. Image classification, SVM, Kernel function, PCA.

1 Introduction

Image classification is a challenging problem in computer vision. With the rapid explosion of images available from various multimedia platforms, effective technologies for organizing, searching and browsing these images are urgently required by common users. Fortunately, image classification can give much help to image indexing and retrieval. In recent years, many methods have been developed for image classification. As the representative work using SVM as classifiers, Chapelle et al. [1] use multiclass classifier framework to train 14 SVM classifiers for 14 image level concepts based on 4096 dimensional HSV histograms. Goh et al. [2] study several ensemble SVM binary classifiers, including one per class (OPC), pairwise coupling (PWC) and error-correction output coding (ECOC) through margin boosting and noise reduction methods to enhance the classification accuracy. Autio and Elomaa [3] focus on indoor image classification by using SVM as the classifier. Yang and Dong [4] propose ASVM-MIL to extend the conventional support vector machine through multiple-instance learning (MIL) so as to train the SVM in a new space. Recently, Anthony et al [5] employ SVM to handle the multiple classification tasks common in remote sensing studies for land cover mapping. And they find that

Z. Shi, D. Leake, and S. Vadera (Eds.): IIP 2012, IFIP AICT 385, pp. 68–76, 2012.

whereas the one-against-all technique is more predisposed to yielding unclassified and mixed pixels, the resulting classification accuracy is not significantly different from one-against-one approach. In addition, Gao and Lin [6] present CGSVM for semantic image retrieval, which utilizes clustering result to select the most informative image samples to be labeled and optimize the penalty coefficient. As a result, CGSVM can get higher search accuracy than conventional SVM for semantic image retrieval. In more recent years, Jiang, He and Guo [7] propose a novel method which adopts learning vector quantization (LVQ) technique to optimize low-level features extracted from given images, and then some representative vectors are selected with LVQ to train support vector machine classifier instead of using all feature data. Agrawal et al. [8] use SVM to implement image classification based on the feature of color image histograms, which benefits from the insensitivity of translation and rotation.

In general, to construct an image classifier includes two important stages: (1) features extracted from a set of training images are used to train the classifier. (2) features extracted from a set of testing images are fed into the built classifier so as to implement the image classification. However, previously related work only focuses on using either global or local (region) image features as the input to train the classifier. In literature [9], Tsai and Lin compare various combinations of feature representation methods including the global and local block-based and region-based features for image database categorization. Then the significant conclusion, i.e. the combined global and block-based feature representation performs best, is drawn in the end. Chow and Rahman [10] present an image classification approach through a tree-structured feature set, in which the image content is organized in a two-level tree, and the root node denotes the whole image features while the child nodes represent the local region-based features. Subsequently, the tree-structured image features are processed by a two-level self-organizing map (SOM) to implement the classification of images. In addition, Lu and Zhang [11] propose a block-based image feature representation in order to reflect the spatial features for a specific concept. For an image with $N \times N$ blocks, partition it by rows with blocks and connect into a block-line. The block lies at the more left side of the block-line represents upper and/or left position in the image. Shi et al. [12] present a square symmetrical local binary pattern texture descriptor, which is a compact symmetrical-invariant variation of local binary pattern, to capture the low-dimensional optimal discriminative features of images, etc. Most of these approaches can achieve state-of-the-art performance and motivate us to explore image classification with the help of their excellent experiences and knowledge. So in this paper, we present a new support vector machine with mixture of kernels (SVM-MK) for image classification. On the one hand, the combined global and local block-based image features are extracted so as to reflect the intrinsic content of images as complete as possible. SVM-MK, on the other hand, is constructed to shoot for better classification performance. Finally, the approach is experimentally tested on the Berg dataset and the precision we obtain is higher than conventional SVM with any single kernels as well as comparative to the state-of-the-art methods.

The rest of the paper is organized as follows. Section 2 elaborates the image classification method proposed in this paper, including the image feature representation and the support vector machine with mixture of kernels (SVM-MK). Section 3 reports the experimental results on the Berg dataset. Finally, some important conclusions and future work are summarized in Section 4.

2   Classification Method

Motivated by the references aforementioned, global and local block-based image feature representation is proposed in this paper. Meanwhile, support vector machine with mixture of kernels (SVM-MK) is constructed as the image classifier. To our best knowledge, this is the first study to apply SVM with mixture of kernels in image classification. Details of them will be described in the following subsections, respectively.

2.1   Visual Feature Representation

(1) Global Feature Representation
To begin with, the image in RGB color space is transformed into HSV (hue, saturation and value) color space since it is more perceptually uniform than RGB. Without loss of generality, color histogram is used to represent the global feature of images in this paper. In details, the histogram of each channel is computed as follows: $h_G = n_G/n_T$, $G=1,2,...,q$, ,where G denotes a quantized level of an HSV color channel, n_G is the total number of pixels in that level, n_T is the total number of pixels, and q is the number of quantized levels. The complete histogram vector can be represented as $H_q = [h_{H1},...,h_{Hq},h_{S1},...,h_{Sq},h_{V1},...,h_{Vq}]$.

(2) Local Feature Representation
The first-order, second-order and third-order color moments of image blocks are calculated for each channel as follows:

$$\mu_i = \frac{1}{N}\sum_{j=1}^{N} f_{ij} \tag{1}$$

$$\sigma_i = \left(\frac{1}{N}\sum_{j=1}^{N}(f_{ij} - \mu_i)^2\right)^{\frac{1}{2}} \tag{2}$$

$$\gamma_i = \left(\frac{1}{N}\sum_{j=1}^{N}(f_{ij} - \mu_i)^3\right)^{\frac{1}{3}} \tag{3}$$

where f_{ij} is the color value of the i-th color component of the j-th image pixel and N is the total number of pixels in the image. Hence, the color moments can be represented as $CM = [\mu_{c1}, \mu_{c2}, \mu_{c3}, \sigma_{c1}, \sigma_{c2}, \sigma_{c3}, \lambda_{c1}, \lambda_{c2}, \lambda_{c3}]$, where $\mu_{ci}, \sigma_{ci}, \lambda_{ci}$ ($i=1,2,3$)

denote the mean, variance and skewness of each channel of an image block, respectively.

Texture is a very useful characterization for a wide range of image. It is generally believed that human visual system uses textures for recognition and interpretation. The common methods for texture feature analysis include Gabor filter, wavelet transform and gray co-occurrence matrices, etc. Among which Gabor filter is the most commonly used method in extracting texture features. Given an image $I(x,y)$, Gabor wavelet transform $G(x,y)$ convolves $I(x,y)$ with a set of Gabor filters of different spatial frequencies and orientations. Assuming the wavelet transform is at u-th scale and v-th orientation, then means and standard deviations of $G_{uv}(x,y)$ can be calculated to represent the Gabor features:

$$\mu_{uv} = \frac{1}{PQ}\sum_x\sum_y|G_{uv}(x,y)|$$

(4)

$$\sigma_{uv} = \frac{1}{PQ}\sqrt{\sum_x\sum_y|G_{uv}(x,y)| - \mu_{uv}}$$

(5)

where $P\times Q$ is the size of image. The complete Gabor feature is then represented as $G_{uv}=[\mu_{00}, \sigma_{00}, \mu_{01}, \sigma_{01},..., \mu_{0(v-1)}, \sigma_{0(v-1)},..., \mu_{(u-1)(v-1)}, \sigma_{(u-1)(v-1)}]$.

In addition, shape feature is also considered, whose purpose is to encode simple geometrical forms such as straight lines in different directions. Since Sobel operator is unsensitive to noise than other edge detectors. Here, the shape features can be extracted by the convolution of 3×3 masks with the image in 4 different directions (horizontal, 45°, vertical and 135°), and the corresponding features can be represented as $E=[e_0,e_{45},e_{90},e_{135}]$. Fig.1 shows several masks used here to compute four specific shapes.

Up to now, the approach for local feature extraction is presented above. The next step is how to partition the image and how to organize the block-based image features. The idea in reference [11] is adopted here, to start with, the image is

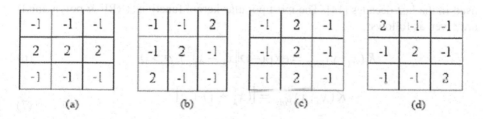

Fig. 1. Mask to detect shape features (a) Horizontal lines. (b) 45° slanted lines. (c) Vertical lines. (d) 135° slanted lines.

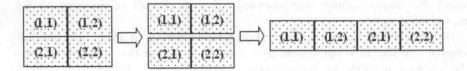

Fig. 2. Partition an image into 2*2 blocks and connect into a block-line

partitioned into regular grids (blocks) and then connect the blocks by left-right and top-down orders. In short, partition the image by rows with blocks and connect into a block-line. For the sake of simplicity, the image is partitioned into 2×2 blocks here, and the basic procedure is illustrated in Fig.2.

2.2 Kernel Function Constructing

Support vector machine (SVM) is a supervised classifier proposed by Vapnik which is based on statistical learning theory [13]. SVM works by mapping the training data into a high dimensional feature space. Then it separates the two classes of data with a hyperplane, and maximizes the distance which is called the margin. By introducing Kernels into the algorithm, it is possible to maximize the margin in the feature space, which is equivalent to nonlinear decision boundaries in the original input space. The algorithm comes with all the theoretical guarantees given by VC theory as well as the convergence properties studied in the statistical mechanics literature. In general case, assume that the labeled training examples $(x_1, y_1), ..., (x_n, y_n)$, where each $x_k \in R^n$ is the $k\text{-}th$ input sample and $y_k \in R$ is the $k\text{-}th$ output pattern. Suppose that each example is assigned a binary label $y_k \in \{+1, -1\}$. In their simplest form, SVM could find out the hyperplanes that separate the training data by a maximal margin. All the vectors lying on one side of the hyperplane are labeled as +1, and all the vectors lying on the other side labeled as -1. The training instances that lie closest to the hyperplane are called support vectors. As we known, the challenging problem in the case of SVM is the choice of the kernel function which is actually a measure of similarity between two vectors. Different choices of kernel functions have been proposed and extensively used in the last decades [14]. The most popular kernel functions (RBF & polynomial) are given as follows:

$$K(x_k, x)_{RBF} = \exp(-\rho\|x_k - x\|^2), \rho > 0 \tag{6}$$

$$K(x_k, x)_{Poly} = [(x_k * x) + c]^q \tag{7}$$

It is reported that kernels used by SVM can be divided into two classes: global and local kernels. In global kernels, points far away from the test point have a great effect on kernel values. While in local kernels, only those close to the test point have a great effect on kernel values. The RBF and polynomial kernel functions are two typical local and global kernels, respectively. Therefore, a kind of SVM modeling method based on mixture of kernels can be constructed as follows:

$$K(x_k, x_l)_{Mix} = \lambda K(x_k, x_l)_{RBF} + (1-\lambda)K(x_k, x_l)_{Poly} \tag{8}$$

where $K(x_k, x_l)_{RBF}$ denotes the RBF kernel function and $K(x_k, x_l)_{Poly}$ denotes polynomial kernel function, $\lambda(0 \leq \lambda \leq 1)$ is the mixed coefficient used to adjust the weight of the two kinds of kernel functions.

3   Experiments and Analysis

For the sake of comparison, we test the proposed image classification scheme on the ten-animal-class dataset of Berg available from [15]. Fig.3 shows some images of Berg dataset[1]. Each image is represented with 668-dimensional visual features since the image representation adopted in this paper consists of global and local block-based segmentation features reduced by principal component analysis (PCA). SVM-MK is implemented by adapting the source code of Libsvm [16] so as to construct the mixture of polynomial kernel and RBF. To determine the optimal classification parameters C and λ, 10-fold cross validation is conducted on the training image dataset and the best parameters are used in testing. Here, the corresponding parameters are predefined as $C=1000$, $c=1$, $q=1$ and the best representative λ value is set to 0.05 by trial and error.

Fig. 3. Example images from Berg dataset

To show the effectiveness of the SVM-MK for image classification, we conduct two experiments in this paper. The first one is the image classification comparison between SVM-MK and the SVM with any single kernels (RBF & polynomial) under usual image feature representation and the image feature representation proposed in

---

[1]   Download from http://tamaraberg.com/animalDataset/index.html

Table 1. Comparison among SVM-Poly., SVM-RBF and SVM-MK under different image feature representations

Classifier	SVM-Poly.	SVM-RBF	SVM-MK
Avg. Prec.(usual feature)	43.6	45.4	46.1
Avg. Prec.(presented feature)	46.8	48.5	50.3

this paper, respectively. The average classification precision over different kinds of images is used to measure the overall quality of SVM-MK.

From table 1, it easily can be seen that the performance of all classifiers with the image feature representation proposed in this paper is better than the corresponding ones with the usual image features. On the other hand, the classification accuracy of SVM-MK is higher than that of SVM-Poly. and SVM-RBF under the same image feature representation, respectively. It is also noticeable that SVM-MK with the proposed image feature representation performs the best, which further demonstrates that the mixture of global and local kernel functions as well as the block-line image feature representation plays a very important role in SVM classifiers. In addition, we also make a comparison between SVM-MK and other state-of-the-art approaches. Here we create two SVM classifiers (SVM-EQ, SVM-TW) with different settings for five categories. SVM-EQ is the SVM created using the same positive and negative data as in SVM-MK; SVM-TW is the SVM created using the ground truth data twice the number of the labeled data in SVM-MK. According to table 2, SVM-MK performs significantly better than the SVM-EQ model. Particularly, the precision of "ant" has been increased from 30 to 56.3. But it is partially outperformed by SVM-TW for the classification of "alligator", "bear" and "leopard".

Table 2. Comparison among SVMs: Precision at top 100 images

	alligator	ant	bear	leopard	penguin
SVM-EQ	58.0	30.0	38.0	52.0	48.0
SVM-TW	77.0	52.0	61.0	67.0	62.0
SVM-MK	73.5	56.3	55.0	61.8	63.6

To further illustrate the effect of SVM-MK proposed in this paper for image classification, examples for the precision of the top 100 images on the Berg dataset by Berg [15] and Schroff [17], together with ours are depicted in Fig.4. From Fig.4, we can clearly see that our approach gives superior precision to both Berg and Schroff over five categories and outperforms [15] on eight categories except "monkey" and "ant". In summary, the performance of our approach is holistically superior to other methods.

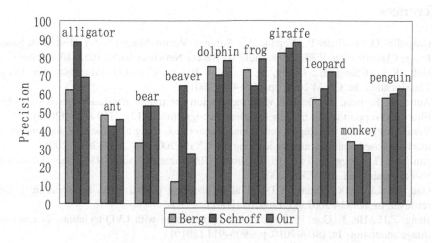

Fig. 4. Comparison with Berg and Schroff

4   Conclusions and Future Work

In this paper, a novel support vector machine with mixture of kernels (SVM-MK) for image classification has been proposed. On the one hand, the combined global and local block-based image features are extracted in order to reflect the intrinsic content of images. SVM-MK, on the other hand, is constructed in this paper as classifiers. Experimental results on Berg dataset show that the proposed image feature representation method together with the constructed image classifier, SVM-MK, can achieve higher classification accuracy than conventional SVM with any single kernels as well as comparable performance with several state-of-the-art approaches.

For the future work, we plan to introduce semi-supervised learning into our approach under the conditions that there are only a few labeled but a large amount of unlabeled images to implement image classification. In addition, we intend to apply other more complicated image datasets to further evaluate the performance of SVM-MK proposed in this paper comprehensively.

Acknowledgements. This work is supported by the National Natural Science Foundation of China (No.61035003, No. 60933004, No. 60903141, No.61072085), the National Program on Key Basic Research Project (973 Program) (No.2007CB311004), the National High-tech R&D Program of China (863 Program) (No.2012AA011003) and the National Science and Technology Support Program (2012BA107B02).

References

1. Chapelle, O., Haffner, P., Vapnik, V.: Support Vector Machines for Histogram-Based Image Classification. IEEE Transactions on Neural Networks 10(5), 1055–1064 (1999)
2. Goh, K.S., Chang, E., Cheng, K.T.: SVM Binary Classifier Ensembles for Image Classification. In: CIKM 2001, pp. 395–402 (2001)
3. Autio, I., Elomaa, T.: Flexible view recognition for indoor navigation based on Gabor filters and support vector machines. Pattern Recognition 36(12), 2769–2779 (2003)
4. Yang, C.B., Dong, M.: Region-based image annotation using asymmetrical support vector machine-based multiple instance learning. In: CVPR 2006, pp. 2057–2063 (2006)
5. Anthony, G., Gregg, H., Tshilidzi, M.: Image Classification Using SVMs: one-against- one Vs one-against-all. In: ACRS (2007)
6. Gao, K., Lin, S.X., Zhang, Y.D., et al.: Clustering guided SVM for semantic image retrieval. In: ICPCA 2007, pp. 199–203 (2007)
7. Jiang, Z.H., He, J., Guo, P.: Feature data optimization with LVQ technique in semantic image annotation. In: ISDA 2010, pp. 906–911 (2010)
8. Agrawal, S., Verma, N.K., Tamrakar, P., et al.: Content Based Color Image Classification using SVM. In: ITNG 2011, pp. 1090–1094 (2011)
9. Tsai, C.F., Lin, W.C.: A Comparative Study of Global and Local Feature Representations in Image Database Categorization. In: NCM 2009, pp. 1563–1566 (2009)
10. Chow, T.W.S., Rahman, M.K.M.: A new image classification technique using tree-structured regional features. Neurocomputing 70(4-6), 1040–1050 (2007)
11. Lu, H., Zheng, Y.B., Xue, X.Y., Zhang, Y.J.: Content and Context-Based Multi-Label Image Annotation. In: CVPRW 2009, pp. 61–68 (2009)
12. Shi, Z.P., Liu, X., Li, Q.Y., He, Q., Shi, Z.Z.: Extracting discriminative features for CBIR. Multimedia Tools and Applications (2011)
13. Vapnik, V.N.: The Nature of Statistical Learning Theory. Springer, New York (1995)
14. Zhu, Y.F., Tian, L.F., Mao, Z.Y., Wei, L.: Mixtures of Kernels for SVM Modeling. Springer, Heidelberg (2005)
15. Berg, T.L., Forsyth, D.A.: Animals on the web. In: CVPR 2006, pp. 1463–1470 (2006)
16. Chang, C.C., Lin, C.J.: Libsvm: a library for support vector machines (2001)
17. Schroff, F., Criminisi, A., Zisserman, A.: Harvesting image databases from the web. In: ICCV 2007, pp. 1–8 (2007)

The BDIP Software Architecture and Running Mechanism for Self-Organizing MAS

Yi Guo[1], Xinjun Mao[1], Fu Hou[1], Cuiyun Hu[1], and Jianming Zhao[2]

[1] Department of Computer Science and Technology,
School of Computer, Nation University of Defence Tehnology
[2] Department of Computer Science and Technology, Zhejiang Normal University
Berniegy@gmail.com

Abstract. As there are huge gaps between the local micro interactions among agents and the global macro emergence of self-organizing system, it is a great challenge to design self-organizing mechanism and develop self-organizing multi-agent system to obtain expected emergence. Policy-based self-organization approach is helpful to deal with the issue, in which policy is the abstraction of self-organizing mechanism and acts as the bridge between the local micro interactions and global macro emergence. This paper focuses on how to develop software agents in policy-based self-organizing multi-agent system and proposes a BDIP architecture of software agent. In our approach, policy is an internal component that encapsulates the self-organizing information and integrates with BDI components. BDIP agent decides its behaviors by complying with the policies and respecting BDI specifications. An implementation model and the running mechanism as well as corresponding decision algorithms for BDIP agents are studied. A case of self-organizing system is studied to illustrate our proposed approach and show its effectiveness.

Keywords: multi-agent system, self-organization, agent architecture.

1   Introduction

Self-organization refers to a process where a system changes its internal structure without explicit external and central control. It often results in emergent behavior in the global system [1].With the pervasiveness of distributed information systems, self-organizing systems become more and more attractive to researchers from different application areas [2]. Agent technology is considered as an appropriate and powerful paradigm to develop large-scale complex systems applications. As a kind of such complex systems, self-organizing systems are usually engineered with agent metaphor, which views the whole system as MAS (Multi-Agent Systems) and using software agents as basic components to construct the systems [4].

However, developing self-organizing MAS in an iterative and effective way is still a great challenge in the literature of software engineering [5]. The obstacle is how to obtain desirable global system characteristic through the local interactions among agents. In self-organizing MAS, there is an absence of centralized control node in the

Z. Shi, D. Leake, and S. Vadera (Eds.): IIP 2012, IFIP AICT 385, pp. 77–86, 2012.

system and the agents only interact with their local environment. This leaves a significant gap between the local interaction and global system characteristic, and brings obstacles which are put in evidence during the development of the systems. To obtain the desirable global characteristic, the developer therefore must adjust the behaviors of agents iteratively. However, self-organizing MAS often consist of large numbers of agents and are deployed in complex environment. Designing and deploying such systems in an iterative way is difficult. How to effectively support the development of such systems is still an open issue [5].

Against the background, we have proposed a policy-based self-organization approach in our previous work [6] which affects the emergences of multi-agent systems by restricting or guiding agents' behaviors in terms of policy. In order to support the development of the PSOMAS (Policy-based Self-Organizing Multi-Agent Systems), this paper proposes a BDIP software architecture in which policies are viewed as component of the agent. Based on the architecture, an implementation model and running mechanisms as well as agent behavior decision algorithms are also provided. The rest of this paper is organized as follows: Section 2 gives a brief introduction of policy-based self-organizing MAS. Section 3 proposes the agent architecture, policy representation, and running mechanism of the software agents. Section 4 discusses the policy-based agent decision algorithms and a case study is illustrated in section 5. Section 6 discusses the related works as well as conclusions and future works are discussed in section 7.

2   Policy Based Approach to Self-organizing Multi-Agent Systems

To intuitively understand the challenges in self-organizing MAS, we firstly introduce an example of a group of self-organizing robots whose aims are to explore and carry ore in a strange environment. Each robot provides functions to randomly walk in the environment to find ore resource, and carry ore back to the base. Furthermore, each robot can broadcast its position to other robots. The system searches and takes ore by using the self-organization of these robots. When the users search for ore depending on these robots in strange environments, they perhaps need to cope with different circumstances, such as various landforms and ore distributing and etc. However, it is impossible that the fixed behaviors of robots can satisfy all scenarios. On the other hand, it is not always feasible that the users redesign and redeploy the robots after they have acquired the new requirements, for example the robots are exploring on the mars. Then we need a new approach to effectively change the behaviors of agent during runtime to meet the variety environment and requirements.

In human society, policies are often used to restrict and guide the behaviors of people. With these policies, human society often presents a self-organizing process and results in different emergent phenomena. For example, the economic policies often result in the changes of macro economic index, which is owing to the people's economic behaviors like stock transaction are affected by such policies. In the literature of self-organizing MAS, in order to facilitate the solving of the issues discussed above, the policies in human society can be used for self-organizing MAS. By this approach, policies give a presentation of the behaviors of agents and the agents must to comply with them at runtime. On one hand, developers need to design the policies as well as the agents in the design phase of such systems. On the other hand, the behaviors of agents are affected by these policies at runtime, and developers can evaluate the system's macro characteristics whether satisfies the requirements or not. If it does not satisfy the requirements, developers can change the policies, which can cause changes of the agents' behaviors and result in the changes of the self-organizing process of the whole system.

3   BDIP Architecture and Running Mechanism

3.1   BDIP Architecture of Software Agent

The BDI architecture of software agents had been accepted for a long time both in academe and industry. The architecture has three components: *Belief, Desire*, and *Intention*. *Belief* means the cognitions of an agent about its environment and internal state. *Desire* means the goals that an agent wants to pursue, and *Intention* means the commitment plans of the agent which is useful for the accomplishment of the goals in *Desire* [7]. We consider that the BDI architecture is useful for analyzing the autonomous behaviors of rational agents and easier to accept the *policy* as a new decision component than other architectures (e.g. reactive architecture). This paper proposes a BDIP (Belief, Desire, Intention and Policy) agent architecture by extending the BDI architecture (see Fig.1).

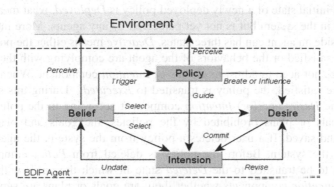

Fig. 1. BDIP architecture of Agent

In PSOMAS, the environment of a BDIP agent consists of policies and other agents [6], and the BDIP agent can perceive the policies from its local environment. The perceived policies are deposited in the *Policy* component, and are triggered by the *policy conditions* (introduced in section 3.2) which are specified in the *Belief* component. Policies can affect both *Desire* component and *Intention* component. For the *Desire* component, a new goal will be created by the policy or some of goals are prohibited. For the *Intention* component, the execution of committed plan will be guided by the policy, for example some actions are forbidden to be executed and some actions are preferential (corresponding algorithms will be introduced in section 4). The relationship among *Belief*, *Desire* and *Intention* is same as [7].

3.2 Representation and Realization of Policy

Policies can be viewed as a set of rules which restrict the behaviors or states of the agents in the system. This paper distinguishes two kinds of policies: *Obligation* and *Prohibition*. The *Obligation* represents the action that an agent need to perform or the state the agent need to keep as well as the *Prohibition* represents the action that an agent must not to perform or the state not to appear. On the other hand, a policy consists of the conditions to be satisfied and the action (state) to be performed (kept) by agents. Formally, it can be described by EBNF as follows:

> Policy:= *Obligation* (*IF* Self-condition *WHEN* Env-event *DO* (Action | State))
> | *Prohibition*(*IF* Self-condition *WHEN* Env-event *DO* (Action | State))

Self-condition specifies the internal state of an agent to be satisfied and *Env-event* represents the happened event of the environment of the agent. *"Obligation(IF* Self-condition *WHEN* Env-event *DO* (Action | State*)"* means that when the *Self-condition* of the agent is satisfied and *Env-event* is happening in the environment, the agent need to perform the *Action* or keep the *State, Prohibition* means the opposite semantics. The *Self-condition* and *Env-event* can be viewed as *policy-condition*.

A policy at the run-time can be in four states: *Deployed, Deactived, Executed,* or *Deleted.* The initial state of a newly deployed policy is *Deployed,* what means that the policy exists in the system but is not yet perceived by any agents. More interesting is the policy inside an agent can has three states. *Deactive* means either the policy conditions are not satisfied or the behaviors of the agent are complying with the policy i.e. the agent does not need to adjust its *Desire* or *Intention* component. When the policy conditions are satisfied, the policy is transited to *Executed.* During this state, agent will adjust the *Desire* and (or) *Intention* component according to the policy. If there are some goals or plans prohibited by the policies, the goals and plans will be suppressed and saved. If a user deletes a policy from the system, the agents will be informed by the system. Before the policy is deleted from *Policy* component, the policy needs to be transited to the *Deleted* state in which the agent will check the *Desire* and *Intention* components whether there are goals or plans are suppressed by this policy and resume them respectively if existing.

3.3 Implementation Model and Running Mechanism of BDIP Software Agent

Fig. 2 depicts the running mechanism of BDIP agents. *B, D, I* and *P* represent Belief set, Goal set, Plan set and the Policy set of the agent respectively. Agenda can be seen as a queue of actions. All executions of plans and other actions must on the agenda.

DE (Deliberation and Execution) is responsible for executing actions on the agenda and adjusting B, D, I, and P. In the DE component, AE (Action Execution) always executes the first action of the agenda. After the execution of an action, there may be a new action need to be added on the agenda. For example, an agent needs to execute action *CreateGoal* to create a sub-goal in the execution of a plan, and AE can directly add these new actions on agenda (direct effects). On the other hand, the execution of actions maybe change the state of belief set, DE will inform the CE (Condition Evaluation) about these belief changes. CE then checks whether to trigger a goal, a plan, or a policy as well as add corresponding actions on the agenda (side effects). Moreover, the event in the environment of an agent may also add actions to the agenda, e.g. messages that have been received from other agents and messages need to be processed (External effects).

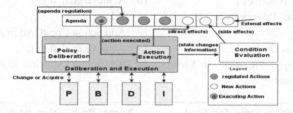

Fig. 2. Implementation Model and Running Mechanism of BDIP Agent

PD (Policy Deliberation) is the component which is used to adjust the D and I component. When the *policy conditions* of a policy are satisfied, CE will add the *TiggerPolicy* on the agenda, the policy will be transited from *Deactived"* to *Executed* state. Moreover AE will add *ExecutePolicy* on the agenda after *TiggerPolicy* executed, and the D and I components will be adjusted by PD after *ExecutePolicy* executed. In the adjusting process, perhaps the actions on the agenda also be regulated by PD (agenda regulation), for example add new actions, delete action, and adjust the sequence of the actions. On the other hand, when a user changes the policies in the system, CE will acquire this message and add *PerceivePolicy* or *DropPolicy* on the agenda.

4   Behavior Decision Algorithms of BDIP Agent

In the agent behaviors adjusting process, PD use different algorithms to cope with the different policy type. Table 1 shows the different algorithms for different policy type, the italic are the names of algorithms and "*Null*" means the regulation of this component is needless. The details of these algorithms are listed in Table 2. Moreover we consider the goal in D component as discussed in [8] is consists of different states at runtime: "Active" state means the goal is currently pursued by agent. "Suspended" state and "Options" state represent the goal is inactive. Therefore regulation of D component is the transition between goal states.

Table 1. Behavior decision Algorithm for Different Policies Type

Policy Type	Object	Algorithm for Intention	Algorithm for Desire
Obligation	State	Null	*Obligation_State_for_Goal*
	Action	*Obligation_Action_for_Plan*	Null
Prohibition	State	*Prohibition_State_for_Plan*	*Prohibition_State_for_Goal*
	Action	*Prohibition_Action_for_Plan*	Null

Table 2. Behavior Decision Algorithms

Name: Obligation_State_for_Goal

Input: Obligated action; goal of *Desire*
Output: goal
01 if (goal(state) is active) {
02 Rise goal(state) Priority in Agenda;
03 } else
04 if(goal(state) is Option){
05 Add action *SuppressContents* in Agenda;
06 Acheve goal(state);
07 } else

08 if(goal(state) is Suspended) {
09 waitfor (satisfyCondition);
10 Achieve goal(state);
11 }else {
12 Add action CreatGoal in Agenda;
13 Achieve goal(state);
14 Add action DropGoal in Agenda;}

Name: Prohibition_State_for_Goal

Input: Prohibited state; goal of *Desire*
Output: goal
01 if (goal(state)is one of achieving goals) {
02 Add action *SuppressContents* in Agenda;
03 Add action *DeliberationNewOption* in Agenda;
04 }else
05 if(goal(state) is Suspended or Option)
06 Add action *SuppressContents* in Agenda;
07 Add action *UpdateGoal* in Agenda;

Name: Prohibition_State_for_Plan

Input: Obligated state; plan of *Intention*
Output: plan
01 if (goal(state) is executing Plan) {
02 Add action *SuppressContents* in Agenda;
03 Add action *ScheduleCandidates* in Agenda;
04 }else
05 if(state is the state of the plan in *Intention* component)
06 Add action *SuppressContents* in Agenda;
07 Add action *UpdateGoal* in Agenda;

Name: Prohibition_Action_for_Plan

Input: Prohibited action; plan of *Intention*
Output: plan
01 if (Action is executing Plan) {
02 Add action *SuppressContents* in Agenda;
03 Add action *ScheduleCandidates* in Agenda;
04 }else
05 if(Action in Plan component)
06 Add action *SuppressContents* in Agenda;
07 Add action *UpdateGoal* in Agenda;

Name: Obligation_Action_for_Plan

Input: Obligated action; plan of *Intention*
Output: plan
01 Add action *CreatPlan* into Agenda;
02 Add action *ExecutePlanStep* in Agenda;
03 Add action *TerminatePlan* in Agenda;

In the algorithm *Obligation_State_for_Goal*, the *goal(state)* means to make the obligation state as a goal. PD searches this goal in *D* component, and upgrades the priority of this goal as soon as the goal is found. If the goal dose not exist in *D* component, PD will create this goal in *D*. *Achieve goal(state)* means to pursuer this goal right now. In the Algorithm *Obligation_Actiion_for_Plan* PD creates a plan which obtains the execution of the obligated action and adds corresponding actions on the agenda. The plan will be deleted as soon as its execution finished. When a certain state of an agent is prohibited by a **Prohibition** policy, PD will search this state both in *D* and in *I* component. In the algorithm *Prohibition_State_for_Goal*, the state will be treated as a goal. PD will suppress this goal and add *DeliberateNewOption* which is used for selecting of another goal to pursue if the goal is "Active" state (line 1-4).

When the prohibited goal is being "Suspended" or "Option" state, PD will suppress it and executes *UpdateGoal* by updating the *D* component. Algorithm *Prohibition_State_for_Planl* is designed to search the prohibited state in *I* component. When the prohibited action belonging to the executing plan, PD will suppress this plan and add *ScheduleCondidates* on the agenda to select another plan. If the state can be achieved by the "Suspended" or "Options" plans, PD will suppress the plan and update *I* component. If the user prohibits a certain action to be executed in terms of **Prohibition** policy, PD will search this action on agenda and in *I* component. When the action is on the agenda, PD will suppress it and call another plan on the agenda (line 1-4). If the action is not executing currently but is contained in some plans in *I* component, PD will suppresses these plans and update the *I* component (line 5-7).

5   Case Study

In this section, we will analyze the case that we introduced in section 2. In this case, the actions of robots should be designed in the design phases. These actions include: random searching, taking ore, sending message, responding message etc. On the other hand, we assume that there are two scenarios need to be considered. The first scenario is that there is only one ore source in the environment. To find resource and carry ore more quickly, it is appropriate to make each robot performs behaviors as follows: 1) randomly works in the environment to find ore resource; 2) carries ore back to base if finds the ore resource; 3) broadcasts position of ore resource if finds it; 4) goes to the position and carries ore to base if having received position information from other robot. To realize these behaviors, the user can deploy the policies as follows:

Obligation(*IF* Searching *WHEN* Others_Send Message *DO* TakeOreFromReceived-
    Position)
Obligation(*IF* Find_Ore_at_Some_Position *WHEN* $ *DO* Broadcast(Position))

In the second scenario, robots will be deployed in an environment which has more than one ore resource. Moreover, the user needs all robots to carry the found ore from near to far. According to this requirement, robots should firstly collect some ore resource positions whatever the position is found by itself or received from others. Then select the nearest one from the base to carry ore. To realize this requirement, user should add some new policies:

> **Obligation** (*IF* Searching *WHEN* Others_Send_Message *DO* storage_message)
> **Obligation** (*IF* Message_number=MAX *WHEN* $ *DO* TakeOre_Nearest)
> **Prohibition** (*IF* Message_number <MAX]*WHEN* $ *DO* RespondMessage)

We have implemented the BDIP robots in a simulation way by using the Jadex platform [10]. The basic actions of agents are implemented as *Plan* of the agent. Some important information such as ore resource positions are implemented as *Belief*, the interaction among agents this case is implemented as the message events of the agent. The *Intention* of the robots are to find the ore resource and carry ore back to base. On the other hand, the *policy conditions* are implanted as either a part of *belief* (e.g. ore resource) or implemented as the message events. When the system is running, if the *policy conditions* are satisfied, the agent will perform its behaviors comply with these policies, the trigger condition of the plans are implemented same as the *policy conditions*.

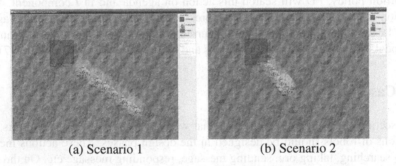

(a) Scenario 1   (b) Scenario 2

Fig. 3. Running snapshot of the Case

Fig. 3 shows the running snap shot of the case. The gray transparent pane is the base. The number below the base is the amount of ore in the base which has been carried by robots. The yellow transparent cycle of a robot presents the scope that the robot can explore for ore. The red points in environment mean the ore resources which have not been found by robots. The gray points mean the ore resource which has been found by robots and the number below the gray points mean the remained reserves of the ore resource. Fig. 3(a) shows the first scenario in which only one ore resource in the environment. From the figure we can see all robots are carrying ore from this resource when one of them has found it. Fig. 3 (b) shows the second scenario which has many ore resources in the environment. In Fig. 3(b), the robots have collected enough positions of the ore resources and carry the ore from the resource which is the nearest to the base among the collected positions.

6   Related Works

Recently, many research efforts have been made on the agent modeling based on norms and policies. However majority of them are engaged in the design of norm-based MAS organization, which propose norms to have an effective influence on agent and agent role, for example organization regimentation [11] and enforcement mechanisms [13][14]. In the agent architecture aspect, the main contributions of norm acceptance of agents are focus on the theory of BDI agent, e.g. [3][9]. These works focus on the theory frameworks e.g. logical expressions to explain how to represent the norms or policies in the agent and how to influence the reasoning process of the agent. The implement of architecture and running mechanisms of BDI or norm-based BDI agent are relatively few. [12] proposes a multi-level agent architecture, in which norms can communicated, adopted and used as meta-goals. [8] proposes a BDI Interpreter architecture for the running mechanisms of BDI agent, which our work can be seen as an extension of it.

7   Conclusion and Future Works

In the development of self-organizing MAS, the great challenge is how to bridge the huge gap between the agent local interaction behaviors and the system macro emergence characteristics. Against this open issue, this paper proposes a BDIP architecture of software agent. In the BDIP architecture, agent can perceive the policies of the system and adjusts its behaviors according to policies. A flexible running mechanism is also proposed based on the architecture, which executing meta-operations from a dynamic agenda structure. The architecture and its running mechanism are flexible enough to support the adjustments of system policies so that the users could control the self-organizing process and result by changed policies in the system. Moreover, the details of policy-based decision algorithms of agents are designed, and a case study of policy-based self-organizing robot system is implemented in a simulation way with the Jadex platform. Through the case study we can see that the changes of policies of the system can alter the self-organizing emergence result, and users can satisfy different system requirements by adjusting the policies.

   In future works we still focus on the development of the policy-based self-organizing multi-agent systems. A PSOMAS developing environment named PSOMASDE is in our current works. The developing environment include agent design platform and running environment, as well as the policies can be represented by the XML files that can be loaded by the system. Besides the implementation of such systems, the development methodology is also in consideration, which is based our previous work named ODAM [15].

Acknowledgement. The research acknowledges financial support from Natural Science Foundation of China under granted No 61070034, Program for New Century Excellent Talents in University, and Opening Fund of Top Key Discipline of Computer Software and Theory in Zhejiang Provincial Colleges at Zhejiang Normal University.

References

1. Giovanna, D.M.S., Marie, P.G., Anthony, K.: Self-Organizsation in MAS. Technology report, Agent Link III Technical Forum Group (2005)
2. Mamei, M., Ronaldo, M., Zambonelli, F.: Case Studies for Self-Organization in Computer Science. Journal of Systems Architecture 52(8), 443–460 (2006)
3. Dignum, F., Kinny, D., Sonenberg, L.: Motivational Attitudes of Agents: On Desires, Obligations, and Norms. In: Dunin-Keplicz, B., Nawarecki, E. (eds.) CEEMAS 2001. LNCS (LNAI), vol. 2296, pp. 83–92. Springer, Heidelberg (2002)
4. Guo, Y., Mao, X., Hu, C.: A Survey of Engineering for Self-Organization Systems. In: 23th Software Engineering & Knowledge Engineering, pp. 527–531. Knowledge Systems Institute Press, USA (2011)
5. Parunak, H.V., Sevn, A.B.: Software Engineering for Self-Organizing Systems. In: 12th International Workshop on Agent-Oriented Software Engineering, AAMAS 2011 (2011)
6. Guo, Y., Mao, X., Hu, C.: Design Pattern for Self-Organization Multi-agent Systems based on Policy. In: 6th International Conference on Frontier of Computer Science and Technology, pp. 1572–1577. IEEE Press, USA (2011)
7. Rao, A.S., Georgeff, M.P.: Modeling Rational Agents within a BDI-Architecture. In: 2nd International Conference on Principles of Knowledge Representation and Reasoning, pp. 473–484. Kaufmann Press, USA (1991)
8. Pokahr, A., Braubach, L., Lamersdorf, W.: A Flexible BDI Architecture Supporting Extensibility. In: 2005 IEEE/WIC/ACM International Conference on Intelligent Agent Technology, pp. 379–385. IEEE Press, USA (2005)
9. Crida, N., Argente, E., Noriega, P., Botti, V.: Towards a Normative BDI Architecture for Norm Compliance. In: 2010 Multi-agent Logics, Languages, and Organisations Federated Worshops, pp. 65–81 (2010)
10. Braubach, L., Pokahr, A., Lamersdorf, W.: Jadex: A BDI Agent System Combining Middleware and Reasoning. In: Rainer, U., Matthias, K., Monique, C. (eds.) Software Agent-Based Applications, Platforms and Development Kits, pp. 143–168. Birkhauser Press (2005)
11. Criado, N., Argente, E., Botti, V.: Thomas: An Agent Platform for Supporting Normative Multi-agent Systems. Journal of Logic and Computation (2011), doi:10.1093/logcom/exr025
12. Castelfranchi, C., Frank, D., Catholijin, M.J., Jan, T.: Deliberative Normative Agents: Principles and Architecture. In: Jennings, N.R. (ed.) ATAL 1999. LNCS, vol. 1757, pp. 364–378. Springer, Heidelberg (2000)
13. Modgil, S., Faci, N., Meneguzzi, F., Oren, N., Miles, S., Luck, M.: A Framework for Monitoring Agent-based Normative Systems. In: 8th International Conference on Autonomous Agents and Multi-agent System, pp. 153–160. ACM Press, USA (2009)
14. Grossi, D., Aldewereld, H., Dignum, F.: *Ubi Lex, Ibi Poena*: Designing Norm Enforcement in E-Institutions. In: Noriega, P., et al. (eds.) COIN 2006. LNCS(LNAI), vol. 4386, pp. 101–114. Springer, Heidelberg (2007)
15. Mao, X., Hu, C., Wang, J.: An Organization-based Approach to Developing Self-Adaptive Multi-Agent Systems. International Transactions on Systems, Science and Applications 5(4), 297–317 (2009)

Optimization of Initial Centroids for K-Means Algorithm Based on Small World Network

Shimo Shen and Zuqiang Meng

College of Computer, Electronics and Information, Guangxi University,
Nanning 530004, China
shenshimo@126.com, mengzuqiang@163.com

Abstract. K-means algorithm is a relatively simple and fast gather clustering algorithm. However, the initial clustering center of the traditional k-means algorithm was generated randomly from the dataset, and the clustering result was unstable. In this paper, we propose a novel method to optimize the selection of initial centroids for k-means algorithm based on the small world network. This paper firstly models a text document set as a network which has small world phenomenon and then use small-world's characteristics to form k initial centroids. Experimental evaluation on documents croups show clustering results (total cohesion, purity, recall) obtained by proposed method comparable with traditional k-means algorithm. The experiments show that results are obtained by the proposed algorithm can be relatively stability and efficiency. Therefore, this method can be considered as an effective application in the domain of text documents, especially in using text clustering for topic detection.

Keywords: k-means, text clustering, small world network, SNN.

1   Introduction

Clustering is useful in a wide range of data analysis fields, including data mining, document retrieval, image segmentation, and pattern classification. The goal of clustering is to group data into clusters so that the similarities among data members within the same cluster are maximal while similarities among data members from different clusters are minimal [1].

Among the different classes of clustering algorithms, the distance-based methods are the most popular methods in a wide variety of applications, while two best-known methods for distance-based clustering are the partition clustering algorithm and hierarchical clustering algorithm. K-means algorithm is one of the most widely used distance-based partitioning algorithms, and that separates data into k mutually excessive groups [2].

K-means algorithm is very popular because of its ability to cluster huge data set and its simplicity. However, K-means algorithm is quite sensitive to the initial cluster centers picked during the clustering, and does not guarantee unique clustering because different results obtained with randomly chosen initial centroids. The final cluster centroids may not be the optimal ones because of the k-means algorithm converges

Z. Shi, D. Leake, and S. Vadera (Eds.): IIP 2012, IFIP AICT 385, pp. 87–96, 2012.

into local optimal solutions. The initial centroids affect the quality of k-means algorithm, especially in documents clustering. Therefore, it is quite important for k-means algorithm to have good initial cluster centroids.

There are several methods to reduce the sensitivity of initial centroids picked during clustering proceeding.

Cutting,etl[3] use group average agglomerative clustering algorithm to select initial centroids.

Likas[4] proposed the global k-means algorithm which is an incremental approach to clustering which dynamically adds one cluster center at a time through a deterministic global search procedure consisting of N (with N being the size of the dataset)executions of the k-means algorithm from suitable initial positions.

Arthur and Vassilvitskii[5] proposed k-means++ algorithm, a specific way of choosing centers for the k-means algorithm, which choose the centers by weighs the data points according to their squared distance from the closest center already chosen, and improves both the speed and the accuracy of k-means. However, the k-means++ clustering method sometimes generates bad clusters because it depends on the selection of the first initial center. The first initial center is chosen uniformly at random from data points set.

Onoda.etl [6] proposed a seeding method based on the independent component analysis for the k-means clustering method. This method can be summarized as two steps. First, k independent components IC_i ($i=1,...,k$) were obtained from given data x. Second, the initial centers were selected according by k independent components. This method is useful for Web corpus.

In the paper, we propose a novel method to optimize the selection of initial centroids for k-means algorithm based on small world network. A novel network which has the small world phenomenon is built by connecting similar documents, and then we use small-world's characteristics to form k initial centroids for k-means algorithm.

The rest of the paper is organized as follows. Section 2 introduces Vector space model and small world network. Selecting initial centroids for k-means algorithm based on small world network is proposed in Section 3. Section 4 is the algorithm of improved k-means clustering algorithm. Section 5 presents extensive experimental results. Conclusion follows in Section 6.

2   Preliminaries

2.1   Text Document Representation

Vector Space Model. The vector space model is used to represent the document as vector with a list of words and a list of weights of each word occurs. The vector of document is defined as:

$$d_i = \{w_{ij}\}, j = 1.2.....m, i = 1.2.....n$$

(1)

Where

w_{ij} is the weight of jth term in ith document

m is the total number of unique terms appearing in the document

n is the number of documents in the document collection.

With the vector space model, the text data is converted into structured data that computer can handle, and the similarity between the two documents is converted into the similarities between the two vectors.

Cosine Similarity Computation between Documents. The cosine similarity is often used to measure the similarity between two documents in text mining. For text matching, the document vectors d_i and d_k are the TF-IDF vectors of the documents. Cosine similarity between d_i and d_k is defined as:

$$SC(d_i, d_k) = \frac{\sum_{j=1}^{t} d_{kj} d_{ij}}{\sqrt{\sum_{j=1}^{t} (d_{ij})^2} \sqrt{\sum_{j=1}^{t} (d_{kj})^2}} \tag{2}$$

Where

d_{ij} (or d_{kj}) is the jth term in d_i (or d_k) document.

The cosine measure gives values between 0 and 1, and the more common words of documents have, the bigger of cosine similarity is.

Definition 1. Cosine Similarity Threshold λ. We suppose that cosine similarity threshold λ to measure the similarity between two documents. While cosine similarity of documents is greater than λ, these documents are considered as the same class, and vice versa. This method is good at handling noise, outliers and reducing the dimension of the network.

Shared Nearest Neighbor Similarity. SNN (shared nearest neighbor similarity) [7], an approach to similarity two documents, with the value is the number of the same points that similar to the two documents.

For two points, x and y, the shared nearest neighbor definition of similarity between in the manner indicated by algorithm as follows.

1) Find the k-nearest neighbors of x and y;
2) If x and y are not among the k-nearest neighbors of each other then
3)   Similarity $(x, y) \leftarrow 0$;
4) Else
5)   Similarity $(x, y) \leftarrow$ number of shared neighbors;
6) End If.

SNN can be any positive integer, the more SNN is, the more relevant to each other the two texts are.

Definition 2. SNN Threshold v. We suppose that SNN threshold v to measure the similarity between two documents.

2.2   Small World Network

The small world phenomenon came from the research of sociologist Milgram carried out in 1967 to trace the shortest path in the U.S. social network.

Watts spent a deep studying on the small world phenomenon in 1998, and proposed that a small world network has characteristics of high concentration and short path [8].

Cluster Coefficient. The cluster coefficient [9] C is defined as follows. Suppose that a vertex i has K_i neighbours; then at most $K_i(K_i - 1)/2$ edges can exist between them (this occurs when every neighbour of i is connected to every other neighbor of i). φ_i denote the actual number of edges that exist between K_i and neighbor nodes. The cluster coefficient of vertex i is C_i:

$$C_i = \frac{\varphi_i}{K_i(K_i - 1)/2}. \tag{3}$$

The cluster coefficient of the whole network is C:

$$C = \frac{1}{N}\sum_{i=1}^{N} C_i. \tag{4}$$

Network with small-world nature have high cluster coefficient feature.

Power Law Distribution of Node Degrees Features. Ferrer's[8] study also showed that: the lexical co-occurrence network also has a scale-free features ,the degree of network node distribution is close to the power law. The probability $P(k)$ of having a node with degree k scales as $P(k)\approx k^{-r}$, where r is a constant.   It reflects that the status of the connection between each node in the network is severe heterogeneity. In this network only a very small number of nodes have many connections with other nodes and become hub nodes, and most of nodes have few connections. Hub nodes play a leading role in the operation of scale-free network. So we use this feature form the k seeds.

3   Selecting Initial Centroids for K-Means Algorithm

The three major steps of the proposed selection of the initial centroids approach are described as follows.

Step 1. Model a document set as document network based on vector space model. Document is represented as the document space vector, and the similarity of documents was calculated by vector way. Two documents which similarity is greater than threshold λ were connected. Form document network and express as $G= \{V, E\}$, where V is a set of vertex and E is a set of edges or connection between documents. In the document network, a vertex corresponds to a document that connected with one document at least. An edge $e = (i, j) \in E$ stand for a connection between vertices i and j.

When drawing a graph of document network, the Fruchterman-Reingold Algorithm [10] was used to. Fig.1shows document network graph obtained after processing about 842 documents.

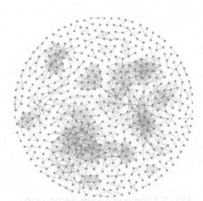

Fig. 1. Document network

In Fig.1, black nodes are documents, and two documents are linked while the similarity of them is greater than threshold λ. The cluster coefficient of the document network is C=0.382, while random network is $C_{random} = 1.55*10^{\wedge(-4)}$. It can be seen that $C >> C_{random}$, and the document network has high cluster coefficient. The distribution of degrees of document network is shown in Table 1.

Table 1. Distribution of nodes' degree

degree	2	3	4	5	6	7	8	9	10	11	12	13	14
number	112	53	37	25	15	26	21	11	5	7	4	1	1
P(k)	0.13	0.06	0.04	0.03	0.017	0.03	0.025	0.013	0.005	0.008	0.004	0.0011	0.001
$k^{-2.3}$	0.2	0.08	0.04	0.03	0.016	0.011	0.008	0.006	0.005	0.004	0.003	0.003	0.002

From Table 1, it shows that the probability P(k) of having a node with degree k scales as $P(k) \approx k^{-2.3}$, and the degree of network node distribution is close to the power law.

Fig.1 and Table 1 illustrate that the document network has a high cluster coefficient characteristic and the node degree distribution is close to the power-law distribution features which is of small world network. According to the distribution of power-law distribution, we can see only a small number of nodes with many connections to other nodes. These nodes can be thought as common documents in the network, and these documents play a leader role in the network. The content of these common documents are the common content that most probably appear in the dataset. So we can use the center of these nodes to represent the initial centroids of k-means algorithm.

Step 2. Split network based on its features with SNN. In order to get these common documents, we can segment network based on its features with the v, and form network as show in Fig.2.

Fig. 2. Common document network

Step 3. Form initial centroids. Compute the centroid of each subgraph and make these centroids as initial centroids of clusters.

4   Proposed Algorithm

We use the method which proposed in section 3 to improve   the k-means algorithm, and the algorithm is described as follows.

1. Set the cosine similarity threshold λ. Make a Link between two documents which similar is greater than λ, and form a network.
2. Initialize the classification sample collection, $M_1 = M_2 = ... = M_k = \{\}$.
3. Set the SNN threshold v.
4. Find the maximum degree of nodes from sample collection and join in set M_1.
5. Identify the nodes with the nearest neighbor similarity is not less than v in the sample set, and added to the collection M_1.
6. Repeat step 4 on the collection of all data points in the M_1 is no longer changes until the collection of M_1.
7. Delete the nodes that in set M_1 from sample collection.
8. Repeat steps 4-6, until M_k is generated.
9. Calculate the centers of $M_1, M_2,...,M_k$ and make these centers as initial centroids of clusters.
10. Form k clusters by assigning each point to its closet centroid. Then recomputed the centroid of each cluster.
11. Repeat step10 until centroids do not change.

5   Experimental Analysis

We use documents datasets that are downloaded from http://www.163.com (November 27, 2011) for evaluation in our experiments. This dataset has 5 categories and conclude 842 documents. The categories come from the http://www.163.com, and contain the following types: News, Reading, Education, Science and Technology.

Experiment compared the traditional k-means and the proposed k-means on various evaluation measures. Details are described as follows.

5.1   Evaluation Measures

To evaluate the proposed approach, we employed three measures of total cohesion, accuracy and recall to evaluate the quality of clusters generated by different methods. The definitions are as follows.

Total Cohesion. It is an objective function, which measures the quality of a clustering, in this experiment our objective is to maximize the similarity of the documents in a cluster to the cluster centroid, this quantity is known as the cohesion of the cluster. The total cohesion is defined as follows:

$$\text{Total Cohesion} = \sum_{i=1}^{k} \sum_{x \in C_i} cosine(x, c_i) \tag{5}$$

where
k is the number of cluster
C_i is the ith cluster
c_i is the centroid of the ith cluster and $\mathbf{x}$ is the node that belong to C_i.

Purity. Purity measure the degree to which each cluster consists of objects of a single. For each cluster, the class distribution of the data is calculated first, i.e., for cluster C_i and class G_j of text .The corresponding purity of C_i and G_j are defined as:

$$precision(C_i, G_j) = n_{ij} / n_i \tag{6}$$

where
n_i is the number of objects in cluster C_i
n_{ij} is the number of objects of class G_j in cluster C_i . The purity of cluster C_i is as:

$$precision(C_i) = \max_{j} precision(C_i, G_j) \tag{7}$$

The overall purity of a clustering is as:

$$precision = \sum_{i=1}^{k} \frac{n_i}{N} precision\,(C_i) \qquad (8)$$

where
N is the total number of text, k is the number of cluster.

Recall. Recall measure the extent to which a cluster contains all objects of a specified class. Similarly as purity, the recall rate of this method can be defined as:

$$recall = \sum_{i=1}^{k} \frac{n_i}{N} recall(C_i) \qquad (9)$$

where
$recall(C_i) = \max_{j} recall(C_i, G_j)$ $recall(C_i, G_j) = n_{ij} / n_i$.

5.2   Setting of Threshold μ and ν

In order to find out more specific value for parameter μ and ν, we perform the proposed algorithm, with different values of μ and ν and the results are shown in Fig.3 and Fig.4.

Fig.3 shows the result of different values of λ when ν=2.

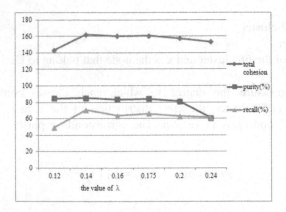

Fig. 3. The result of different values of λ

Fig.3 shows that better results are gotten when the value of λ is between 0.14 and 0.2. Specially, the three evaluation values have the highest value when λ=0.14. May be 0.14 is the suitable value to measure the similarity between two documents.

Fig.4 shows that the result of different values of v when λ=0.14. The best results are gotten when the value of v is 2.

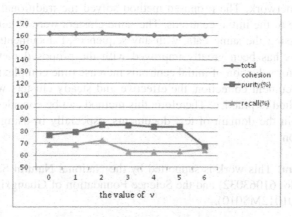

Fig. 4. The result of different values of v

5.3   Evaluation of Clustering Efficiency

To further evaluate the proposed approach, the Comparison of traditional k-means algorithm [2] and proposed algorithm with the best parameters on total cohesion, purity and recall have done. The traditional k-means algorithm is run 50 times, and the three better results are selected from all of the run results and are listed in Table 2. The proposed algorithm is run only one time because it can get a unique initial seeding. It is observed that proposed algorithm has high values in every evaluation.

Table 2. Cluster quality of different cluster algorithms

The name of algorithm	Totalcohesion	purity(%)	Recall(%)
First of Basic k-means algorithm	142.54	47.41	50.37
Second of Basic K-means algorithm	146.7	63.3	53.7
Third of Basic K-means algorithm	145.55	53.5	48.4
Proposed algorithm	162.56	85.14	70.5

There are two aspects which can prove the stable results can be obtained by the proposed algorithm. On the one hand, theoretically, for the same text corpus, with any sequence of corpus and at any time to run, the document network which the corpus are model as to is constant, and then the common document networks are stable. Thus the center of these common document network (the initial centroids for k-means clustering algorithm) are stable, Therefore, the results are obtained by this method is stable. On the other hand, the experiment shows the result of proposed method that runs several times on the same corpus is the same.

6   Conclusion

In this paper, we improve the k-means algorithm in selection of initial centroids based on small world network. The proposed method solved the traditional k-means algorithm is sensitive to the initial centroids. The same clusters were obtained by the proposed method, using the same data, with any sequence of documents. And also the purity of clusters has been greatly improved with the proposed method. Although additional work for selection of initial centroids increase time complexity to $O(n^{\wedge 2})$, n is the size of document collection, the effective and steady clusters with $O(n^{\wedge 2})$ remain can be applied to practice. Therefore, this method can be considered as an effective application in the domain of text documents, especially in using text clustering for topic detection.

Acknowledgment. This work is supported by the National Natural Science Foundation of China (No. 61063032) and the Science Foundation of Guangxi Education Department (No. 201012MS010).

References

1. Feldman, R., Sanger, J.: The text mining handbook, pp. 82–92. Posts & Telecom Press, Beijing (2009)
2. Aggarwal, C., Zhai, C.: A survey of text clustering algorithms, pp. 77–128. Springer (2012)
3. Cutting, D., Karger, D., Pedersen, J., Scatter/Gather, J.: A Cluster-based Approach to Browsing Large Document Collections. In: ACM SIGIR Conference (1992)
4. Likas, A., Vlassis, N., Jakob, J.V.: The global k-means algorithm algorithm. Pattern Recognition 36(2), 451–461 (2003)
5. Arthur, D., Vassilvitskii, S.: K-means++: the advantages of careful seeding. In: ACM-SIAM Symposium (2007)
6. Onoda, T., Sakai, M., Yamada, S.: Independent Component Analysis based Seeding method for k-means Clustering. In: IEEE/WIC/ACM Conference (2011)
7. Tan, P., Steinbach, M., Kumar, V.: Introduction to Data Mining, pp. 385–387. Posts & Telecom Press (2011)
8. Cancho, R.F., Sole, R.V.: The small world of human language. The Royal Society of London, Biological Sciences(Series B) 268(1482), 2261–2265 (2001)
9. Wars, D.J., Strogatz, S.H.: Collective dynamics of small-world networks. Nature 393(6684), 440–442 (1998)
10. Thomas, M.J.F., Edward, M.R.: Graph Drawing by Force-directed Placement. Software: Practice and Experience 21(11), 1129–1164 (1991)

ECCO: A New Evolutionary Classifier with Cost Optimisation

Adam Omielan and Sunil Vadera

School of Computing, Science and Engineering,
University of Salford, Salford M5 4WT, UK

Abstract. Decision tree learning algorithms and their application represent one the major successes of AI. Early research on these algorithms aimed to produce classification trees that were accurate. More recently, there has been recognition that in many applications, aiming to maximize accuracy alone is not adequate since the cost of misclassification may not be symmetric and that obtaining the data for classification may have an associated cost. This has led to significant research on the development of cost-sensitive decision tree induction algorithms. One of the seminal studies in this field has been the use of genetic algorithms to develop an algorithm known as ICET. Empirical trials have shown that ICET produces some of the best results for cost-sensitive decision tree induction. A key feature of ICET is that it uses a pool that consists of genes that represent biases and parameters. These biases and parameters are then passed to a decision tree learner known as EG2 to generate the trees. That is, it does not use a direct encoding of trees. This paper develops a new algorithm called ECCO (Evolutionary Classifier with Cost Optimization) that is based on the hypothesis that a direct representation of trees in a genetic pool leads to improvements over ICET. The paper includes an empirical evaluation of this hypothesis on four data sets and the results show that, in general, ECCO is more cost-sensitive and effective than ICET when test costs and misclassifications costs are considered.

Keywords: Decision Tree Induction, Cost-Sensitive Learning.

1 Background

Decision tree learning algorithms have been widely studied since Quinlan first developed the ID3 algorithm [1], with significant impetus given by developments such as C4.5 [2], and its availability in the Weka system [3]. These decision tree learning algorithms aim to take a table of examples as input and produce a decision tree of the kind shown in Figure 1, where there are decision nodes such as Test A, Test B and Test C, and outcomes of the tests decision nodes that are labeled on the links, such as Improve, Same and Deteriorate.

Z. Shi, D. Leake, and S. Vadera (Eds.): IIP 2012, IFIP AICT 385, pp. 97–105, 2012.
© IFIP International Federation for Information Processing 2012

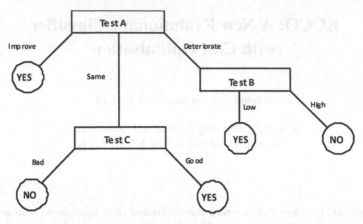

Fig. 1. A decision tree

Early decision tree learning algorithms, such as CART [4], and ID3 [1] aimed to learn such trees from a set of training data. They built decision trees in a greedy fashion, where a test is first chosen as a root. The training examples are then divided into subsets associated with each possible value of the chosen test. The process is then applied recursively until the examples in the subsets all have the similar outcomes, resulting in leaf or decision nodes. A significant step in this greedy algorithm is the criteria for selecting the next test. Most of the early algorithms adopted information theoretic measures that selected tests on the basis of the amount of information gained towards the final classification. The primary aim of these greedy algorithms was to produce accurate decision trees, where accuracy was estimated by the proportion of cases for which the data in a testing set were correctly classified.

However, several authors recognized that maximizing accuracy is not adequate for many real world applications and that costs need to be taken into account (e.g., [4,5,6]). There are several types of costs but the main ones include the costs of misclassifying an example and the cost of acquiring information [7]. For example, in a medical application misclassifying a person as healthy when they are ill can be higher than misclassifying them as ill, and the cost of carrying out an MRI scan can be higher than a blood test.

Research on development of cost-sensitive decision tree algorithms can be broadly classified into the following main categories:

1. Algorithms that adopt the Greedy algorithm but that adapt the information theoretic selection measures to include costs. The key difference amongst algorithms in this category is how the information gain measure is adapted to include costs and whether they take account of just the costs of the tests, or also take cost of misclassification. Algorithms that take account of just costs of attributes include CS-ID3 [8], IDX [9], EG2 [10] , CSGain [11]. Algorithms that also adapt the information theoretic measure to include costs of misclassification include PM [12], and CS-4.5 [13].

2. Algorithms that utilise bagging and boosting methods that generate alternative trees and combine them in a way that reduces cost. For example, the MetaCost system [14] resamples the data several times and applies a base learner (such as C4.5) to each sample to generate alternative decision trees. The decisions made on each example by the alternative trees are combined to predict the class of each example that minimizes the cost and the examples relabeled. The relabeled examples are then processed by the base learner, resulting in a cost-sensitive decision tree. Other examples of systems that adopt bagging include B-PET & B-LOT [15], and examples of systems that use boosting methods include AdaCost [16] and Lp-CSB [17].

3. The use of Genetic Algorithms (GAs) to generate and evolve cost-sensitive trees. The idea with these methods is to begin with a genetic pool, select the fittest, apply evolution operators such as mutation and crossover to generate a new pool and repeat the evolution cycles, resulting in improved pools. The main algorithm in this category is Turney's ICET [5] (Inexpensive Classification with Expensive Tests). ICET begins by dividing the training set of examples into two random but equal parts: a sub-training set and a sub-testing set. An initial population is created consisting of individuals with random values of CA_i, ω, and CF, which are parameters required by C4.5 and EG2 . C4.5, with the EG2's cost function, is then used to generate a decision tree for each individual. These decision trees are then passed to a fitness function to determine fitness. This is measured by calculating the average cost of classification on the sub-testing set. The next generation is then obtained by using the roulette wheel selection scheme, which selects individuals with a probability proportional to their fitness. Mutation and crossover are used on the new generation and passed through the whole procedure again. After a fixed number of generations the best decision tree is selected.

A comprehensive survey of these algorithms that includes a framework and timeline covering over 50 algorithms can be found in Lomax and Vadera [18]. This paper focuses on this last category: algorithms that use genetic algorithms for generating cost-sensitive decision tree algorithms. By far the most widely known and cited algorithm in this category is Turney's ICET [5]. In our previous empirical evaluations, ICET produced some of the best results in comparison to algorithms from other categories [19]. As mentioned above, a key feature of ICET is that the population of individuals consists of biases that are utilized when C4.5 is used to generate decision trees. That is the individuals in the population are not direct encodings of trees.

This paper presents a new algorithm called ECCO that is based on the hypothesis that direct encoding of trees in a pool could improve upon the results of Turney's seminal   system ICET.

The next section of the paper describes how ECCO is developed using a GA and is followed by a section that presents an empirical evaluation of the hypothesis that a direct representation of trees leads to improvements over the ICET system.

2   The ECCO Algorithm

To develop a cost-sensitive decision tree algorithm that is based on the use of a genetic algorithm we need to address the following questions:

- How can the trees be represented as genes?
- What mutation and cross-over operators are appropriate?
- What sort of fitness function is appropriate?

The following subsections describe how these questions are addressed, leading to the ECCO algorithm.

2.1   Encoding the Tests and Trees as Genes

The genes in GAs are represented as bit strings. To represent a tree as a bit string, we need to first code each test. Each test is given a unique identification number, from 1 to n, where n is the total number of tests. Each identifier is then converted to its binary equivalent. To code a tree as a fixed length binary string, we need to assume a fixed size tree. The maximal size of a tree is determined by the number of possible values of the tests. For example, a test such as 'It has wings' has only two outcomes, true or false. More complex tests give rise to more outcomes, e.g. A test such as 'How many eggs?' when classifying a recipe may well have '0, 1, 2, 3' eggs possible, and so four child nodes are needed. Given that we know the tests and the maximal number of possible outcomes of tests, we can compute the maximal size of a tree. This can then be used to compute the length of the bit string required to represent a tree. Figure 2 illustrates a mapping for tree of depth 3 assuming a maximal of two outcomes per test (the idea extends to n-ary trees).

It's worth noting that continuous variables are handled by using the process used in systems such as C4.5 where they discretized to axis parallel binary tests (See Quinlan C4.5 [2] for details).

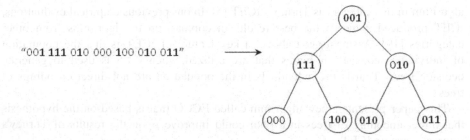

"001 111 010 000 100 010 011"   ⟶

Fig. 2. An illustration of how trees are represented genes

In general, a bit string may not correspond to a fully populated tree and, so such maximal trees need interpretation. Even if we begin with fully constituted trees in the initial population, such trees may arise because of the evolution process described in the next section. The interpretation used is to assume that the highest parent node that does not correspond to a test is assumed to be a leaf node (i.e. a classification denoting an outcome).

2.2   Evolving the Population

Having developed a representation of trees, the next question is: how can the populations evolve? This is done by using the standard crossover and mutation operations. The crossover process takes two of the fittest genes of a population, and combines the first part of one, and the second part of the other to create a new gene. Figure 3 illustrates the crossover operator on two genes representing decision trees. The position of the split is at any random point in the string, but must be on a boundary between nodes. The genes that are chosen for the crossover operation include a random mix of the genes, with greater weighting given to the fittest genes.

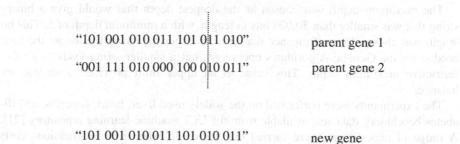

"101 001 010 011 101 011 010"   parent gene 1

"001 111 010 000 100 010 011"   parent gene 2

"101 001 010 011 101 010 011"   new gene

Fig. 3. The crossover operation on representations of trees

The mutation operation changes up to 2% of the nodes. Thus if there are fifty nodes at most in a string, up to one of them will be changed into a different node. This helps prevent the tree from becoming stuck if the fittest genes become identical, and to evolve further to test other slightly different trees.

2.3   Fitness Function

The fitness function has to assess the overall cost of the tree. First the tree needs to be trained to set the classification at each leaf node. That is, a proportion of the data provided is processed through the tree, marking at each node that it has been accessed, and at the classification node the class that row of data belongs to, is appropriately marked. The class chosen at a leaf node is then selected as the one that minimizes the cost of misclassification based on the user provided costs of misclassification.

Once a tree is trained, it is possible to traverse the tree with each example and compute its cost as the sum of the tests used and the cost of any misclassification. The average cost over the examples in the training set is then the fitness function.

3   Empirical Comparison with ICET

This section presents an empirical evaluation of ECCO. The primary idea being explored is that utilizing a direct encoding of decision trees for a GA instead of utilizing an intermediate decision tree generator will result in better performance. The ECCO algorithm was implemented by utilizing the GENSIS [20] genetic algorithm and a version of ICET implemented as part of previous work [6] was utilized.

The experiments were designed to test the performance of ECCO under a variety of conditions, and to make comparisons to the performance of ICET. To understand how the algorithms are affected, experiments were first conducted where the cost of misclassification of one class over another were set to: 10, 50, 100, 500, 1000, 5000, 10000. All the experiments were performed with ten iterations using randomly selected training sets consisting of two-thirds of the data and the remaining thirds used as testing data to compute the averages. The depth of trees for ECCO was also varied from 5 to 11.

The maximum depth was chosen as the deepest depth that would give a binary string that was smaller than 50,000 bits in length, with a minimum depth of 5. This bit length was chosen for performance reasons; the larger the string the longer the time needed for the Genetic Algorithm's operations, but a smaller string leads to a more restrictive maximum depth. This value as an upper limit provided a satisfactory balance.

The experiments were performed on the widely used liver, heart, hepatitis, and diabetes benchmark data sets available from the UCI machine learning repository [21]. A range of experiments were carried out, including when misclassifications costs alone are considered, and when both test and misclassification costs are considered. The results had a common pattern for all the data sets and occurs irrespective of whether or not test costs were included. This common pattern is illustrated by Figure 4, which shows the performance of ICET and ECCO (with a depth of 5, with pruning and including both test costs) on the Hepatitis data set. As this illustrates, as the cost of misclassification increases, ECCO is more sensitive and performs better than ICET. The scale of improvement is similar on the other data sets.

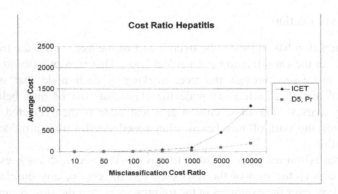

Fig. 4. Average cost as misclassification costs increase for hepatitis data

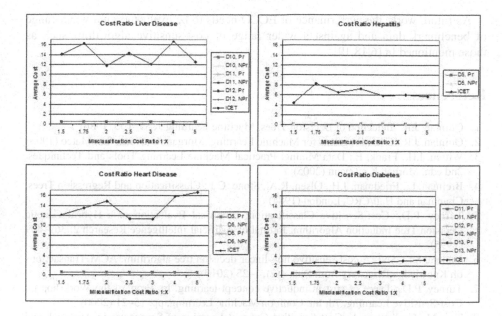

Fig. 5. Results with misclassification costs in the range 1.5 to 5

Given these results, the obvious question to explore is the behaviour of the two approaches when the costs of misclassification are smaller. Figure 5, below, presents the results when the misclassifications cost ratios are varied from 1.5 to 5 and the test costs available in the benchmark data sets are used. The Dn indicates that a tree of depth n is used and results with pruning are indicated with a Pr while results without pruning are indicated with a NPr. These results also show: (a) that ECCO converges to minima more quickly than ICET, (b) that ICET is not very sensitive to increases in misclassification costs, (c) in this range, there isn't much variation in the results whether or not ECCO uses pruning (indeed the differences are not visible in Figure 5 where the lines combine).

4   Conclusion

This paper has introduced a new cost-sensitive decision tree learner, ECCO, that is based on the use genetic algorithms to evolve trees. The algorithm utilizes a more direct coding of trees as genes than the widely cited ICET algorithm that uses biases as genes and utilizes C4.5 and EG2 to generate the trees from the biases. Empirical trials were conducted on four benchmark data sets and the results show that ECCO is more cost-sensitive than ICET to increases in the ratio of costs of misclassification and ECCO produces trees that enable more cost-effective decision making than ICET.

As future work, the performance of ECCO needs to be evaluated on a wider range of benchmark data and against a wider range of cost-sensitive algorithms such as those mentioned in [6,18,19].

References

1. Quinlan, J.R.: Induction of Decision Trees. Machine Learning 1, 81–106 (1986)
2. Quinlan, J.R.: C4.5: Programs for Machine Learning. Morgan Kaufman, San Mateo (1993)
3. Witten, I.H., Frank, E.: Data Mining: Practical Machine Learning Tools and Techniques, 2nd edn. Morgan Kaufmann (2005)
4. Breiman, L., Friedman, J.H., Olsen, R.A., Stone, C.J.: Classification and Regression Trees. Chapman and Hall/CRC, London (1984)
5. Turney, P.D.: Cost-Sensitive Classification: Empirical Evaluation of a Hybrid Genetic Decision Tree Induction Algorithm. Journal of Artificial Intelligence Research 2, 369–409 (1995)
6. Vadera, S.: CSNL A cost-sensitive non-linear decision tree algorithm. ACM Transactions on Knowledge Discovery from Data 4(2), 1–25 (2010)
7. Turney, P.D.: Types of cost in inductive concept learning. In: Proc. of the Workshop on Cost-Sensitive Learning, 7th Int. Conf. on Machine Learning, pp. 15–21 (2000)
8. Tan, M., Schlimmer, J.: Cost-Sensitive Concept Learning of Sensor use in Approach and Recognition. In: Proceedings of the 6th International Workshop on Machine Learning. ML 1989, Ithaca, New York, pp. 392–395 (1989)
9. Norton, S.W.: Generating Better Decision Trees. In: Proceedings of the Eleventh International Joint Conference on Artificial Intelligence, IJCAI 1989, Detroit, Michigan, USA, pp. 800–805 (August 1989)
10. Núnez, M.: The Use of Background Knowledge in Decision Tree Induction. In: Machine Learning, vol. 6, pp. 231–250. Kluwer Academic Publishers, Boston (1991)
11. Davis, J.V., Ha, J., Rossbach, C.J., Ramadan, H.E., Witchel, E.: Cost-Sensitive Decision Tree Learning for Forensic Classification. In: Fürnkranz, J., Scheffer, T., Spiliopoulou, M. (eds.) ECML 2006. LNCS (LNAI), vol. 4212, pp. 622–629. Springer, Heidelberg (2006)
12. Liu, X.: A New Cost-Sensitive Decision Tree with Missing Values. Asian Journal of Information Technology 6(11), 1083–1090 (2007)
13. Freitas, A., Costa-Pereira, A., Brazdil, P.: Cost-Sensitive Decision Trees Applied to Medical Data. In: Song, I. Y., Eder, J., Nguyen, T.M. (eds.) DaWaK 2007. LNCS, vol. 4654, pp. 303–312. Springer, Heidelberg (2007)
14. Domingos, P.: MetaCost: A general method for making classifiers cost-sensitive. In: Proceedings of the fifth ACM SIGKDD International Conference on Knowledge Discovery and Data Mining, pp. 155–164. ACM, New York (1999)
15. Moret, S., Langford, W., Margineantu, D.: Learning to predict channel stability using biogeomorphic features. Ecological Modelling 191(1), 47–57 (2006)
16. Fan, W., Stolfo, S.J., Zhang, J., Chan, P.K.: AdaCost: misclassification cost-sensitive boosting. In: 16th International Conference on Machine Learning, Bled, Slovenia, June 27-30, pp. 97–105 (1999)
17. Lozano, A.C., Abe, N.: Multi-class cost-sensitive boosting with p-norm loss functions. In: Proceeding of the 14th ACM SIGKDD International Conference on Knowledge Discovery and Data Mining, KDD 2008, Las Vegas, USA, August 24-24 (2008)

18. Lomax, S., Vadera, S.: A Survey of Cost-Sensitive Decision Tree Induction Algorithms. To Appear in ACM Computing Surveys 45(2) (2013)
19. Lomax, S., Vadera, S.: An Empirical Comparison of Cost-Sensitive Decision Tree Induction Algorithms. Expert Systems The Journal of Knowledge Engineering 28(3), 227–268 (2011)
20. Grefenstette, J.: Optimization of control parameters for genetic algorithms. IEEE Transactions on Systems, Man, and Cybernetics 16, 122–128 (1986)
21. Blake, C., Merz, C.: UCI Repository of Machine Learning Databases. University of California, Department of Information and Computer Science, Irvine, CA (1998), http://www.ics.uci.edu/mlearn/MLRepository.html

Reasoning Theory for D3L
with Compositional Bridge Rules

Xiaofei Zhao[1], Dongping Tian[1], Limin Chen[2], and Zhongzhi Shi[1]

[1] Key Laboratory of Intelligent Information Processing, Institute of Computing Technology,
Chinese Academy of Sciences, Beijing, China
zhaoxf@ics.ict.ac.cn
[2] Research Institute of China Unicom, Beijing, China

Abstract. The semantic mapping in Distributed Dynamic Description Logics (D3L) allows knowledge to propagate from one ontology to another. The current research for knowledge propagation in D3L is only for a simplified case when only two ontologies are involved. In this paper we study knowledge propagation in more complex cases. We find in the case when more than two ontologies are involved and bridge rules form chains, knowledge does not always propagate along chains of bridge rules even if we would expect it. Inspired by Package-based description Logics, we extend the original semantics of D3L by imposing so called compositional consistency condition on domain relations in D3L interpretations. Under this semantics knowledge propagates along chains of bridge rules correctly. Furthermore we provide a distributed Tableaux reasoning algorithm for deciding satisfiability of concepts which is decidable in D3L under compositional consistency. Compared with original one, the extended D3L provides more reasonable logic foundation for distributed, dynamic system such as the information integration system and the Semantic Web.

Keywords: Distributed Dynamic Description Logics(D3L), Knowledge Propagation, Compositional Consistency, Bridge Rule Chain.

1   Introduction

As the distributed extension of traditional dynamic description logics(DDL)[1][2], distributed dynamic description logics(D3L)[3][4] enables reasoning with multiple heterogeneous DDL ontologies interconnected by directional semantic mapping. D3L captures the idea of importing and reusing knowledge between several ontologies. This idea combines well with the basic assumption of the distributed and dynamic system such as the Semantic Web and the heterogeneous information integration system.

An important innovation of D3L is to describe the semantic mapping between heterogeneous ontologies through bridge rules[3]. Bridge rules can assert that the concept/relation/action, local to ontology T_1, is mapped to an independent ontology T_2 as a subconcept/relation/action or superconcept/relation/action. Bridge rules are directed, and hence if there is a bridge rule with direction from T_1 to T_2, then T_2 reuses knowledge from T_1 but not necessarily the other way around. Bridge rules lead to the im-

Z. Shi, D. Leake, and S. Vadera (Eds.): IIP 2012, IFIP AICT 385, pp. 106–115, 2012.

portant characteristic--knowledge dissemination which makes D3L different from DDL. Knowledge propagation theory of D3L has been studied in our early work, but only for a simplied case when only two DDL ontologies are involved and bridge rules doesn't construct chain. Because D3L is presented primarily as the logical foundation for complex distributed dynamic ontology, the research for knowledge propagation in more complex cases has important theoretical value and practical value.

In this paper we study subsumption propagation theory in more complex cases, when bridge rules in multiple DDL ontologies construct chains. We discover that original semantics of D3L can't assure knowledge propagate correctly in that situations even if we would expect it. Inspired by Package-based description Logics, we extend the original semantics of D3L by imposing so called compositional consistency condition on domain relations in D3L interpretations. Under this semantics knowledge propagates along chains of bridge rules correctly. Then we study the general laws of knowledge propagation under the new semantics. At last, we study the decidability for reasoning satisfiability of concepts in D3L under compositional consistency and provide a distributed Tableaux reasoning algorithm for deciding it. The algorithm can reason with acyclic D3L knowledge bases build on top of D-ALC as the local language.

2   Demonstrations of the Problem

By two examples, this section will illustrate the problem which occurs in D3L knowledge propagation in the case of bridge rule chains exist.

Example 1. In DDL ontology T_1 there are action *BuyVehicle* and its subaction *BuyCar*. There is also another DDL ontology T_2 with action *Buy* and *BuyFord*. An into-action bridge rule maps from *BuyVehicle* to *Buy* and states that *BuyVehicle* is a subaction of *Buy*. An onto-action bridge rule maps from *BuyCar* to *BuyFord* and states that *BuyCar* is a superaction of *BuyFord*. Thanks to action subsumption propagation theorem[4], the knowledge that *BuyCar* is a subaction of *BuyVehicle* propagates to T_2 and we infer that *BuyFord* is a subaction of *Buy* and other inferences such as action realizability.

Consider a change in the situation, T_1 is divided into T_1 and $T_1{}'$, as depicted in Figure 1, $T_1{}'$ contains *BuyCar*, and there is an into-action bridge rule from $T_1{}'$ to T_1 stating that *BuyCar* is a subaction of *BuyVehicle*. Actions in T_2 are unchanged, but the two bridge rules now map, first: from T_1 to T_2 asserting *BuyVehicle* a subaction of *Buy*, and second from $T_1{}'$ to T_2: asserting *BuyCar* a superaction of *BuyFord*. In D3L we no longer infer that *BuyFord* is a subaction of *Buy*.

Example 2. There are three DDL ontologies T_1, T_2, T_3 in D3L knowledge base, as depicted in Figure 2(a), by applying concept subsumption propagation theorem repeatedly we can infer that *MyMastiff* is a kind of *DangerousAnimal*. This situation can be extended to another one where there are many similar intermediate ontologies as T_2 between T_1 and T_3. In such cases knowledge can be correctly propagated along bridge rule chains. However, the restriction condition in above cases is too strict, if

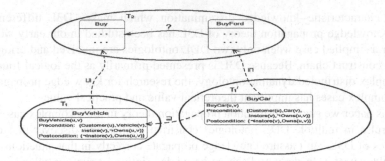

Fig. 1. Depiction of distributed ontologies in example 1

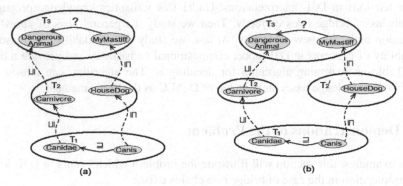

Fig. 2. Depiction of distributed ontologies in example 2

conditions are violated, *Carnivore* and *HouseDog* are located in different ontology, as depicted in Figure 2(b), we can't infer that *MyMastiff* is a subconcept of *DangerousAnimal* although the superconcept-subconcept relation between *Canidae* and *Canis* should be propegated to *Carnivore* and *HouseDog*, and then to *DangerousAnimal* and *MyMastiff* again.

From above two examples we can see, the traditional semantics of D3L does not guarantee the knowledge that we hope to import into the target ontology spreads along the bridge rule chains correctly. In section 4, we will give a detailed analysis of the reasons, and extend D3L to solve this problem.

3   Distributed Dynamic Description Logics

A basic form of D3L knowledge base dubbed distributed *Tbox*, includes a set of local *Tboxes* each over its own DDL language and a set of bridge rules that provides mappings between local *Tboxes*. In this work, we use the formal definition as follows.

Definition 1. Assume a DDL language DDL_X, a non-empty index set I, a set of concept names $N_C = \cup_{i \in I} N_{ci}$, a set of relation names $N_R = \cup_{i \in I} N_{Ri}$ and a set of action names $N_A = \cup_{i \in I} N_{Ai}$, A Distributed Tbox over DDL_X is a pair $<\{T_i\}_{i \in I},\ BR>$ such that: each local *Tbox* T_i is a collection of general inclusions over N_{Ci}, N_{Ri} and N_{Ai} in the local language of T_i, a sub-language of DDL$_X$; the set of bridge rules BR divides into sets of bridge rules $BR=\cup_{i,\ j\in I,\ i\neq j}\ BR_{ij}$. Each BR_{ij} is a collection of bridge rules in direction from T_i to T_j which are of four forms:

$i : C \xrightarrow{\ \sqsubseteq\ } j : E$ (into-concept(relation) bridge rule);

$i : C \xrightarrow{\ \sqsupseteq\ } j : E$ (onto- concept(relation) bridge rule);

$i : \alpha \xrightarrow{\ \sqsubseteq\ } j : \beta$ (into-action bridge rule);

$i : \alpha \xrightarrow{\ \sqsupseteq\ } j : \beta$ (onto-action bridge rule).

As we have already mentioned in the introduction, the direction of bridge rules matters and hence B_{ij} and B_{ji} are possibly and expectedly distinct. The bridge graph $G_{DTB}=<V,\ E>$ of a distributed *Tbox DTB* is defined as follows: $V=I$ and $<i,\ j>\in E$ if $BR_{ij}\neq\emptyset$. We say that *DTB* is acyclic if G_{DTB} is acyclic.

Given a *Tbox T* , a hole is an interpretation $I^\epsilon=<\emptyset,\ \cdot^\epsilon>$ with empty domain. Holes are used for fighting propagation of inconsistency. A distributed interpretation $DI=<\{I_i\}_{i\in I},\ \{r_{ij}\}_{i,\ j\in I,\ i\neq j}>$ of a distributed *Tbox DTB* consists of a set of local interpretations $\{I_i\}_{i\in I}$ and a set of domain relations $\{r_{ij}\}_{i,\ j\in I,\ i\neq j}$. For each $i\in I$, either $I_i=\ (\Delta^{I_i},\ \cdot^{I_i})$ is an interpretation of local Tbox or $I_i=I^\epsilon$ is a hole. Each domain relation r_{ij} is a subset of $\Delta^{I_i}\times\Delta^{I_j}$. We denote by $r_{ij}\ (d)$ the set $\{d'\ |<d,\ d'>\in r_{ij}\}$ and by $r_{ij}\ (D)$ the set $\cup_{d\in D}\ r_{ij}\ (d)$.

Definition 2. For every $i,\ j\in I$, a distributed interpretation DI satisfies the elements of a distributed *Tbox DTB* (denoted by $DI\vDash_d\bullet$) according to the following clauses:

1. $DI\vDash_d i : C \xrightarrow{\ \sqsubseteq\ } j : E$ if $r_{ij}\ (C^{I_i})\ \subseteq E^{I_j}$;
2. $DI\vDash_d i : C \xrightarrow{\ \sqsupseteq\ } j : E$ if $r_{ij}\ (C^{I_i})\ \supseteq E^{I_j}$;
3. $DI\vDash_d i : \alpha \xrightarrow{\ \sqsubseteq\ } j : \beta$ if $r_{ij}\ (\alpha^{I_i})\ \subseteq \beta^{I_j}$;
4. $DI\vDash_d i : \alpha \xrightarrow{\ \sqsupseteq\ } j : \beta$ if $r_{ij}\ (\alpha^{I_i})\ \supseteq \beta^{I_j}$;
5. $DI\vDash_d i : C\sqsubseteq E$ if $I_i\vDash C\sqsubseteq E$;
6. $DI\vDash_d i : \alpha\sqsubseteq\beta$ if $I_i\vDash\alpha\sqsubseteq\beta$;
7. $DI\vDash_d T_i$ if $I_i\vDash T_i$;
8. $DI\vDash_d BR$ if $DI\vDash_d br$ for every $br\in BR$;
9. $DI\vDash_d DTB$ if $DI\vDash_d T_i$ and $DI\vDash_d BR$ for every $i\in I$;
10. $DTB\vDash_d i : C\sqsubseteq E$ if $DI\vDash_d DTB$ implies $DI\vDash_d i : C\sqsubseteq E$ for every DI;
11. $DTB\vDash_d i : \alpha\sqsubseteq\beta$ if $DI\vDash_d DTB$ implies $DI\vDash_d i : \alpha\sqsubseteq\beta$ for every DI.

If $DI\vDash_d DTB$ then we say that DI is a distributed model of *DTB*.

Among the properties of D3L we find the characterization of subsumption propagation, which formally describes the mechanism of knowledge reuse of D3L. Theorem 1 constitutes the most basic version of this property: thanks to a pair of one into-bridge rule and one onto-bridge rule local subsumption relationship is propagated from the source ontology of these bridge rules to the target ontology.

Theorem 1. In each distributed *Tbox DTB*, if $i : C \xrightarrow{\sqsupseteq} j : G \in BR$ and $i : D \xrightarrow{\sqsubseteq} j : H \in BR$, then the following holds: $DTB \vDash_d i : C \sqsubseteq D \Rightarrow DTB \vDash_d j : G \sqsubseteq H$; if $i : \alpha \xrightarrow{\sqsupseteq} j : \beta \in BR$ and $i : \pi \xrightarrow{\sqsubseteq} j : \rho \in BR$, then the following holds: $DTB \vDash_d i : \alpha \sqsubseteq \pi \Rightarrow DTB \vDash_d j : \beta \sqsubseteq \rho$.   (Subsumption propagation theorem)

Theorem 1 states subsumption propagation property, one of the D3L properties. So far subsumption propagation laws in more complex cases have been studied. However all these cases do not extend the case captured by Theorem 1 in that aspect that only two local DDL ontologies that are directly connected with bridge rules are studied. The reader is kindly redirected to reference [4] for all details and discussion.

4   Knowledge Propagation in D3L under Compositional Consistency

Why does not the knowledge propagate along bridge rule chains as we would expect in the examples shown in section 2? In the following we will give the detailed analysis from semantic perspective. There are three bridge rules in Example 1: $1' : BuyCar \xrightarrow{\sqsupseteq} 2 : BuyFord \in BR$, $1 : BuyVehicle \xrightarrow{\sqsubseteq} 2 : Buy \in Br$ and $1' : BuyCar \xrightarrow{\sqsubseteq} 1 : BuyVehicle \in BR$. Given a model of *DTB*, we get some semantic constraints within Δ^{I2}, particularly $BuyFord^{I2} \sqsubseteq r_{1'2} (BuyCar^{I1'})$ and $r_{12} (BuyVehicle^{I1}) \sqsubseteq Buy^{I2}$ thanks to two of the bridge rules. However, the third bridge rule does not help anyhow to establish relation between $r_{1'2} (BuyCar^{I1'})$ and $r_{12} (BuyVehicle^{I1})$. In Example 2, in each distributed model *DI* of *DTB* the interpretations of the two concepts "on the way"-- $HouseDog^{I2}$ and $Carnivore^{I2}$ --are totally unrelated. By composition of the bridge rules that are available here we derive the inclusions: $MyMastill^{I3} \sqsubseteq r_{2'3} (HouseDog^{I2'}) \sqsubseteq r_{2'3} (r_{12'} (Canis^{I1}))$ and $r_{23} (r_{12} (Canidae^{I1})) \sqsubseteq r_{23} (Carnivore^{I2}) \sqsubseteq DangerousAnimal^{I3}$. However $r_{2'3} (r_{12'} (Canis^{I1}))$ and $r_{23} (r_{12} (Canidae^{I1}))$ are not related in Δ^{I3}, it does not help that $Canis^{I1} \sqsubseteq Canidae^{I1}$ in Δ^{I1}.

From above analysis we can see the main reason why the subsumption does not always propagate to remote ontologies in more complex cases as we would intuitively expect is all the semantic constraints generated by remote bridge rules do not propagate to remote parts of the system. In Package-based Description Logics (P-DL)[5][8], so called compositional consistency semantics is imposed on the importing relation in a distributed ontology environment so that knowledge can be propagated along importing relation chains uninterruptedly. The problem in our work is similar to that in P-DL. Inspired by this, we apply compositional consistency semantics on the D3L framework to try to solve the problem.

Definition 3. Given a distributed interpretation DI, we say that domain relation r(and also DI) satisfies compositional consistency if for each $i,\ j,\ k \in I$ and for each $x \in \Delta^{li}$ with $r_{ij}\ (x)\ = d$ we have $r_{jk}\ (d)\ = r_{ik}\ (x)$.

We say that D3L is under compositional consistency if only domain relations that satisfy the compositional consistency condition are allowed in distributed interpretations. We denote by D3L $_{(ccy)}$ D3L under compositional consistency. Accordingly, D3L under original semantics is denoted by D3L $_{(orl)}$.

Let us now reconsider what happens in the examples shown in section 2 if the compositional consistency semantics is enforced. In Example 2, by composition of the bridge rules that are available in the example we derive the inclusions: $MyMastill^{l3} \sqsubseteq r_{2'}$ $_3$ $(HouseDog^{l2'})\ \sqsubseteq r_{2'3}\ (r_{12'}\ (Canis^{l1}))$ and $r_{23}\ (r_{12}\ (Canidae^{l1}))\ \sqsubseteq$ $r_{23}\ (Carnivore^{l2})\ \sqsubseteq DangerousAnimal^{l3}$. Compositional consistency implies $r_{2'3}\ (r_{12'}$ $(Canis^{l1}))\ = r_{13}\ (Canis^{l1})$ and $r_{13}\ (Canidae^{l1})\ = r_{23}\ (r_{12}\ (Canidae^{l1}))$. Finally, since $DTB \vDash_d l : Canis \sqsubseteq Canidae$ we are now able to derive $r_{13}\ (Canis^{l1})\ \sqsubseteq$ $r_{13}\ (Canidae^{l1})$ and $MyMastill^{l3} \sqsubseteq DangerousAnimal^{l3}$, and hence we can derive that $MyMastill$ is a kind of $DangerousAnimal$ according to Definition 2. Similar case also occurrs in Example 1, due to space limitations we omit the discussion for it. From above analysis we can see the problem of subsumption propagation between ontologies which are connected indirectly can be solved by our extension. In the following we will give a in-depth theoretical analysis and provement for our extension.

Theorem 2. Given a distributed *Tbox DTB* and a subsumption formula ϕ, if $DTB \vDash_d$ ϕ in D3L $_{(orl)}$, then $DTB \vDash_d \phi$ also holds in D3L $_{(ccy)}$, i.e. the subsumption propagation property is satisfied in D3L $_{(ccy)}$.

Proof. This follows from the fact that each model that satisfies compositional consistency is also a model in the original D3L semantics. If a formula ϕ is satisfied in each model from the set of models of DTB according to the original semantics, it is also satisfied by each model from its subset--the set of models of DTB that we obtain under compositional consistency. Hence we can derive that if $DTB \vDash_d \phi$ in D3L $_{(orl)}$, then $DTB \vDash_d \phi$ also holds in D3L $_{(ccy)}$.   □

Subsumption relations that can be propagated in D3L $_{(orl)}$ can also be propagated in D3L $_{(ccy)}$. The only difference is that in the adjusted semantics possibly some more subsumption relations which can not be propagated along bridge rule chains in the former can be propagated correctly in addition, thus Theorem 1 must be extended for D3L $_{(ccy)}$. Theorem 3 gives the general form of subsumption propagation in D3L $_{(ccy)}$ (take action subsumption for example):

Theorem 3. Given a distributed *Tbox DTB*=<$\{T_0, T_1, ..., T_n\}$, BR>, action π, $\rho \in T_0$, $\alpha_i, \beta_i \in T_i$, $1 \leq i \leq n$, $1 \leq k \leq n$ such that

1. $DTB \vDash_d i : \alpha_i \sqsubseteq \beta_i, \ 1 \leq i \leq n$,

2. $i+1 : \alpha_{i+1} \xrightarrow{\quad \sqsupseteq \quad} i : \beta_i \in BR, \ 1 \leq i < k$,

3. $i : \beta_i \xrightarrow{\quad \sqsubseteq \quad} i+1 : \alpha_{i+1} \in BR, \ k \leq i < n$,

4. $1 : \alpha_1 \xrightarrow{\quad \sqsupseteq \quad} 0 : \pi \in BR$ and $n : \beta_n \xrightarrow{\quad \sqsubseteq \quad} 0 : \rho \in BR$.

In D3L $_{(ccy)}$ it follows that $DTB \vDash_d 0 : \pi \sqsubseteq \rho$.

Proof. The chain of bridge rules $i+1 : \alpha_{i+1} \xrightarrow{\quad \sqsupseteq \quad} i : \beta_i \in BR, \ 1 \leq i < k$ together with the local subsumptions in T_1, T_2, T_3, ..., T_k in the assumptions allows us to relate $r_{10} (\alpha_1^{I1}) \subseteq r_{10} (r_{21} (...r_{kk-1} (\alpha_k^{Ik}) ...))$. From compositional consistency we derive $r_{10} (r_{21} (...r_{kk-1} (\alpha_k^{Ik}) ...)) = r_{k0} (\alpha_k^{Ik})$. For the other chain of bridge rules alike. Now it is easy to see that $\pi^{I0} \subseteq r_{10} (\alpha_1^{I1}) \subseteq r_{k0} (\alpha_k^{Ik}) \subseteq r_{k0} (\beta_k^{Ik}) \subseteq r_{10} (\beta_1^{I1}) \subseteq \rho^{I0}$, hence we can derive $DTB \vDash_d 0 : \pi \sqsubseteq \rho$. $\square$

Extending Theorem 3 to the case of concept (relation) bridge rule chains, similarly we can draw concept (relation) subsumption propagation theorem, due to space limitations we omit the discussion for it.

5   Distributed Tableaux Algorithm for D3L $_{(ccy)}$

In this section, we introduce a distributed Tableaux algorithm for deciding satisfiability of concepts with respect to an acyclic distributed *Tbox* for D3L $_{(ccy)}$ over D-ALC. We use the formal definition as follows:

Definition 4. Assume a distributed *Tbox* $DTB=<\{T_i\}_{i \in I}, BR>$ over D-ALC with index set I, concept names $N_C = \cup_{i \in I} N_{ci}$, relation names $N_R = \cup_{i \in I} N_{Ri}$ and action names $N_A = \cup_{i \in I} N_{Ai}$. Let CC_i be the set of all (atomic and complex) concepts over N_{ci}, N_{Ri} and N_{Ai} in negation normal form. A distributed completion tree $T=\{T_i\}_{i \in I}$ is a set of labeled trees $Ti=<V_i, E_i, L_i, r_i>$, such that for each $i \in I$:

1. the members of $\{V_i\}_{i \in I}$ are mutually disjoint;
2. the members of $\{E_i\}_{i \in I}$ are mutually disjoint;
3. the labeling function L_i labels each node $x \in V_i$ with $L (x) \subseteq 2^{CCi}$ and each edge $<x, y> \in E_i$ with $L (<x, y>) \in N_{Ri}$;
4. the labeling function r_i labels each node $x \in V_i$ with a set of references to its r-images $r_i (x) \subseteq \{j : y | j \in I \wedge y \in V_j\}$.

During the run of the tableaux algorithm, tableaux expansion rules are applied on the completion tree and the tree is expanded by each rule application. If no more rules are applicable any more, we say that the tree is complete. There is a clash in the completion tree T if for some $x \in V_i$ and for some $C \in N_{ci}$ we have $\{C, \neg C\} \subseteq L_i (x)$. Given a distributed completion tree $T = \{T_i\}_{i \in I}$, a node $x \in V_i$ is blocked, if it has an ancestor $y \in V_i$ such that $L_i (x) \subseteq L_i (y)$. In such a case we also say that x is blocked by y. Node $y \in V_i$ is said to be an R-successor of $x \in V_i$, if $<x, y> \in E_i$ and $L_i (<x, y>) = R$.

Taking a distributed *Tbox DTB* and a concept C in NNF as its inputs, the distributed tableaux algorithm continues in three steps:

(1) Create new distributed tree $T = \{T_j\}_{j \in I}$ such that $T_j = <\{s_0\}, \varnothing, \{s_0 \mapsto \{C\}\}, \varnothing>$ for $j = i$ and $T_j = <\varnothing, \varnothing, \varnothing, \varnothing>$ for $j \neq i$.

(2) Apply the following tableaux expansion rules exhaustingly:

(a) $\sqcap$-rule: If $C_1 \sqcap C_2 \in L_i (x)$ for some $x \in V_i$ and $\{C_1, C_2\} \not\subseteq L_i (x)$, and x is not blocked, then set $L_i (x) = L_i (x) \cup \{C_1, C_2\}$.

(b) $\sqcup$-rule: If $C_1 \sqcup C_2 \in L_i (x)$ for some $x \in V_i$ and $\{C_1, C_2\} \cap L_i (x) = \varnothing$, and x is not blocked, then either set $L_i (x) = L_i (x) \cup \{C_1\}$ or set $L_i (x) = L_i (x) \cup \{C_2\}$.

(c) $\exists$-rule: If $\exists R.C \in L_i (x)$ for some $x \in V_i$ with no R-successor y s.t. $C \in L_i (y)$, and x is not blocked, then add new node z to V_i, add the edge $<x, z>$ to E_i, and set $L_i (z) = \{C\}$ and $L_i (<x, z>) = \{R\}$.

(d) $\forall$-rule: If $\forall R.C \in L_i (x)$ for some $x \in V_i$, and there is R-successor y of x s.t. $C \notin L_i (y)$, and x is not blocked, then set $L_i (y) = L_i (y) \cup \{C\}$.

(e) Action α-rule: If $[\alpha]C \in L_i (x)$ for some $x \in V_i$, $\alpha = (P_\alpha, E_\alpha)$, then set $L_i (x) = L_i (x) \setminus \{$ all conditions in $P_\alpha\} \cup \{$all conclusions in $E_\alpha\} \cup \{C\}$

(f) $\xrightarrow{\sqsupseteq}$-rule: If G (or β) $\in L_j (x)$ for some $x \in V_j$, $i : C$ (or α) $\xrightarrow{\sqsupseteq} j$: G (or β) $\in BR$, and there is no $y \in V_i$ s.t. C (or α) $\in L_i (y)$ and $j : x \in r_i (y)$, and x is not blocked, then add new node y to V_i and set $L_i (y) = \{C$ (or α) $\}$, and set $r_i (y) = \{j : x\} \cup r_j (x)$.

(g) $\xrightarrow{\sqsubseteq}$-rule: If D (or π) $\in L_i (x)$ for some $x \in V_i$, $i : D$ (or π) $\xrightarrow{\sqsubseteq} j : H$ (or ρ) $\in BR$ and there is $y \in V_i$ s.t. $j : y \in r_i (x)$ and H (or ρ) $\notin L_j (y)$, then set $L_j (y) = L_j (y) \cup \{H$ (or ρ) $\}$.

(3) If none of the tableaux expansion rules is applicable any more (i.e , the distributed tree is now complete), answer "C is satisfiable" if a clash-free completion tree has been constructed. Answer "C is unsatisfiable" otherwise.

Theorem 4. The satisfiability problem of concept in D3L$_{(ccy)}$ is decidable.

Proof. To determine whether the concept is satisfiable is to check whether the completion tree includes clash. In D3L$_{(ccy)}$, a distributed tree can be expanded by applying tableaux rules, as described in above algorithm. Now we ought to prove that the algorithm, once starts with input, eventually terminates. We will consider each step of the algorithm separately:

It is obvious that the establishment of a new tree in the first step eventually terminates for both $j=i$ and $j \neq i$.

In the second step, tree is expanded by applying seven tableaux rules. Only ⊔-rule is uncertain, but at worst complete binary tree is constructed which can be finished in exponential time. ⊓-rule, ∃-rule and ∀-rule which can be finished in polynomial time are all terminable. Action α-rule has no effect on the decidability of D3L$_{(ccy)}$ too, for detailed discussion please refer to [1]. The terminability of $\xrightarrow{\sqsupseteq}$ -rule and $\xrightarrow{\sqsubseteq}$ -rule is discussed as follows: If there are onto-bridge rules ingoing into T_i then possibly in some $x \in V_i$ such that C *(or α)* $\in L_i$ *(x)* and C *(or α)* appears on a right hand side of an onto-bridge rule, say $k : D$ *(or π)* $\xrightarrow{\sqsupseteq} i : C$ *(or α)* , then the $\xrightarrow{\sqsupseteq}$ -rule is applied and computation is triggered in T_k. By structural subsumption we assume that this computation eventually terminates (the length of the longest incoming path of into-bridge rules decreased for T_k compared to T_i, and all such paths are finite because of acyclicity). During this process we possibly get some new concepts in L_i *(x)* that are introduced thanks to incoming into-bridge rules that trigger the $\xrightarrow{\sqsubseteq}$ -rule--but only finitely many. The computation now continues in x and its descendants and possibly the $\xrightarrow{\sqsupseteq}$ -rule is triggered again in some $y \in V_i$, a descendant of x. But thanks to subset blocking, this happens only finitely many times. Hence $\xrightarrow{\sqsupseteq}$ -rule and $\xrightarrow{\sqsubseteq}$ -rule eventually terminate.

Clash checking process of distributed completion tree in the third step is terminable. Either concept is unsatisfiable because there are clashes in the tree, or concept is satisfiable because there are not clashes. Hence the satisfiability problem of concept in D3L$_{(ccy)}$ is decidable.   □

6   Conclusion and Future Work

Although knowledge propagation theory has been studied, only simple cases with two DDL ontologies were involved. In this paper we focused on the general cases with arbitrary number of ontologies and bridge rules chains. We found original semantics of D3L could not ensure correct knowledge propagation because domain relations were not transitive. Inspired by Package-based description Logics, we extended the original semantics of D3L by imposing so called compositional consistency condition on domain relations. Under this semantics knowledge propagates along chains of bridge rules correctly. At last, we studied the decidability of satisfiability problem of concepts in D3L under compositional consistency and provided a distributed Tableaux algorithm for deciding it.

Extending presented algorithm for expressive D3L, analysing computational complexity of the algorithm and developing corresponding reasoners are left for future work.

Acknowledgment. This work is supported by the Chinese National Natural Science Foundation(No. 61035003, No. 61072085, No. 60970088).

References

1. Shi, Z., Dong, M., Jiang, Y., et al.: A Logic Foundation for the Semantic Web. Science in China, Series F Information Sciences 48(2), 161–178 (2005)
2. Shi, Z., Jiang, Y., Zhang, H., et al.: Agent service matchmaking based on description logic. Chinese Journal of Computers 27(5), 625–635 (2004) (in Chinese)
3. Jiang, Y., Shi, Z., Tang, Y., et al.: A Distributed Dynamic Description Logic. Journal of Computer Research and Development 43(9), 1603–1608 (2006) (in Chinese)
4. Wang, Z.: Distributed Information Retrieval Oriented Automatic Reasoning. Doctor's thesis. Institute of Computing Technology. Chinese Academy of Sciences, Beijing (2010)
5. Bao, J., Caragea, D., Honavar, V.G.: On the Semantics of Linking and Importing in Modular Ontologies. In: Cruz, I., Decker, S., Allemang, D., Preist, C., Schwabe, D., Mika, P., Uschold, M., Aroyo, L.M. (eds.) ISWC 2006. LNCS, vol. 4273, pp. 72–86. Springer, Heidelberg (2006)
6. Franz, B., Diego, C., Deborah, M., et al.: The description logic handbook: Theory, implementation, and applications, 2nd edn. Cambridge University Press, Cambridge (2007)
7. d'Aquin, M., Schlicht, A., Stuckenschmidt, H., Sabou, M.: Criteria and Evaluation for Ontology Modularization Techniques. In: Stuckenschmidt, H., Parent, C., Spaccapietra, S. (eds.) Modular Ontologies. LNCS, vol. 5445, pp. 67–89. Springer, Heidelberg (2009)
8. Bao, J., Caragea, D., Honavar, V.G.: A distributed tableau algorithm for package-based description logics. In: 2nd International Workshop on Context Representation and Reasoning. IOS Press, Italy (2008)
9. Grau, B.C., Parsia, B., Sirin, E.: Combining OWL ontologies using $\mathcal{E}$-connections. Journal of Web Semantics 4(1), 40–59 (2006)
10. Cuenca Grau, B., Parsia, B., Sirin, E.: Ontology Integration Using ε-Connections. In: Stuckenschmidt, H., Parent, C., Spaccapietra, S. (eds.) Modular Ontologies. LNCS, vol. 5445, pp. 293–320. Springer, Heidelberg (2009)
11. Kutz, O., Lutz, C., Wolter, F., et al.: $\mathcal{E}$-connections of abstract description systems. Artificial Intelligence 156(1), 1–73 (2004)
12. Bao, J., Voutsadakis, G., Slutzki, G., Honavar, V.: Package-Based Description Logics. In: Stuckenschmidt, H., Parent, C., Spaccapietra, S. (eds.) Modular Ontologies. LNCS, vol. 5445, pp. 349–371. Springer, Heidelberg (2009)
13. Lutz, C., Walther, D., Wolter, F.: Conservative extensions in expressive description logics. In: Twentieth International Joint Conference on Artificial Intelligence (IJCAI 2007), pp. 453–458. AAAI Press, California (2007)
14. Serafini, L., Borgida, A., Tamilin, A.: Aspects of Distributed and Modular Ontology Reasoning. In: Tenth International Joint Conference on Artificial Intelligence (IJCAI 2005), pp. 570–575. AAAI Press, California (2005)

Semantic Keyword Expansion: A Logical Approach

Limin Chen

China Unicom Research Institute, Beijing 100032, China
chenlm49@chinaunicom.cn

Abstract. Keyword search is the primary way for ordinary users to access the Web content. It is essentially a syntax match between users' key-ins and the index structure of information systems with a relevance-ranking way to sort the hitted documents. The syntax match can rarely satisfy users' information need when the keywords and index are not syntically similar while share a lot semantically. This paper proposed a way of semantic keyword expansion with a re-ranking to handle this problem. Experimental results show that our method helps in improving the quality of keyword search and particularly in the cases of keywords with widely-used synonyms or parasynonyms.

Keywords: keyword search, semantic expansion, probabilistic uncertainty, re-rank.

1 Introduction

Web Search has changed radically with the advent of *Semantic Web [1]*. The Semantic Web technology helps in better understanding of users' information need and more complex queries by interpreting Web search queries and resources relative to one or more underlying ontologies, describing some background domain knowledge, in particular, by connecting the Web resources to semantic annotations, or by extracting semantic knowledge from Web resources [2]. However, for naïve users with no familiarity with such technologies, accessing the information in need is not a trivial job and keyword query is still the primary way.

Adding semantics to keyword query is not a new topic and there are many literatures in this aspect. Fazzinga and Lukasiewicz give a comprehensive overview of state-of-the-art approaches [2]. Generally speaking, these approaches employ the Semantic Web technologies to augment or refine the results of traditional keyword search with the help of knowledgebase outside, such as domain-specific ontologies, Wiki, Wordnet, and etc [3-6].

The main drawback in such approaches is their lack of quantitative specification of the overlap of keywords, which is of ever-increasing importance due to the fact that two keywords are rarely logically related via a subsumption or disjointness relationship, and that they certainly show a certain degree of overlap.

This paper presented a logical approach for keyword semantic expansion with a re-ranking to handle this problem. It employs PD_ALCO@, a formalism proposed in [7] to represent and maintain a knowledgebase on the overlap degrees of keywords,

Z. Shi, D. Leake, and S. Vadera (Eds.): IIP 2012, IFIP AICT 385, pp. 116–124, 2012.

which is used to expand the user-formulated keyword to some highly-overlapped keywords. With the original and the expanded keywords, our approach performs traditional keyword search and re-ranks these results to get the overall result.

Paper Outline. We first give a brief overview of the formalism in Sec.2, including its syntax (Sec.2.1), semantics (Sec.2.2) and some reasoning tasks (Sec.2.3). Sec.3 details our semantic keyword expansion with experimental results. Finally, we conclude the paper in Sec.4.

2   The Formalism

PD_ALCO@ may be seemed as a PDL [8]-like extension of description logics with probabilistic uncertainty admitting dynamic reasoning under probability uncertainty. It employs conditional constraints [9] to express interval restrictions for conditional probabilities over concepts, and lexicographic entailment for probabilistic reasoning.

2.1   Syntax

Primary alphabets of D-ALCO@ include: i) N_R for role names; ii) N_C for concept names; iii) N_I for individual names; and iv) N_A for atomic action names. The concepts and roles are the same as that in *ALCO* with "$C,D \rightarrow C_i \mid \{o\} \mid \neg C \mid (C \sqcap D) \mid @_o C$ $(C \sqcap D) \mid \exists R.\ C$" & "$R \rightarrow P$", where $C_i \in N_C$, $o \in N_I$, $P \in N_R$. We use $\bot$, $\top$, $(C \sqcup D)$, and $\forall R.\ C$ to short $(C \sqcap \neg C)$, $\neg \bot$, $\neg(\neg C \sqcap \neg D)$, and $\neg \exists R.\ \neg C$, resp. Roles in PD_ALCO@ are built up with the same syntax rules as that for roles and concepts in ALCO@, resp. The definitions of axioms, TBox and ABox in PD_ALCO@ are also the same as those corresponding definitions in ALCO@. The set N_I of individuals in PD_ALCO@ is divided into two disjoint sets: the set I_C of classical individuals and the set I_P of probabilistic individuals which are those individuals in N_I related to which we store some probabilistic knowledge.

A *conditional constraint* is an expression of the form $(C|D)[l,u]$, where C, D are concepts free of probabilistic individuals, and l, u are reals in $[0,1]$. The conditional constraint $(C|D)[l,u]$ encodes an interval restriction for conditional probabilities over concepts C and D : for a randomly chosen individual o, if $D(o)$ holds, then the probability of $C(o)$ lies in $[l,u]$.

A PTBox $PT =(T,P)$ consists of a TBox T and a finite set of conditional constraints P. A PABox $PA=P_o$ for $o \in I_P$ is a finite set of conditional constraints that are specific probabilistic knowledge about o.

An atomic action in D-ALCO@ is defined as $\alpha \equiv (Pre, Eff)$, where i) $\alpha \in N_A$ is the name of the atomic action; ii) *Pre* is a finite set conditional constraints (generally or specific to some individuals) specifying the action's preconditions; and iii) *Eff* is a finite set of possibly negated primitive ALCO@-assertions.

Actions are built with: π, $\pi' \rightarrow a \mid \varphi? \mid \pi \cup \pi' \mid \pi$; $\pi' \mid \pi^*$, where a is an atomic action, and φ is a possibly negated ALCO@-assertion or a conditional constraint.

Dynamic conditional constraints (dynamic c-constraints for short) are more complex than conditional constraints and built up with $f \rightarrow cc \mid <\pi>f$, where π is an action and cc is a conditional constraint.

A dynamic probabilistic knowledge base in KB=$(T,P,\{P_o\}_{o \in Ip}, A_C)$ consists of a PTBox (T,P), a PABox P_o for each o$\in I_P$, and an A_C. Informally, a dynamic probabilistic knowledge base extends a probabilistic knowledge base in [9] by an ActionBox which encodes the dynamic aspects of the domain.

2.2   Semantics

A PD_ALCO@ interpretation is a pair $Pr = (M, \mu)$ consisting of a D_ALCO@ interpretation $M = (\Delta, W, I)$ and a probability function over Δ, i.e., $\mu: \Delta \rightarrow [0,1]$ subject to for each o$\in \Delta$, $\mu(o) \geqq 0$ and $\Sigma \mu(o)=1$.

Pr interprets concepts and roles at w$\in W$ as $I(w)$ does: 1) $A^{Pr,w} = A^{M,w} = A^{I(w)} \subseteq \Delta$; 2) $P^{Pr,w} = P^{I(w)} \subseteq \Delta \times \Delta$; 3) $(\neg C)^{Pr,w} = \Delta \backslash C^{Pr,w}$; 4) $(C \sqcap D)^{Pr,w} = C^{Pr,w} \cap D^{Pr,w}$; 5) $\{o\}^{Pr,w} = \{o\}$; 6) $(@_o C)^{Pr,w} = \Delta$ if o$\in C^{Pr,w}$ and $= \varnothing$ o.w.; 7) $(\forall R.C)^{Pr,w} = (\forall R.C)^{I(w)} = \{ x \mid \forall y \in \Delta$ subject to $(x, y) \in R^{I(w)}$ implies y$\in C^{I(w)}\}$.

The probability of concept C in $Pr = (\Delta, W, I, \mu)$ at w$\in W$, noted $Pr_w(C)$, is defined as

$$Pr_w(C) = \Sigma \mu(o), \text{ for each } o \in C^{Pr,w} \tag{1}$$

We abbreviate $Pr_w(C \sqcap D)/Pr_w(D)$ as $Pr_w(C \mid D)$ when $Pr_w(D) \neq 0$.

Pr satisfies $(C \mid D)[l,u]$ at possible world w, noted $Pr, w \models (C \mid D)[l,u]$ iff $Pr_w(D)=0$ or $Pr_w(C \mid D) \in [l,u]$. Pr satisfies a set P of conditional constraints at w, noted $Pr, w \models P$, iff $Pr, w \models p$ for each $p \in P$.

Actions are still interpreted as accessibility between possible worlds in Pr: 1) $\alpha^{Pr} = (Pre, Eff)^{Pr} = \{ (w,w') \mid w,w' \in W$ such that $Pr, w \models Pre$ and $I(w) \rightarrow_a I(w') \}$; 2) $(\varphi?)^{Pr} = \{ (w,w) \mid (w,w) \in W$ such that $I(w) \models \varphi \}$; 3) $(\pi \cup \pi')^{Pr} = (\pi)^{Pr} \cup (\pi')^{Pr}$; 4) $(\pi ; \pi')^{Pr} = \{ (w,w') \mid \exists w_t \in W$ such that $(w,w_t) \in (\pi)^{Pr}$ and $(w_t,w) \in (\pi')^{Pr} \}$; 5) $(\pi^*)^{Pr} =$ the reflexive and transitive closure of $(\pi)^{Pr}$.

The updated probability of concept C from w in $Pr=(\Delta, W, I, \mu)$ w.r.t. atomic action a, noted $Pr_{a,w}(C)$, is defined as $Pr_{a,w}(C) = Pr_v(C)$, where $(w,v) \in \alpha^{Pr}$.

Pr satisfies a dynamic c-constraint $<\pi>f$ at w, noted $Pr, w \models <\pi>f$, iff there exists v$\in$ such that $(w,v) \in (\pi)^{Pr}$ and $Pr, v \models f$. Pr satisfies a finite set F of dynamic c-constraints at w, noted $Pr, w \models F$, iff $Pr, w \models f$ for each f$\in F$. A dynamic c-constraint f is satisfiable iff there exists a Pr subject to $\exists w$ such that $Pr, w \models f$.

Pr verifies $(C \mid D)[l,u]$ at w iff $Pr_w(D)=1$ and $Pr, w \models (C \mid D)[l,u]$. Pr falsifies $(C \mid D)[l,u]$ at w iff $Pr_w(D)=1$ and $Pr, w \not\models (C \mid D)[l,u]$. A finite set F of conditional constraints tolerates a conditional constraint f iff there exists a $Pr= (\Delta, W, I, \mu)$ subject to $\exists w \in W$, such that $Pr, w \models F$ and Pr verifies f at w.

PTBox $PT=(T,P)$ is *consistent* iff i) T is satisfiable, and ii) there exists an ordered partition $(P_0,...,P_k)$ of P such that each P_i with $i \in \{0,...,k\}$ is the set of all conditional constraints that are tolerated w.r.t. T by $P\backslash(P_0 \cup ... \cup P_{i-1})$.

A knowledge base $KB=(T,P,\{P_o\}_{o \in Ip}, A_C)$ is *consistent* iff i) (T,P) is consistent, ii) $T \cup P_o$ is satisfiable for each $o \in I_P$, and iii) $T \cup Eff$ is satisfiable for each atomic action $\alpha= (Pre, Eff)$ in A_C.

The notions of *lexicographical preference* and *lexicographical entailment* can be generalized to the dynamic setting as follows. First we use the *z-partition* $P_0,...,P_k)$ of P to define a lexicographic preference relation on probabilistic dynamic interpreta- tions. For PD_ALCO@ interpretations $Pr=(\Delta, W, I, \mu)$ and $Pr'=(\Delta', W, I', \mu')$, we say Pr at w is *lexicographically preferable* (or *lex-preferable*) to Pr' w' iff there exists $i \in \{0,...,k\}$ such that $|\{F \in P_i | Pr, w \models F\}| > |\{F \in P_i | Pr', w \models F\}|$ and $|\{F \in P_j | Pr, w \models F\}| = |\{F \in P_j | Pr', w \models F\}|$ for all $i<j \leq k$.

For a TBox I and a set F of conditional constraints, an interpretation Pr at w is a *lexicographically minimal* (or *lex-minimal*) model of $T \cup F$ iff no interpretation Pr' at $w' \models T \cup F$ and is lex-preferable to Pr at w.

$(C|D)[l,u]$ is a *lexicographic consequence* (or *lex-consequence*) of a set F of condi- tional constraints w.r.t. $PT=(T, P)$, $F \models^{lex} (C|D)[l,u]$ w.r.t. PT, iff $Pr_w(C) \in [l,u]$ for every lex-minimal model Pr at w of $T \cup F \cup \{(D|\top)[1,1]\}$. $(C|D)[l,u]$ is a *tight lexico- graphic consequence* (or *tight lex-consequence*) of F w.r.t. PT, denoted $F \models^{lex}_{tight}$ $(C|D)[l,u]$ w.r.t. PT, iff l (resp., u) is the infimum (resp., supremum) of $Pr_w(C)$ for all lex-minimal models Pr at w of $T \cup F \cup \{(D|\top)[1,1]\}$. Note that $[l,u]=[1,0]$ (where $[1,0]$ represents the empty interval when no such model exists.

Given a TBox T and a set F of conditional constraints, $T \cup F$ is satisfiable iff there exists an interpretation Pr that satisfies $T \cup F$ at some w. A conditional constraint $(C|D)[l,u]$ is a logical consequence of $T \cup F$, denoted $T \cup F \models (C|D)[l,u]$, iff each Pr at w that models $T \cup F$ also models $(C|D)[l,u]$; $(C|D)[l,u]$ is a tight logical consequence of $T \cup F$, denoted $T \cup F \models_{tight} (C|D)[l,u]$, iff l (resp., u) is the infimum (resp., supremum) of $Pr_w(C|D)$ subject to each Pr at w models $T \cup F$ with $Pr_w(D)>0$.

A dynamic c-constraint $<\pi> (C|D)[l,u]$ w.r.t. PT is a *lexicographic consequence* (or *lex-consequence*) of F w.r.t. PT, denoted $F \models^{lex} <\pi> (C|D)[l,u]$ w.r.t. PT, iff $Pr_{w'}(C) \in [l,u]$ for every lex-minimal model Pr at w of $T \cup F \cup \{(D|\top)[1,1]\}$, where $(w,w') \in \pi^{Pr}$. $<\pi> (C|D)[l,u]$ is a *tight lexicographic consequence* (or *tight lex-consequence*) of F w.r.t. PT, denoted $F \models^{lex}_{tight} <\pi> (C|D)[l,u]$ w.r.t. PT, iff l (resp., u) is the infimum (resp., supremum) of $Pr_{w'}(C)$ for all lex-minimal models Pr at w of $T \cup F \cup \{(D|\top)[1,1]\}$ and $(w,w') \in \pi^{Pr}$. Note that $[l,u]=[1,0]$ (where $[1,0]$ represents the empty interval when no such model exists.

A (dynamic) conditional constraint f is a *lex-consequence* of PT, denoted $PT \models^{lex} f$, iff $\emptyset \models^{lex} (C|D)[l,u]$ w.r.t. PT,; and f is a tight lex-consequence of PT, denoted $PT \models^{lex}_{tight} f$, iff $\emptyset \models^{lex}_{tight} f$, w.r.t. PT. A (dynamic) conditional constraint f about a probabilistic individual $o \in I_P$ is a *lex-consequence* of a $KB=(T,P,\{P_o\}_{o \in Ip}, A_C)$, denoted $KB \models^{lex} f$, iff

$P_o \models^{lex} f$ w.r.t (T,P), and f is a tight lex-consequence of KB, denoted $K \models^{lex}_{tight} f$, iff $P_o \models^{lex}_{tight} f$ w.r.t (T,P).

2.3   Reasoning Tasks

The main reasoning tasks in PD_ALCO@ include: i) PTBox Consistency (*PTCon*): Decide whether a given PTBox $PT = (T,P)$ is consistent; ii) Probabilistic Dynamic Knowledge Base Consistency (*PDKBCon*): Decide whether a probabilistic dynamic knowledge base $KB=(T,P,\{P_o\}_{o \in Ip}, A_C)$ is consistent; iii) Tight Lex-Entailment (*TLexEnt*): Given a $KB=(T,P,\{P_o\}_{o \in Ip}, A_C)$, a finite set F of conditional constraints, for concepts free of probabilistic individuals C and D, and action π from A_C, compute the rational numbers $l, u \in [0,1]$ such that $F \models^{lex}_{tight} (C|D)[l,u]$ w.r.t. *PT*.

As shown in [7], the above tasks can be reduced to the following two problems, which can be reduced to deciding the satisfiability of classical DL-knowledgebase, deciding the solvability of linear constraints and computing the optimal value of linear programs:

a)   Satisfiability (*SAT*): decide whether $T \cup F$ is satisfiable, where T is a TBox and F is a set of conditional constraints;

b)   Tight Logical Entailment (*TLogEnt*): Given a TBox T, a finite set F of conditional constraints, concepts C,D free of probabilistic individuals, compute the rational numbers compute the rational numbers $l, u \in [0,1]$ such that $T \cup F$ $\models_{tight} (C|D)[l,u]$.

We refer the interested readers to [7] for further technical details.

3   Semantic Keyword Expansion

3.1   The Big Picture

The keyword search is essentially a syntax match between users' key-ins and the index structure of information systems with a relevance-ranking to sort the hitted documents. In many cases, there may not be an exact syntax match even the users' keywords and the index share a lot semantically. For example, a user wants to get some papers in "logic programming", and uses "logic programming" as the keyword to claim his information need. In traditional keyword search, it will lose the papers indexed with "deductive databases" while the two keywords are closely related.

The formalism in Sec. 2 provides a way to represent the degrees of overlap between keywords and maintain such kind of knowledge w.r.t the evolution of the Web, i.e., some documents are no longer classified as a certain topic while others are new comers to the topic. To put it in another way, documents in the Web can be seemed as individuals, the keywords as concepts, and document classifications as concept memberships. Then the degrees of overlap between the keywords provide a means of deriving a probabilistic membership to the related keywords and so an estimation of the relevance to the original keyword, which helps in the re-ranking process.

3.2   The Keyword Expansion

To semantically expand users' keywords, our approach employs a knowledgebase about the degrees of the overlap between keywords, which can be constructed 1) with the help of existing search engines, such a Yahoo!, Google, Baidu, and etc.; or 2) with the help of domain experts.

For example, we need to figure the overlap degree between the aforementioned "logic programming" and "deductive databases". One way is to turn to some expert in this field for help and the expert gives an interval as his estimation, say [0.9, 0.98]. So this piece of knowledge about the overlap degree can be encoded in the following conditional constraints:

"(logic programming | deductive databases) [0.9, 0.98]"   (2)

Due to the semantics of conditional constraints, in "(logic programming | deductive databases) [l, u]", the [l, u] is the constraint on the conditional probability of a document being "deductive databases" also in the category of "logic programming". So, the interval can be assessed with the search results of corresponding keywords on some search engines by the following formula:

$$(C|D) [l, u] = [mid(C|D)-\xi, mid(C|D)+ \xi]   (3)$$

where mid(C|D) =(C⊓D)/D, i.e., the conditional probability of document being D over the document being C and D, and   is a positive number as the error.

Take the above example, we search on some engines with "deductive databases" and Take the above example, we search on some engines with "deductive databases" and "logic programming" respectively. The first hits 10000 documents while the second hits 12000 documents with 9000 in the first results. So the mid(…) in this case are 0.9, if the pre-set error is 0.05, then our assessment about the degree of overlap by above search is:

"(logic programming | deductive databases) [0.85, 0.95]"   (4)

Using the basic facts constructed in the above methods, we can compute the overlap degree with the reasoning mechanism in Sec.2. With those knowledge, when a user formulates a keyword "C" to claim his information need, our method expands the original keyword "C" with the most closely related keyword "D", i.e., the keywords in conditional constraints "(D| C) [l, u]" with the maximum l.

Other than searching with "C" directly on some engines, our methods search with the original keyword and the expanded keyword respectively, and re-rank the two search results with the help of keywords' overlap-degree.

As for re-rank, we have fixed a re-ranking function as follows:

$$re - rank(r) = \begin{cases} \log(rank_1(r) + 1) & \text{if } r \in R1 \text{ and } r \notin R2 \\ \frac{2}{l+u}\log(rank_2(r) + 1) & \text{if } r \in R1 \text{ and } r \notin R2 \\ \log\left(rank_{1(r)} + \frac{rank_2(r)}{(l+u)}\right) & \text{if } r \in R1 \text{ and } r \in R2 \end{cases} \qquad (5)$$

where $rank_i(r)$ is the sequence number in the corresponding searching result Ri.

Let us recap, our semantic keyword expansion can be depicted in Fig.1.

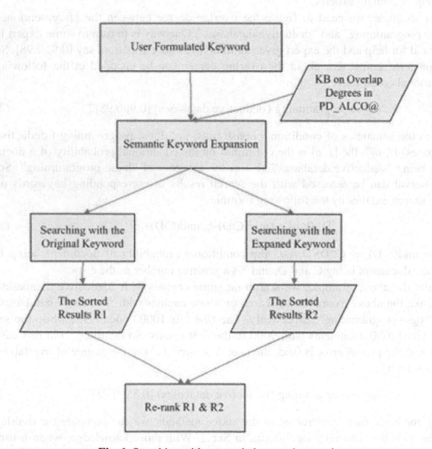

Fig. 1. Searching with semantic keyword expansion

3.3   Experiment and Evaluation

We construct a preliminary knowledge base with the aforementioned two ways, i.e., the assessment of keywords overlap-degree is given either by some experts or by the index structure of some search engines. For example, by Equation (3) with some search engine such as Yahoo!, we may get a conditional constraint as "(deductive databases |logic programming) [0.87, 0.97]", while "(PC| Laptop) [1, 1]" is some expert's belief.

Twenty-five different keywords are selected and for each keyword the volunteer formulated it is asked to label the relevance of the top 20 returned pages from Yahoo!, Google, Youdao, and Baidu respectively. The MAP of the returned pages can be calculated with the following formula:

$$MAP = \frac{\sum_1^4 AveP(q)}{4}, \text{ where } AveP = \frac{\sum_{r=1}^{20}(P(r)*rel(r))}{20}. \tag{6}$$

In this paper, we choose the relevance function rel(r) as rel(r) = 1/rank(r), and in the formula (6), P(r) is the number of pages listed before the page r.

Figure 2 shows the MAP comparison of both the original and the expanded search. We can see most of (more than 4/5) the expanded results become better.

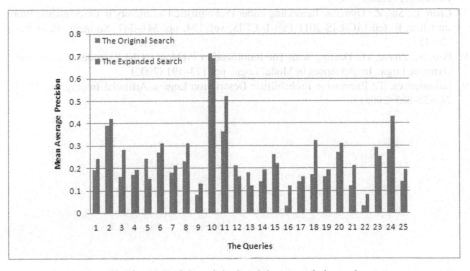

Fig. 2. MAP of the original and the expanded search

4   Conclusion

In this paper, we proposed a method of semantic keyword expansion based on a DL-based formalism admitting dynamic reasoning under probability uncertainty. The main motivation behind this work is to overcome the drawback in keyword search: the user-formulated keywords may not be exact syntax matches with the index structures in the main search engines while share a lot semantically. It employs conditional constraints to encode the overlap-degrees of keywords and dynamic reasoning under uncertainty to compute or maintain such kind of knowledge. We also designed a re-ranking function to re-rank the traditional search results with the original and the expanded keywords. The preliminary results certified the benefits of the approach. We think it will be of some interest to researchers in the field.

References

1. Berners-Lee, T., Hendler, J., Lassila, O.: The Semantic Web. Scientific American (2001)
2. Fazzinga, B., Lukasiewicz, T.: Semantic Search on the Web. Semantic Web 1, 89–96 (2010)
3. Buitelaar, P., Eigner, T., Declerck, T.: OntoSelect: A dynamic ontology library with support for ontology selection. In: Proc. Demo Session at ISWC 2004 (2004)
4. Guha, R.V., McCool, R., Miller, E.: Semantic search. In: Proc. WWW 2003, pp. 700–709. ACM Press (2003)
5. Cheng, G., Ge, W., Qu, Y.: Falcons: Searching and browsing entities on the Semantic Web. In: WWW 2008, pp. 1101–1102 (2008)
6. Lei, Y., Uren, V.S., Motta, E.: SemSearch: A Search Engine for the Semantic Web. In: Staab, S., Svátek, V. (eds.) EKAW 2006. LNCS (LNAI), vol. 4248, pp. 238–245. Springer, Heidelberg (2006)
7. Chen, L., Shi, Z.: Dynamic Reasoning under Probabilistic Uncertainty in the Semantic Web. In: Chen, R. (ed.) ICICIS 2011 Part I. CCIS, vol. 134, pp. 341–347. Springer, Heidelberg (2011)
8. Foo, N., Zhang, D.: Dealing with The Ramification Problem in the Extended Propositional Dynamic Logic. In: Advances in Modal Logic, pp. 173–191 (2002)
9. Lukasiewicz, T.: Expressive Probabilistic Description Logics. Artificial Intelligence 172(6-7), 852–883 (2008)

An ABox Abduction Algorithm
for the Description Logic ALCI

Yanwei Ma, Tianlong Gu, Binbin Xu, and Liang Chang

Guangxi Key Laboratory of Trusted Software, Guilin University of Electronic Technology,
Guilin 541004, China
liberty_myw520@163.com, cctlgu@guet.edu.cn, changl@guet.edu.cn

Abstract. ABox abduction is the foundation of abductive reasoning in description logics. It finds the minimal sets of ABox axioms which could be added to a background knowledge base to enforce the entailment of certain ABox assertions. In this paper, an abductive reasoning algorithm for the description logic ALCI is presented. The algorithm is an extension of an existing ABox abduction algorithm for the description logic ALC, with the feature that it is based on the Tableau of ALCI directly and do not need to use arguments and Skolem terms. It firstly transforms the ABox abduction problem into the consistency problem of knowledge base; then traditional Tableau construction process for ALCI is expanded to deal with this problem; finally the solution of the abduction problem is constructed by a process of backtracking.

Keywords: abductive reasoning, ABox abduction problem, Tableau, consistency of knowledge base.

1 Introduction

In recent decades abduction has gained considerable attention in fields such as logic, artificial intelligence and computer science. It has been widely recognized that the style of reasoning, usually illustrated as an inference from puzzling observations to explanatory hypotheses, is in fact inherent in a vast majority of problem-solving and knowledge acquisition tasks. In the face of such a widespread and diverse application interest, researchers proposed many algorithms for abduction problem in the formalisms of propositional logic, first-order logic and modal logics [1,2].

In recent years, abductive reasoning in description logics become an important topic, since description logics is the logic foundation of the Web Ontology Language OWL and is playing an important role in the semantic Web. In paper [3], based on a combination of the tableau-based reasoning mechanism and the resolution-based reasoning mechanism of both the first-order logic and the modal logic, Klarman et.al. proposed an abduction algorithm for the description logic ALC. In paper [4], Elsenbroich et al. presented several application scenarios in which various forms of abduction would be useful; they also developed some formal apparatus needed to employ abductive inference in expressive description logics. In paper [5], Davenport et al. developed a cognitive model and presented an algorithm for

Z. Shi, D. Leake, and S. Vadera (Eds.): IIP 2012, IFIP AICT 385, pp. 125–130, 2012.

abductive reasoning over instances of a triples ontology. In paper [6], Jianfeng Du et al. proposed a new ABox abduction method based on reducing the original problem to an abduction problem in the logic programming.

In this paper, an abductive reasoning algorithm for the description logic ALCI is presented. This algorithm is an extension of the algorithm proposed by Klarman et.al. for the description logic ALC [3]. Besides the mechanisms for dealing with inverse roles occurring in ALCI, compared with Klarman et.al.'s work, a feature of the algorithm presented here is that the abduction process is based on the Tableau of ALCI and do not need to use a large number of arguments and Skolem terms.

2   Abduction Problem in the Description Logic ALCI

Due to the space limitation, here we omit the introduction of the description logic ALCI. The syntax and semantics of ALCI can be found in [7].

Definition 1. Let $K=<T, A>$ be a knowledge base of ALCI, and let Q be a concept assertion. If $K \nvDash Q$, then we call the pair $<K, Q>$ as an ABox abduction problem of ALCI.

Definition 2. Let $<K, Q>$ be an ABox abduction problem of ALCI. If there exists an ABox E such that not only $<T, A \cup E> \vDash Q$ but also the following conditions holds, then we say that E is a solution for the ABox abduction problem:

(1) E is consistent with the problem, i.e., $<T, A \cup E> \nvDash \bot$;

(2) E is relevant with the problem, i.e., $E \nvDash Q$;

(3) E is minimal, i.e., if there is some ABox B with $K \cup B \vDash Q$, $K \cup B \nvDash \bot$, $B \nvDash Q$ and $A \vDash B$, then it must be $B \vDash A$.

3   Abductive Reasoning Algorithm for the Problem

Let $<K, Q>$ be an ABox abduction problem, where $K=<T, A>$, Q is an ABox concept assertion, and $K \nvDash Q$. Intuition of the algorithm is that the ABox abduction problem will be transformed into the consistency problem of knowledge base, then traditional Tableau construction process can be extended correspondingly to deal with the abduction problem. More precisely, since $K \nvDash Q$, the knowledge base $K'=<T, A \cup \{\neg Q\}>$ must be consistent. The order of the algorithm is to find a set E of ABox assertions such that the knowledge base $K''=<T, A \cup \{\neg Q\} \cup E>$ is inconsistent, and the set E satisfies all the three conditions listed in Def. 2.

For the ease of presentation, we firstly transform the TBox into a set T' of concepts according to the following three steps:

(1) transform every concept definition $C \equiv D$ into two GCIs $C \sqsubseteq D$ and $D \sqsubseteq C$;

(2) transform every GCI $C \sqsubseteq D$ into $\top \sqsubseteq \neg C \sqcup D$ if C is not the Top concept $\top$;

(3) abbreviate the GCI $\top \sqsubseteq \neg C \sqcup D$ as $\neg C \sqcup D$.

For every concept occurring in the knowledge base $K'=<T', A \cup \{\neg Q\}>$, transform it into a normal form according to the following steps:

(1) transform it into an NNF (i.e., negation normal form) such that negation signs occur only in front of concept names;

(2) transform the NNF further into a conjunction normal form by the function τ_n:

① $\tau_n(\sqcup_{1 \leq i \leq n} C_i) = \sqcup_{1 \leq i \leq n} C_i$, where $C_i \in \{C, \exists r.C, \forall r.C\}$;

② $\tau_n(B \sqcup (C \sqcap D) \sqcup E) = \tau_n(B \sqcup C \sqcup E) \sqcap \tau_n(B \sqcup D \sqcup E)$;

③ $\tau_n(\exists r.C) = \exists r.\tau_n(C)$, and $\tau_n(\forall r.C) = \forall r.\tau_n(C)$;

④ $\tau_n(\exists r^-.C) = \exists r^-.\tau_n(C)$, and $\tau_n(\forall r^-.C) = \forall r^-.\tau_n(C)$.

(3) eliminate redundancy by the function τ_{eli}:

① $\tau_{eli}(\forall r.\top) = \top$, $\tau_{eli}(C \sqcap \top) = C$, $\tau_{eli}(C \sqcap \neg C) = \bot$;

② $\tau_{eli}(\exists r.\bot) = \bot$, $\tau_{eli}(C \sqcap \bot) = \bot$, $\tau_{eli}(C \sqcap C) = C$;

③ $\tau_{eli}(\forall r.C \sqcap \forall r.D) = \forall r.(C \sqcap D)$, $\tau_{eli}(\exists r.C \sqcap \exists r.(C \sqcap D)) = \exists r.(C \sqcap D)$;

④ $\tau_{eli}(\exists r.C \sqcap \forall r.\bot) = \bot$, $\tau_{eli}(\exists r.\top \sqcap \exists r.C) = \exists r.C$.

(4) for each resulted conjunction normal form $(B_1 \sqcup B_2 \sqcup \cdots \sqcup B_n) \sqcap (C_1 \sqcup C_2 \sqcup \cdots \sqcup C_n) \sqcap \cdots (D_1 \sqcup D_2 \sqcup \cdots \sqcup D_n)$ which is contained in T', split it into the disjunction terms of $B_1 \sqcup B_2 \sqcup \cdots \sqcup B_n$, $C_1 \sqcup C_2 \sqcup \cdots \sqcup C_n$, $\cdots$, and $D_1 \sqcup D_2 \sqcup \cdots \sqcup D_n$.

Let $K''=<T'', A'' \cup \{\neg Q''\}>$ be the resulted knowledge base. Let $J=T'' \cup A'' \cup \{\neg Q''\}$, and let N_C^K be all the individuals occurring in K''.

Now, we will construct a Tableau for J. Starting from the initial node $\{\neg Q''\}$, the Tableau is constructed by using rules introduced in the following paragraphs. During the construction process, we will use four set $Role_\exists$, $Role_\forall$, $Role_{\exists^-}$ and $Role_{\forall^-}$ which are initialized as empty sets to record the relationship between individuals.

All the rules used for constructing the Tableau are divided into two groups: node evolution rules, and branch expansion rules.

There are six node evolution rules.

(1)$\sqcap$-rule: for any assertion of the form $(C_1 \sqcap C_2 \cdots \sqcap C_n)(x)$ in the current leaf, generate n Tableau and the assertion in current leaf changes to $C_1(x)$, $C_2(x)$, $\cdots$, $C_n(x)$ respectively;

(2)$\sqcup$-rule: for any assertion of the form $(C_1 \sqcup C_2 \cdots \sqcup C_n)(x)$ in the current leaf, attach n new successor node $\{C_1(x)\}$, $\{C_2(x)\}$, $\cdots$, $\{C_n(x)\}$;

(3)$\exists$-rule: for any assertion of the form $\exists r.C(x)$ in the current leaf, generate a new individual b, put $r(x, b)$ in the set $Role_\exists$, and replace $\exists r.C(x)$ with $C(b)$;

(4)$\exists^-$-rule: for any assertion of the form $\exists r^-.C(x)$ in the current leaf, generate a new individual c, put $r(c, x)$ in the set $Role_{\exists^-}$, and replace $\exists r^-.C(x)$ with $C(c)$;

(5)$\forall$-rule: for any assertion of the form $\forall r.D(x)$ in the current leaf, generate a new individual argument y, put $r(x, y)$ in the set $Role_\forall$, and replace $\forall r.D(x)$ with $D(y)$;

(6)$\forall^-$-rule: for any assertion of the form $\forall r^-.D(x)$ in the current leaf, generate a new individual argument z, put $r(z, x)$ in the set $\text{Role}_{\forall}^-$, and replace $\forall r^-.D(x)$ with $D(z)$.

Expanding a branch needs to select a concept from J according to the assertion in the current leaf node, e.g. $C(z)$ is the assertion in the current leaf node. In order to investigate whether z has a relationship in $\text{Role}_{\exists}$, $\text{Role}_{\forall}$, $\text{Role}_{\exists}^-$ or $\text{Role}_{\forall}^-$, we use a transition function **Translate** to map $C(z)$ into the initial value of $C(z)$. The recursive definition of transition function **Translate** is as follows:

If the function is **Translate**$(C(z))$, then Retrieving $r(y, z)$ in $\text{Role}_{\exists}$ and $\text{Role}_{\forall}$ and Retrieving $r(z, y)$ in $\text{Role}_{\exists}^-$ and $\text{Role}_{\forall}^-$:

① $r(y, z)$ is in $\text{Role}_{\exists}$, execute **Translate**$(\exists r.C(y))$;

② $r(z, y)$ is in $\text{Role}_{\exists}^-$, execute **Translate**$(\exists r^-.C(y))$;

③ $r(y, z)$ is in $\text{Role}_{\forall}$, execute **Translate**$(\forall r.C(y))$;

④ $r(z, y)$ is in $\text{Role}_{\forall}^-$, execute **Translate**$(\forall r^-.C(y))$;

⑤ or else **Translate**$(C(z))=C(z)$.

For example, suppose $\text{Role}_{\exists}=\{r_3(y, z)\}$, $\text{Role}_{\forall}=\{r_1(a, x)\}$, $\text{Role}_{\exists}^-=\{r_2(y, x)\}$ and $a\in N_C^K$. By the procedure **Translate**$(C(z))$ we get the assertion $\exists r_3.C(y)$ since $r_3(y, z)$ $\in \text{Role}_{\exists}$, and then will execute **Translate**$(\exists r_3.C(y))$; since $r_2(y, x)\in\text{Role}_{\exists}^-$, we will get the assertion $\forall r_2^-.\exists r_3.C(x)$ and then will execute **Translate**$(\exists r_2^-.\exists r_3.C(x))$; since $r_1(a,x)\in\text{Role}_{\forall}$, then we will get the assertion $\forall r_1.\exists r_2^-.\exists r_3.C(a)$ and therefore will execute **Translate**$(\forall r_1.\exists r_2^-.\exists r_3.C(a))$; finally we have **Translate**$(\forall r_1.\exists r_2^-.\exists r_3.C(a))$ $=\forall r_1.\exists r_2^-.\exists r_3.C(a)$ and the procedure terminates.

Expanding a branch needs to select a concept from J according to the assertion in the current leaf node. The concept selected should satisfy the following conditions:

(1)Condition 1: when the assertion in the current leaf is $C(x)$, if $D(y)=$**Translate** $(C(x))$, $\neg D$ must be a part of the concept selected from J;

(2)Condition 2: when expanding a branch, if $E(x)$ is any assertion in the node of the current branch, and $F(y)=$**Translate**$(E(x))$, then F should not be a part of the concept selected from J.

Let $E(x)$ be the assertion in the current leaf, and $C_1\sqcup C_2\cdots\sqcup C_n$ be the concept selected from J. In that case, attach n new successor node $\{C_1(x)\}$, $\{C_2(x)\},\cdots,\{C_n(x)\}$ to $E(x)$ when we expand the current branch.

If on a branch in a Tableau appears both $C(x)=$**Translate**$(D(y))$ and $\neg C(x)$ $=$**Translate**$(E(z))$, then the branch is called closed. A Tableau is called closed if every branch of it is closed.

A Tableau T is called saturated if it no rule can be applied to it anymore.

Finally, given an abduction problem $<K, Q>$ of ALCI, the complete procedure of the algorithm is as follows:

(1) Construct a knowledge base $K'=<T, A\cup\{\neg Q\}>$ and transform it into the knowledge base $K''=<T'', A''\cup\{\neg Q''\}>$, get the set $J=T''\cup A''\cup\{\neg Q''\}$;

(2) Construct the Tableau of J, get the assertions that can make the unclosed branches close for every unclosed Tableau modal, and put these assertions in set E; for example, if $C(x_n)$ is the last node of an unclosed branch in the Tableau, then put $\neg\textbf{\textit{Translate}}(C(x_n))$ into E;

(3) Check every element of E according to Def. 2, remove the elements that don't satisfy the requirement of consistency, relevance or minimality from the set E;

(4) The result set E is a solution for $<K, Q>$.

At the end of this section, we illustrate the procedure of the algorithm with a small example. Consider an ALCI abduction problem $<K, Q>$, where $Q=HAPPY(John)$ and K is shown in Table 1.

After the pretreatment for $T\cup A\cup\neg Q$, we will get the set $J=\{\neg NIHILIST\sqcup\exists$ $loves^-.\forall owns.(\neg DOG\sqcup\neg BOOK)\sqcup HAPPY, \neg OPTIMIST\sqcup\forall loves^-.\exists owns.DOG, NI$

Table 1. The knowledge base K

TBox	ABox
$NIHILIST\sqcap\forall loves^-.\exists owns.(DOG\sqcap BOOK))\sqsubseteq HAPPY$	$NIHILIST(John)$
$OPTIMIST\sqsubseteq\forall loves^-.\exists owns.DOG$	$DOG(Snoopy)$

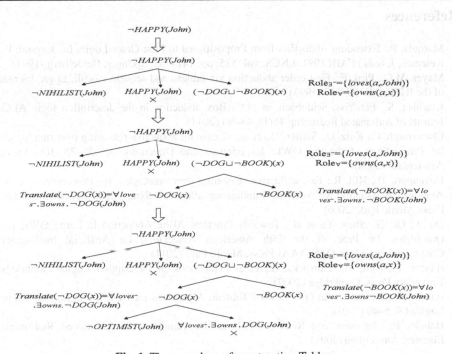

Fig. 1. The procedure of constructing Tableau

HILIST(John), *DOG(Snoopy)*, ¬*HAPPY(John)*}. The procedure for constructing th e Tableau is illustrate in Fig. 1. According to Fig. 1 and the transition function *T ranslate*, we will get *E*={*NIHILIST(john)*, *OPTIMIST(john)*, ∃*owns.BOOK(john)*}. According to Definition 2, we get a solution {*NIHILIST(john)*, *OPTIMIST(john)*, ∀*loves⁻*.∃*owns.BOOK(john)*}.

4   Conclusion

In this paper, the ABox abduction problem is transformed into the consistency problem of knowledge base, and the traditional Tableau construction process of the description logic ALCI is extended and altered to deal with this problem. As a result, the ABox abduction problem can be solved directly in description logics and do not need to use arguments and Skolem terms of first order logics. One of our future work is to improve the algorithm to deal with role assertions. Another work is to expend the algorithm for more expressive description logics.

Acknowledgements. This work is supported by the National Natural Science Foundation of China (Nos. 60903079, 60963010), the Natural Science Foundation of Guangxi Province (No.2012GXNSFBA053169), and the Innovation Project of Guangxi Graduate Education (No. 2011105950812M21).

References

1. Marquis, P.: Extending Abduction from Propositional to First-Order Logic. In: Jorrand, P., Kelemen, J. (eds.) FAIR 1991. LNCS, vol. 535, pp. 141–155. Springer, Heidelberg (1991)
2. Mayer, M.C., Pirri, F.: First order abduction via tableau and sequent calculi. Logic Journal of the IGPL 1(1), 99–117 (1993)
3. Klarman, S., Eendriss, Schlobach, et al.: ABox abduction in the description logic ALC. Journal of Automated Reasoning 46(1), 43–80 (2011)
4. Elsenvroich, C., Kutz, O., Sattler, U., et al.: A case for abductive reasoning over ontologies. In: Proceedings of the 9th OWL: Experiences and Directions, pp. 56–75. IOS Press, Amsterdam (2006)
5. Davenport, D., Hill, R.: Fast abductive reasoning over ontologies. In: Proceedings of the American Association for Artificial Intelligence 2006 Fall Symposium, pp. 84–97. AAAI Press, Menlo Park (2006)
6. Du, J., Qi, G., Shen, Y., et al.: Towards Practical ABox Abduction in Large OWL DL Ontologies. In: Proc. of the 25th American Association for Artificial Intelligence Conference, pp. 1160–1165. AAAI Press, Menlo Park (2011)
7. Baader, F., Nutt, W., Horrocks, L., et al.: Handbook of Description Logic. Cambridge University Press, Cambridge (2007)
8. Baader, F., Sattler, U.: An Overview of Tableau Algorithms for Description Logics. Studia Logica 69, 5–40 (2001)
9. Hahnle, R.: Tableaux and Related Methods. In: Handbook of Automated Reasoning. Elsevier, Amsterdam (2001)

Reasoning about Assembly Sequences
Based on Description Logic and Rule

Yu Meng[1,2], Tianlong Gu[2], and Liang Chang[2]

[1] School of Electronic Engineering, Xidian University, Xi'an 710071, China
[2] Guangxi Key Laboratory of Trusted Software, Guilin University of Electronic Technology,
Guilin 541004, China
mengyucoco@163.com, cctlgu@guet.edu.cn, changl@guet.edu.cn

Abstract. Reasoning about assembly sequences is useful for identifying the feasibility of assembly sequences according to the assembly knowledge. Technologies used for reasoning about assembly sequences have crucial impacts on the efficiency and automation of assembly sequence planning. Description Logic (DL) is well-known for representing and reasoning about knowledge of static application domains; it offers considerable expressive power going far beyond propositional logic while reasoning is still decidable. In this paper, we bring the power and character of description logic into reasoning about assembly sequences. Assembly knowledge is firstly described by a description logic enhanced with some rules. Then, the feasibility of assembly operations is decided by utilizing the reasoning services provided by description logics and rules. An example has been provided to demonstrate the usefulness and executability of the proposed approach.

Keywords: assembly sequence, knowledge representation, reasoning, description logic, rule.

1 Introduction

Related researches show that 40%~50% of manufacturing cost is spent on assembly, and 20%~70% of all the manufacturing work is assembly[1]. Assembly sequence is the most important part of an assembly plan. Reasoning about assembly sequences is useful for identifying the feasibility of assembly sequences according to the assembly knowledge. Technologies used for reasoning about assembly sequences have crucial impacts on the efficiency and automation of assembly sequence planning.

Bourjault[2] and De Fazio and Whitney[3] have developed the structured methodologies in which a series of Yes-or-No questions must be answered to generate the feasible sequences. In both cases, it will become a difficult and error-prone process for all but simplest assemblies. While Homem de Mello et al.[4] presented the cut-set method to finding the feasible assembly operations. This method suffers from the fact that there may be an exponential number of candidates partitioning to test, even though few satisfy physical constraints. Gottipolu and Ghosh[5] developed an algorithmic procedure to generate all feasible assembly sequences by representing the contact and translation function into truth tables and then applying the Boolean

Z. Shi, D. Leake, and S. Vadera (Eds.): IIP 2012, IFIP AICT 385, pp. 131–136, 2012.

algebra principles. It does not preclude that the amount of effort required also increases dramatically with the number of parts in product.

In recent years, due to the strong expression power and the decidability of reasoning, Description Logic (DL)[6] has been emphasized by more and more researchers on knowledge representation and reasoning. But it is hard for DL to define multiple, concurrent conditions rules. However, the multiple, concurrent conditions rules must be represented to implement reasoning in some application domains. For example, Fiorentini et al.[7] and Zhu et al.[8] have proposed the approaches based on DL and rule in the product assembly domain in order to satisfy design tasks such as verifying conditions for design completeness, product qualification and requirements.

In reasoning about assembly sequences, multiple constraints (e.g. connectivity constraints, precedence constraints) must be considered to verify the feasibility of assembly. In other words, the rules for judging the feasibility of assembly are multiple, concurrent conditions rules. In order to improve the level of reasoning automation about assembly sequences, this paper proposes an approach based on DL and rule. In this approach, we bring the power and character of description logic enhanced with some rules into reasoning about assembly sequences.

2   Description Logic and Rule Representation

Description Logic (DL)[6] is a knowledge representation formalism and it is a decidable subset of first-order logic (FOL). In many of the formal methods for representing knowledge, DL has received particular attention in the recent years. The cause is mainly that it is highly effective for concept hierarchy and providing reasoning service. The knowledge base (KB) based on DL is partitioned into an assertional part (ABox) and a terminological part (TBox). In DL, a distinction is drawn between TBox and ABox. In general, TBox is a set defining concepts, relationships among concepts, and relationships among relationships, which is an axiom set describing domain structure. ABox describes which concept an individual belongs to and what relationship one individual have with another individual.

However, there is the scarcity of DL in reasoning rule representation, that is, it is hard for DL to define multiple, concurrent conditions rule. For example, the follow rule which states that x1 is the father of x2 and x3 is the brother of x1, then there exists that x3 is the uncle of x2, can't be expressed in DL.

$$hasFather(x2, x1) \land hasBrother(x1, x3) \rightarrow hasUncle(x2, x3)$$

To offer sophisticated representation and reasoning capabilities, the integration of DL knowledge bases and rule expression representation is necessary. It is one of the methods that the rule-based systems use vocabulary specified in DL knowledge bases.

3   The DL Representation of Assembly Knowledge

Assembly sequence planning begins with representing the assembly knowledge that can be extracted directly from the CAD model of assembly. Gottipolu and Ghosh [5] represented the assembly knowledge as two types of uni-directional matrices, which are called the contact and the translation functions. The contact function

$C_{ab} = (C_1, C_2, C_3, C_4, C_5, C_6)$ is a 1×6 binary function representing contacts between the parts a and b. It can be defined as $C_{ab} : C_i \rightarrow \{0, 1\}$, $i = 1\sim6$, where $C_i = 1$ indicates presence of contact in the direction i, i.e. part b is in contact with part a in the direction i; $C_i = 0$ indicates absence of contact in that direction. The freedom of translation motion between two parts in an assembly can be defined by a 1×6 binary translation function $T_{ab} = (T_1, T_2, T_3, T_4, T_5, T_6)$ or $T_{ab} : T_i \rightarrow \{0, 1\}$, $i = 1\sim6$. If the part b has the freedom of translation motion with respect to the part a in the direction i, then $T_i = 1$, else $T_i = 0$. Here, direction 1, 2 and 3 indicate the positive sense of X, Y and Z axes (X+, Y+ and Z+) respectively, whereas direction 4, 5 and 6 correspond to the negative sense of X, Y and Z axes (X−, Y− and Z−) respectively.

Given an assembly and its contact and translational functions, we can represent the assembly model by defining concepts, roles and creating assertions in DL. The related concepts and roles in DL are defined as follows:

(1) Each assembly is made up of several parts. The **Part** is an atomic concept, the concept **Part** will be defined to represent parts in the assembly.

(2) For the contact functions, we define six atomic roles **DC₁**, **DC₂**, **DC₃**, **DC₄**, **DC₅** and **DC₆**, called contact roles, corresponding to C_1, C_2, C_3, C_4, C_5 and C_6 in C_{ab}.

(3) We represent the translation functions as six atomic roles **DT₁**, **DT₂**, **DT₃**, **DT₄**, **DT₅** and **DT₆**, called translation roles.

For example, if a and b are individual names, then the assertion Part(a) means that a is a part, and Part(b) means that b is a part. For the above part a and part b, if b is in contact with a in the direction i ($i = 1\sim6$), there exist the assertion $DC_i (a, b)$. In the same way, if b has the freedom of translational motion with respect to a in the direction i ($i = 1\sim6$), there exist the assertion $DT_i(a, b)$.

4   The Representation of Reasoning Rule of Assembly

To verify the feasibility assembly, two types of constraints, connectivity constraints and precedence constraints, must be considered. We verify these constraints by representing the contact and translational relations into DL roles and then applying reasoning rules. These reasoning rules are expressed as multiple, concurrent conditions rules. The related inference will use vocabulary specified in DL knowledge bases.

For representing reasoning rules, some concepts and roles will be defined in DL. These concepts include **ComponentSeq**, **Subassembly**, while these roles include **FeasibleAssemblyRole**, **hasLeftComponent** and **hasRightComponent**. **ComponentSeq** and **Subassembly** denote the sequence of components and subassembly respectively. **FeasibleAssemblyRole** means that two components can be assembled. **hasLeftComponent** and **hasRightComponent** denote which components compose the sequence of components or subassembly. There are concept assertions such as

Part(a), Part(b), Part(c),

ComponentSeq($s1$), hasLeftComponent ($s1$, a), hasRightComponent ($s1$, b).

stating that the individuals named a, b and c are parts; $s1$ is a sequence of components and assembled by a and b.

According to the above concepts and roles, the reasoning rule of subassembly is presented as follow.

ComponentSeq(x) $\wedge$ hasLeftComponent(x, xL) $\wedge$ hasRightComponent(x, xR) $\wedge$ FeasibleAssemblyRole(xL, xR) $\rightarrow$ Subassembly(x)

In the rest of this section, we represent all reasoning rules for determining the feasibility of assembly sequences as multiple, concurrent conditions rules.

Firstly, the feasibility of two part subassemblies can be verified from the roles DC_i and DT_i ($i = 1\sim6$) of that pair. For any pair (a, b) of parts, at least one of DC_i (a, b)($i = 1\sim6$) assertions must exist and at least one of DT_i (a, b)($i = 1\sim6$) assertions must exist to make that pair (a, b) a feasible subassembly. For describing these conditions, two roles **DRC** and **DRT** will be defined in DL knowledge bases, where $DC_i \sqsubseteq DRC$ and $DT_i \sqsubseteq DRT$ ($i = 1\sim6$) hold. The reasoning rule for verifying the feasibility of two component subassemblies is described as

DRC($p1$, $p2$) $\wedge$ DRT($p1$, $p2$) $\rightarrow$ FeasibleAssemblyRole($p1$, $p2$)

Secondly, the feasibility of subassemblies with more than two parts, both roles DC_i and DT_i ($i = 1\sim6$) of the involved ordered pairs must be used. For example, the feasibility of assembly of the subassembly (a, b) and the part c is verified as explained as follows.

Step 1. Consider the subassembly (a, b) as a set $\{a, b\}$ and the part c to be added as another set $\{c\}$. The Cartesian product of these two sets ($\{a, b\}\times\{c\}$) gives a set of ordered pairs $\{(a, c), (b, c)\}$. For the pair (a, c) and (b, c), if the assertion DRC(a, c) or DRC(b, c) hold, then there is a contact between c and (a, b). So we define the role **DRTC** in DL knowledge bases and describe this condition as rule expressions such as

Subassembly(s) $\wedge$ hasLeftComponent(s, sL) $\wedge$ DRC(sL, p) $\rightarrow$ DRTC(s, p)
Subassembly(s) $\wedge$ hasRightComponent(s, sR) $\wedge$ DRC(sR, p) $\rightarrow$ DRTC(s, p)

Step 2. The presence of contact above provides only the necessary condition, but it does not guarantee the feasibility of assembly operation due to the precedence constraints. To consider the precedence constraints, the translation roles DT_i ($i = 1\sim6$) of the Cartesian ordered pairs must be used.

Considering the assembly of part c to the subassembly (a, b), there are two pairs (a, c) and (b, c). Firstly, if the DT_i(a, c) and DT_i(b, c) assertions hold, then the component c has collision-free disassembly in the directions i with respect to the subassembly (a, b). Secondly, if one of the six directions is a collision-free disassembly direction of the component c with respect to the subassembly (a, b), then the subassembly (a, b, c) is feasible. In this regard, the role **DRTT** is defined in DL knowledge bases and the above precedence conditions is presented as

Subassembly(s) $\wedge$ hasLeftComponent(s, sL) $\wedge$ hasRightComponent(s, sR) $\wedge$ DT_i(sL, p) $\wedge$ DT_i(sR, p) $\rightarrow$ DRTT(s, p) , $i = 1\sim6$.

The same logic can be applied to the subassembly including multiple components and the single component.

Step 3. If the rules of **Step 1** and **Step 2** hold, then we can obtain the following reasoning rule for judging the feasibility of assembly.

DRTC(s, p) $\wedge$ DRTT(s, p) $\rightarrow$ FeasibleAssemblyRole(s, p)

5   The Reasoning of Assembly Sequence

To demonstrate the usefulness of the proposed approach, we described the assembly knowledge and reasoning rules of the assembly shown in Fig. 1 by the languages OWL DL and SWRL with the help of the editing tool Protégé, and the reasoning is carried out by the JESS reasoning engine [9]. This example assembly includes four parts, a, b, c and d are the name of those parts. The process of generating assembly sequences for the example shown in Fig. 1 is explained as follows.

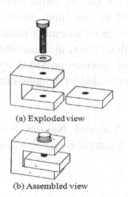

(a) Exploded view

(b) Assembled view

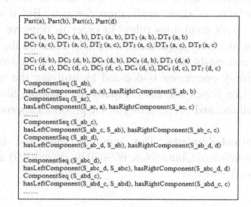

```
Part(a), Part(b), Part(c), Part(d)

DC₄ (a, b), DC₂ (a, b), DT₁ (a, b), DT₃ (a, b), DT₆ (a, b)
DC₅ (a, c), DT₁ (a, c), DT₂ (a, c), DT₃ (a, c), DT₄ (a, c), DT₆ (a, c)
......
DC₁ (d, b), DC₃ (d, b), DC₄ (d, b), DC₆ (d, b), DT₂ (d, a)
DC₁ (d, c), DC₂ (d, c), DC₃ (d, c), DC₄ (d, c), DC₆ (d, c), DT₂ (d, c)

ComponentSeq (S_ab),
hasLeftComponent(S_ab, a), hasRightComponent(S_ab, b)
ComponentSeq (S_ac),
hasLeftComponent(S_ac, a), hasRightComponent(S_ac, c)
......
ComponentSeq (S_ab_c),
hasLeftComponent(S_ab_c, S_ab), hasRightComponent(S_ab_c, c)
ComponentSeq (S_ab_d),
hasLeftComponent(S_ab_d, S_ab), hasRightComponent(S_ab_d, d)

ComponentSeq (S_abc_d),
hasLeftComponent(S_abc_d, S_abc), hasRightComponent(S_abc_d, d)
ComponentSeq (S_abd_c),
hasLeftComponent(S_abd_c, S_abd), hasRightComponent(S_abd_c, c)
......
```

Fig. 1. An example assembly   **Fig. 2.** The ABox of the example shown in Fig. 1

According to the representation of assembly knowledge and reasoning rules given in Section 3 and 4, the ABox of the example shown in Fig. 1 is derived as Fig. 2. After using JESS engine to make inference about OWL individuals of Fig. 2 over the total reasoning rules, **Subassembly** class stores SWRL inference results, that is **Subassembly** instances are generated accordingly. These **Subassembly** instances include:

(1) S_ab, S_ac, S_ad, S_bd and S_cd, which denote the feasible assembly tasks are $\{(a), (b)\}$, $\{(a), (c)\}$, $\{(a), (d)\}$, $\{(b), (d)\}$, $\{(c), (d)\}$, and the corresponding feasible two part subassemblies are (a, b), (a, c), (a, d), (b, d), and (c, d);

(2) S_ab_c, S_ac_b, S_ab_d, S_ac_d, S_cd_a and S_cd_b, which denote the feasible assembly tasks are $\{(a, b), (c)\}$, $\{(a, c), (b)\}$, $\{(a, b), (d)\}$, $\{(a, c), (d)\}$, $\{(c, d), (a)\}$ and $\{(c, d), (b)\}$, and the corresponding feasible three part subassemblies are (a, b, c), (a, b, d), (a, c, d) and (b, c, d).

(3) S_abc_d and S_acb_d, which denote the feasible assembly task is $\{(a, b, c), (d)\}$, and the corresponding feasible four part subassembly is (a, b, c, d).

Through analyzing the above results, the two part subassemblies (a, d) and (b, d) are not used in any tasks in the subsequent assembly, that is, no further higher order subassemblies can be generated from these subassemblies, so (a, d) and (b, d) are invalid subassemblies and should be deleted. Similarly, the (a, b, d), (a, c, d) and (b, c, d) should be deleted. After deleting all the invalid subassemblies, the resulting feasible subassemblies are (a, b), (a, c), (c, d), (a, b, c), (a, b, c, d), and the corresponding feasible assembly tasks are $\{(a), (b)\}$, $\{(a), (c)\}$, $\{(c), (d)\}$, $\{(a, b), (c)\}$, $\{(a, c), (b)\}$ and $\{(a, b, c), (d)\}$. Then all the feasible assembly sequences are obtained as follows:

(1) ({{a} {b} {c} {d }} {{a, b} {c} {d }} {{a, b, c} {d }} {{a, b, c, d }});
(2) ({{a} {b} {c} {d }} {{a, c} {b} {d }} {{a, b, c} {d }} {{a, b, c, d }});
(3) ({{a} {b} {c} {d }} {{a, b} {c} {d }} {{a, b} {c, d }} {{a, b, c, d }});
(4) ({{a} {b} {c} {d }} {{a} {b} {c, d }} {{a, b} {c, d }} {{a, b, c, d }}).

6   Conclusion

Description Logic (DL) is well-known for representing and reasoning about knowledge of static application domains. But it is hard for DL to define multiple, concurrent conditions rules. In order to improve the level of reasoning automation about assembly sequences, we firstly describe the knowledge on assembly model by DL and establish a rule set of the assembly reasoning rules. Then, the feasibility of assembly operations is decided by utilizing the reasoning services provided by description logics and rules. Finally, an example has been provided to demonstrate the usefulness and executability of the proposed approach.

Acknowledgments. This work is supported by the National Natural Science Foundation of China (Nos. 60903079, 60963010) and the Guangxi Key Laboratory of Trusted Software.

References

1. Wang, J.F., Li, S.Q., Liu, J.H., et al.: Computer aided assembly planning: a survey. Journal of Engineering Graphics 26(2), 1–6 (2005)
2. Bourjault, A.: Contribution à une approche méthodologique de l'assemblage automatisé: élaboration automatique des séquences opératoires. Thèse d'État, Université de Franche-Comté, Besancon, France (1984)
3. De Fazio, T.L., Whitney, D.E.: Simplified generation of all mechanical assembly sequences. IEEE Journal of Robotics and Automation RA-3(6), 640–658 (1987)
4. Homem de Mello, L.S., Sanderson, A.C.: A correct and complete algorithm for the generation of mechanical assembly sequences. IEEE Transactions on Robotics and Automation 7(2), 228–240 (1991)
5. Gottipolu, R.B., Ghosh, K.: A simplified and efficient representation for evaluation and selection of assembly sequences. Computers in Industry 50(3), 251–264 (2003)
6. Baader, F., McGuinness, D., et al.: The description logic handbook: theory, implementation and applications. Cambridge University Press (2003)
7. Fiorentini, X., Rachuri, S., et al.: An Evaluation of Description Logic for the Development of Product Models (2008)
8. Zhu, L., Jayaram, U., Kim, O.: Semantic Applications Enabling Reasoning in Product Assembly Ontologies—Moving Past Modeling. Journal of Computing and Information Science in Engineering 12(1) (2012)
9. Dizza, B.: Using OWL and SWRL to represent and reason with situation-based access control policies. Data and Knowledge Engineering 70(6), 596–615 (2011)

Dynamic Logic for the Semantic Web

Liang Chang[1], Qicheng Zhang[1], Tianlong Gu[1], and Zhongzhi Shi[2]

[1] Guangxi Key Laboratory of Trusted Software,
Guilin University of Electronic Technology, Guilin 541004, China
[2] Key Laboratory of Intelligent Information Processing, Institute of Computing Technology,
Chinese Academy of Sciences, Beijing 100190, China
changl@guet.edu.cn, cctlgu@guet.edu.cn, shizz@ics.ict.ac.cn

Abstract. The propositional dynamic logic PDL is one of the most successful variants of modal logic; it plays an important role in many fields of computer science and artificial intelligence. As a logical basis for the W3C-recommended Web ontology language OWL, description logic provides considerable expressive power going far beyond propositional logic as while as the reasoning is still decidable. In this paper, we bring the power and character of description logic into PDL and present a dynamic logic *ALC*-DL for the semantic Web. The logic *ALC*-DL inherits the knowledge representation ability of both the description logic ALC and the logic PDL. With an approach based on Buchi tree automaton, we prove that the satisfiability problem of *ALC*-DL formulas is still decidable and is EXPTIME-complete. The logic *ALC*-DL is suitable for modeling and reasoning about dynamic knowledge in the semantic Web environment.

Keywords: dynamic logic, description logic, satisfiability problem, Buchi tree automaton, semantic Web.

1 Introduction

The propositional dynamic logic PDL is one of the most successful variants of modal logic [1]. It is widely used in many fields of computer science and artificial intelligence. In order to enhance the knowledge representation capability of PDL and correspondingly extend the application domain, many extended logics were proposed by introducing action constructors into PDL [2-4].

Description logic is a family of languages for describing and reasoning about the knowledge of static domains. It plays an important role in the semantic Web, acting as the logic basis of the W3C-recommended Web ontology language OWL. A feature of description logic is that it provides considerable expressive power going far beyond propositional logic, while reasoning is still decidable.

In this paper, we bring the power and character of description logic into dynamic logic. This work is motivated by the fact that, in order to model Web services, actions, or intelligent agents in the semantic Web, one has to deal with lots of knowledge represented as ontologies. Since the Web ontology language OWL is based on description logic, the combination of description logic and dynamic logic will provide

Z. Shi, D. Leake, and S. Vadera (Eds.): IIP 2012, IFIP AICT 385, pp. 137–146, 2012.

the capability to deal with both the static knowledge of ontology and the dynamic knowledge on actions and services.

For the simplicity of presentation, this paper take the typical logic ALC as an example of description logics and adopt it to extend the propositional dynamic logic PDL. As a result, an extended dynamic logic named ALC-DL is presented, which inherits the knowledge representation capability of both the description logic ALC and the propositional dynamic logic PDL. With the help of the Buchi tree automaton, the satisfiability problem of ALC-DL formulas is investigated and demonstrated to be EXPTIME-complete.

2   Dynamic Logic ALC-DL

The dynamic logic ALC-DL is constructed by embracing the description logic ALC into the propositional dynamic logic PDL. From the point of view of syntax, ALC-DL is constructed by replacing all the atomic propositions of PDL with ABox assertions and TBox axioms of ALC.

Primitive symbols of ALC-DL are a set N_C of concept names, a set N_R of role names, and a set N_I of individual names. Starting from these symbols, with the help of a set of constructors, the concepts, formulas and actions of ALC-DL can be constructed inductively.

Definition 1. Concepts of ALC-DL are formed according to the following syntax rule:
$$C, D ::= C_i \mid \neg C \mid C \sqcup D \mid \forall R.\, C$$
where $C_i \in N_C$, $R \in N_R$. Furthermore, concepts of the forms $C \sqcap D$ and $\exists R.\, C$ can also be introduced as abbreviations of concepts $\neg(\neg C \sqcup \neg D)$ and $\neg(\forall R.\, \neg C)$ respectively.

A *general concept inclusion axiom* is an expression of the form $C \sqsubseteq D$, where C, D are any concepts. A *concept assertion* (resp., *role assertion*) is of the form $C(p)$ (resp., $R(p,q)$), where C is a concept, $R \in N_R$, and $p,q \in N_I$.

Definition 2. Formulae of ALC-DL are formed according to the following syntax rule:
$$\varphi, \psi ::= C \sqsubseteq D \mid C(p) \mid R(p,q) \mid <\pi>\varphi \mid \neg\varphi \mid \varphi \wedge \psi$$
where $p, q \in N_I$, $R \in N_R$, C, D are concepts, and π is an action.

Formulas of the forms $[\pi]\varphi$, $\varphi \vee \psi$, $\varphi \rightarrow \psi$, *false* and *true* can also be introduced as abbreviations of formulas $\neg<\pi>\neg\varphi$, $\neg(\neg\varphi \wedge \neg\psi)$, $\neg(\varphi \wedge \neg\psi)$, $\varphi \wedge \neg\varphi$ and $\neg(\varphi \wedge \neg\varphi)$ respectively. Formulas of the form $<\pi>\varphi$ are called *action existence assertions*.

Definition 3. Actions of ALC-DL are formed according to the following syntax rule:
$$\pi, \pi' ::= \alpha \mid \varphi? \mid \pi \cup \pi' \mid \pi\,;\pi' \mid \pi^*$$
where $\alpha \in N_A$, and φ is a formula. Actions of the forms α, $\varphi?$, $\pi \cup \pi'$, $\pi;\pi'$ and π^* are respectively called *atomic actions*, *test actions*, *sequential actions* and *iterated actions*.

The interpretation structure of ALC-DL is a combination of the interpretation structure of PDL and the interpretation of ALC. In such a structure, each action is interpreted as a binary relation between states, while each state is mapped to a classical interpretation of description logic.

Definition 4. A *ALC*-DL interpretation structure is of the form $M=(W,T,\Delta,I)$, where,

(1) W is a non-empty finite set of states;

(2) T is a function which maps each action name $\alpha_i \in N_A$ to a binary relation $T(\alpha_i) \subseteq W \times W$;

(3) Δ is a non-empty set of individuals; and

(4) I is a function which associates with each state $w \in W$ a description logic interpretation $I(w)=(\Delta, \cdot^{I(w)})$, where the function $\cdot^{I(w)}$

(i) maps each concept name $C_i \in N_C$ to a set $C_i^{I(w)} \subseteq \Delta$,

(ii) maps each role name $R_i \in N_R$ to a binary relation $R_i^{I(w)} \subseteq \Delta \times \Delta$, and

(iii) maps each individual name $p_i \in N_I$ to an individual $p_i^{I(w)} \in \Delta$, with the constraints that $p_i^{I(w)} = p_i^{I(w')}$ for any state $w' \in W$. The interpretation $p_i^{I(w)}$ is also represented as p_i^I, since it is not effected by the state w.

Definition 5. Given an interpretation structure $M=(W,T,\Delta,I)$, the semantics of concepts, formulas and actions of *ALC*-DL are defined inductively as follows.

Firstly, for each state $w \in W$, a concept C is interpreted as a set $C^{I(w)} \subseteq \Delta$ which is defined inductively as follows:

(1) $(\neg C)^{I(w)} := \Delta_M \setminus C^{I(w)}$;

(2) $(C \sqcup D)^{I(w)} := C^{I(w)} \cup D^{I(w)}$;

(3) $(\forall R.C)^{I(w)} := \{\, x \mid$ for every $y \in \Delta$: if $(x, y) \in R^{I(w)}$, then $y \in C^{I(w)} \}$.

Secondly, for each state $w \in W$, the satisfaction-relation $(M,w) \vDash \varphi$ for any formula φ is defined as follows:

(4) $(M,w) \vDash C(p)$ iff $p^I \in C^{I(w)}$;

(5) $(M,w) \vDash R(p,q)$ iff $(p^I,q^I) \in R^{I(w)}$;

(6) $(M,w) \vDash C \sqsubseteq D$ iff $C^{I(w)} \subseteq D^{I(w)}$;

(7) $(M,w) \vDash \langle\pi\rangle\varphi$ iff some state $w' \in W$ exists with $(w,w') \in T(\pi)$ and $(M,w') \vDash \varphi$;

(8) $(M,w) \vDash \neg\varphi$ iff it is not the case that $(M,w) \vDash \varphi$;

(9) $(M,w) \vDash \varphi \wedge \psi$ iff $(M,w) \vDash \varphi$ and $(M,w) \vDash \psi$.

Finally, each action π is interpreted as a binary relation $T(\pi) \subseteq W \times W$ according to the following definitions:

(10) $T(\varphi?) := \{(w,w) \in W \times W \mid (M,w) \vdash \psi\}$;

(11) $T(\pi \cup \pi') := T(\pi) \cup T(\pi')$;

(12) $T(\pi \; ; \; \pi') := \{(w,w') \in W \times W \mid$ there is some state $u \in W$ with $(w,u) \in T(\pi)$ and $(u,w') \in T(\pi')\}$;

(13) $T(\pi^*) :=$ reflexive and transitive closure of $T(\pi)$.

Definition 6. An *ALC*-DL formula φ is *satisfiable* if and only if there is an interpretation structure $M=(W,T,\Delta,I)$ and a state $w \in W$ such that $(M,w) \vDash \varphi$.

The satisfiability problem of formulae is a primary inference problem for *ALC*-DL. Many other inference problems which we might concern can be reduced to this problem. For example, given a knowledge base $\mathcal{K}=(\mathcal{T}, \mathcal{A})$ of the description logic *ALC*, we want to know whether it is consistent or not. Since the TBox $\mathcal{T}$ is a finite set composed of general concept inclusion axioms, and the ABox $\mathcal{A}$ is a finite set composed of concept assertions, negations of concept assertions, role assertions, and negations of role assertions, it is obvious that both $\mathcal{T}$ and $\mathcal{A}$ are formula sets of the logic *ALC*-DL, and therefore the consistence problem of the knowledge base $\mathcal{K}$ can be decided by checking whether the conjunction of all the formulas contained in $\mathcal{T}$ and $\mathcal{A}$ is satisfiable or not.

If an *ALC*-DL formula is just a Boolean connection of general concept inclusion axioms, concept assertions and role assertions, then it is also called a *Boolean knowledge base* of the description logic *ALC* [5]. A Boolean knowledge base $\mathcal{B}$ is consistent if and only if there is an interpretation structure $M=(W,T,\Delta,I)$ and a state $w \in W$ such that $(M,w) \vDash \varphi_i$ for every $\varphi_i \in \mathcal{B}$. It is EXPTIME-complete to decide whether a Boolean knowledge base of the description logic *ALC* is consistent or not [5].

3   An Automata-Based Variant of *ALC*-DL

In order to adopt Buchi tree automaton for studying the satisfiability problem of *ALC*-DL formulae, be similar with the approach used in [6], we use automata on finite words rather than regular expressions to describe complex actions of *ALC*-DL.

Firstly, for any *ALC*-DL action π, taking it as a regular expression, then the language $L(\pi)$ represented by it is defined as following: ①$L(\pi):=\{\pi\}$ if π is an atomic action or a test action, ②$L(\pi \cup \pi'):=L(\pi) \cup L(\pi')$, ③$L(\pi$; $\pi'):=\{l_1 l_2 \mid l_1 \in L(\pi)$ and $l_2 \in L(\pi')\}$, and ④$L(\pi^*):=L(\pi^0) \cup L(\pi^1) \cup L(\pi^2) \cup ...$, where $L(\pi^0)$ is the empty word ε, and $L(\pi^j):= L(\pi^{j-1};\pi)$ for every $i \geq 1$. Obviously, all the possible executions of π are captured by the words in $L(\pi)$.

Secondly, we introduce some notations on nondeterministic finite automata.

Definition 7. A nondeterministic finite automaton (NFA) $A=(Q,\Sigma, \rho, q_0, F)$ consists of a finite set of states Q, a finite input alphabet Σ, a transition function $\rho:Q \times \Sigma \to 2^Q$, an initial state $q_0 \in Q$, and a set of final states $F \subseteq Q$.

The transition function ρ can also be inductively extended to a function ρ': $Q \times \Sigma^* \to 2^Q$, such that $\rho'(q,\varepsilon)=q$ and $\rho'(q, \omega a) = \rho(\rho'(q, \omega), a)$ for any $q \in Q$, $\omega \in \Sigma^*$ and $a \in \Sigma$, where ε is the empty word.

Given a NFA $A=(Q,\Sigma, \rho, q_0, F)$, a word $\omega \in \Sigma^*$ is accepted by it if and only if $\rho'(q_0, \omega) \in F$. The language accepted by A is the set of all the words accepted by A, and is denoted as $L(A)$.

For the simplicity of presentation, for any NFA $A=(Q,\Sigma, \rho, q_0, F)$, we use $Q_A, \Sigma_A, \rho_A, q_A$ and F_A to denote the set of states, the alphabet, the transition function, the initial state, and the set of final states of this automaton respectively. Furthermore, for any NFA A and any state $q \in Q_A$, we use A_q to denote the automaton which is constructed by setting the initial state of A as q, i.e., $A_q=(Q_A,\Sigma_A, \rho_A, q, F_A)$.

Now, for any *ALC*-DL action π, there must be a NFA A such that $L(A)=L(\pi)$ [7]. On the contrary, given any NFA $A=(Q,\Sigma, \rho, q_0, F)$ whose input alphabet Σ is composed of atomic actions and test actions of *ALC*-DL, an *ALC*-DL action π can also be constructed such that $L(\pi)=L(A)$. Based on this fact, we can replace each action of *ALC*-DL by a corresponding NFA, and combine Definition 2 and Definition 3 to the following definition.

Definition 8. Let Φ^0 be a set composed of all the general concept inclusion axioms, concept assertions and role assertions of *ALC*-DL. Use Φ^A (resp., Π^A) to denote the sets of all the formulae of *ALC*-DL (resp., all the NFAs of *ALC*-DL). Then, Φ^A and Π^A are the smallest sets satisfying the following conditions:

   (1) $\Phi^0 \subseteq \Phi^A$;

   (2) if φ, $\psi \in \Phi^A$, then $\{\neg\varphi, \varphi\wedge\psi\}\subseteq\Phi^A$;

   (3) if $A\in \Pi^A$ and $\varphi\in \Phi^A$, then $<A>\varphi\in \Phi^A$;

   (4) if $A=(Q,\Sigma, \rho, q_0, F)$ is a NFA with $\Sigma=N_A\cup\{\varphi? \mid \varphi\in \Phi^A\}$, then $A\in \Pi^A$.

We also call formulas of the form $<A>\varphi$ as *action existence assertions*.

Correspondingly, the semantics defined by Definition 5 should be updated. It can be realized by two steps.

Firstly, rules (11), (12) and (13) of Definition 5 are replaced by the following single rule:

   (11') $T(A):= \{(w,w')\in W\times W \mid$ there exist some positive integer m$\geq$0, some word $l_1...l_m\in L(A)$ and m+1 states $u_0, u_1, ..., u_m\in W$, such that $u_0=w$, $u_m=w'$ and $(u_{i-1},u_i)\in l_i^T$ for every $1\leq i\leq m\}$.

Secondly, rule (7) of Definition 5 is updated as follows:

   (7') $(M,w)\vDash<A>\varphi$ iff there is some state $w'\in W$ such that $(w,w')\in A^T$ and $(M,w')\vDash\varphi$.

Obviously, from the point of view of representation and reasoning ability, the automata-based variant of *ALC*-DL presented here is equivalent with the logic defined in former section. Furthermore, for any *ALC*-DL action π in a regular expression form, the NFA A satisfying $L(A)=L(\pi)$ can be constructed in polynomial time, with a property that the number of states in A is linearly bounded by the size of π [7]. Therefore, the satisfiability problem of *ALC*-DL formulae can be studied based on the variant presented in this section, and all the results from this study are also suitable for the logic defined in former section.

4   Buchi Tree Automaton for *ALC*-DL Formulae

Taken a similar approach used in the literatures of [3] and [6], we study the satisfiability problem of *ALC*-DL formulae by Buchi tree automaton. Given any *ALC*-DL formula φ, we will construct a Buchi tree automaton $B\varphi$ such that there exists a one-to-one correspondence between the models of φ and the language accepted by $B\varphi$, and by which the satisfiability problem of φ can be decided by checking whether the language accepted by B_φ is empty or not. Before going to more details, we firstly introduce some notations.

Definition 9. For any finite set Σ and any natural number k, let [k] denote the set $\{1,...,k\}$. A function T: $[k]^* \rightarrow \Sigma$ is called a k-ary Σ-tree. The empty word ε is called the root of the tree. Furthermore, for any $x \in [k]^*$ and $i \in [k]$, the node xi is called the ith-child of the node x.

For any infinite non-empty word $\gamma \in [k]^\omega$, we use $\gamma[0]$ to denote the empty word ε, and use $\gamma[n]$ (where $n \geq 1$) to denote the word formed by the former n characters of the word γ. The word γ is also called a path of a k-ary Σ-tree. It is obvious that the path γ starts from the root ε, and sequentially goes through notes $\gamma[1]$, $\gamma[2]$, $\gamma[3]$ and so on.

Definition 10. A Buchi tree automaton B on k-ary Σ-trees is a quintuple $B=(Q,\Sigma, \rho$, I, F), where Q is a finite set of states, Σ is a finite alphabet, $\rho \subseteq Q \times \Sigma \times Q^k$ is the transition relation, $I \subseteq Q$ is the set of initial states, and $F \subseteq Q$ is the set of accepting states.

Let T be a k-ary Σ-tree. Then, a run of a Buchi tree automaton B on T is a k-ary Q-tree $r:[k]^* \rightarrow Q$ such that $r(\varepsilon) \in I$, and $(r(x), T(x), r(x1), ..., r(xk)) \in \rho$ for all nodes $x \in [k]^*$. A run r of B on T is accepting if and only if $inf(r, \gamma) \cap F \neq \varnothing$ for every path $\gamma \in [k]^\omega$, where $inf(r, \gamma)$ is a set composed of the states in Q that occur infinitely often along the path γ, i.e., $inf(r, \gamma) = \{q \in Q \mid$ for every $n \geq 0$, there always exists some $m \geq n$ with $\gamma[m]=q\}$.

Let B be a Buchi tree automaton on k-ary Σ-trees. Then, a k-ary Σ-tree T is accepted by B if and only if there exists an accepting run of B on T. The language accepted by B, denoted as $L(B)$, is the set of all the k-ary Σ-trees accepted by B.

Let φ be an *ALC*-DL formula for deciding whether it is satisfiable or not. We will construct a Buchi tree automaton B_φ for it. Some notations used in the construction is defined as follows.

Firstly, for any *ALC*-DL formula ψ, let sub(ψ) be a set composed of all the sub-formulae of ψ. More precisely, sub(ψ) is defined inductively as follows:

(1) if ψ is a general concept inclusion axiom, concept assertion or role assertion, then sub(ψ):=$\{\psi\}$;

(2) if ψ is of the forms $\neg\psi'$ or $<\pi>\psi'$, then sub(ψ):=$\{\psi\}\cup$sub(ψ');

(3) if ψ is of the form $\psi'\wedge\psi''$, then sub(ψ):=$\{\psi\}\cup$sub(ψ')$\cup$sub(ψ'').

Secondly, let cl(φ) be the smallest set satisfying the following conditions:

(1) $\varphi \in$ cl(φ);
(2) if $\psi \in$ cl(φ), then sub(ψ)$\subseteq$cl(φ);
(3) if $\psi \in$ cl(φ), then $\neg\psi \in$ cl(φ);
(4) if $<A>\psi \in$ cl(φ), then $\psi \in$ cl(φ) for each test action $\psi? \in \Sigma_A$;
(5) if $<A>\psi \in$ cl(φ), then $<A_q>\psi \in$ cl(φ) for each state $q \in Q_A$.

Finally, for any set $h \subseteq$ cl(φ), it is called an *ALC-Hintikka set* of φ if it satisfies the following five conditions:

(H1) for every formula $\psi \in \text{sub}(\varphi)$: $\psi \in h$ if and only if $\neg\psi \notin h$;

(H2) if $\psi \wedge \psi \in h$, then both $\psi \in h$ and $\psi \in h$;

(H3) if $\neg(\psi \wedge \psi) \in h$, then either $\neg\psi \in h$ or $\neg\psi \in h$;

(H4) if $\neg<A>\psi \in h$ and $q_A \in F_A$, then $\neg\psi \in h$;

(H5) if $\neg<A>\psi \in h$, then for any state $q \in Q_A$ and any test action $\psi? \in \Sigma_A$: it must be $\neg\psi \in h$ or $\neg<A_q>\psi \in h$ whenever $q \in \rho_A(q_A, \psi?)$.

Algorithm 1. Given any ALC-DL formula φ, let $\mathcal{H}_\varphi$ be a set composed of all the ALC-Hintikka sets of φ, let π_φ be a set composed of all the atomic actions occurring in φ, let $\epsilon_1, \epsilon_2, \ldots, \epsilon_k$ be all the action existence assertions contained in $\text{cl}(\varphi)$, and let $\Lambda_\varphi = \mathcal{H}_\varphi \times (\pi_\varphi \cup \{\bot\}) \times \{0,\ldots,k\}$. Then, the corresponding Buchi tree automaton $B_\varphi = (Q, \Lambda_\varphi, \rho, I, F)$ is a Buchi tree automaton on k-ary Λ_φ-trees, and is constructed as follows:

(1) $Q := \Lambda_\varphi \times \{\varnothing, \uparrow\}$;

(2) $I := \{ ((h, \pi, l), d) \in Q \mid \varphi \in h \text{ and } d=\varnothing \}$;

(3) $F := \{ ((h, \pi, l), d) \in Q \mid d=\varnothing \}$; and

(4) $(((h_0, \pi_0, l_0), d_0), (h, \pi, l), ((h_1, \pi_1, l_1), d_1), \ldots, ((h_k, \pi_k, l_k), d_k)) \in \rho$ if and only if

(i) $(h_0, \pi_0, l_0) = (h, \pi, l)$;

(ii) let T be a conjunction of all the (positive or negative) general concept inclusion axioms, (positive or negative) concept assertions, and (positive or negative) role assertions contained in h, then, as a Boolean knowledge base, T is consistent;

(iii) for each $1 \leq i \leq k$, if $\epsilon_i \in h_0$ and ϵ_i is of the form $<A>\psi$, then there exist some integer $n \geq 0$, some word $w=\psi_1?...\psi_n? \in \Sigma_A^*$ composed of test actions, and some state $q_1 \in Q_A$ such that $\{\psi_1,...,\psi_n\} \subseteq h_0$, $q_1 \in \rho_A(q_A, w)$ and either ①$q_1 \in F_A$, $\psi \in h_0$, $\pi_i=\bot$, $l_i=0$, or ②there is some atomic action $\alpha \in \Sigma_A$ and some state $q_2 \in Q_A$ with $q_2 \in \rho_A(q_1, \alpha)$, $\epsilon_j=<A_{q2}>\psi \in h_i$, $\pi_i=\alpha$ and $l_i=j$;

(iv) for each $1 \leq i \leq k$, if there is some formula $\neg<A>\psi \in h_0$, some state $q \in Q_A$ and some atomic action $\alpha \in \Sigma_A$ with $q \in \rho_A(q_A, \alpha)$ and $\pi_i=\alpha$, then it must be $\neg<A_q>\psi \in h_i$;

(v) for each $1 \leq i \leq k$, $d_i=\uparrow$ if and only if either ①$d_0=\varnothing$, $l_i \neq 0$ and $\epsilon_i \in h$, or ②$d_0=\uparrow$, $l_i \neq 0$ and $l_0=i$.

The following property holds for the Buchi tree automaton constructed above:

Theorem 1. An ALC-DL formula φ is satisfiable if and only if the language accepted by the corresponding Buchi tree automaton B_φ is not empty.

This theorem can be proved with a similar process presented in [3]. Due to space limitations, we omitted the proof here.

Let n be the length of φ. In the rest of this section, we give an analyses for the complexity of the satisfiability problem.

Firstly, according to the constructions presented above, the cardinality of $\text{cl}(\varphi)$ is polynomial in n, and the cardinality of $\mathcal{H}_\varphi$ is at most exponential in n. Therefore, for the Buchi tree automaton $B_\varphi = (Q, \Lambda\varphi, \rho, I, F)$, cardinalities of both $\Lambda\varphi$ and Q are at most exponential in n. At the same time, since the number of action existence assertions contained in $cl(\varphi)$ is polynomial in n, the cardinality of the relationship ρ is at most exponential in n. So, the size of the Buchi tree automaton B_φ is at most exponential in n.

Secondly, during the process of constructing the Buchi tree automaton B_φ, majority of the time is spent on step (4) for checking condition (ii). Since the cardinality of each ALC-Hintikka set h is polynomial in n, the size of the Boolean knowledge base T is also polynomial in n. So, the time for checking condition (ii) on step (4) is at most exponential in n. Since the cardinality of the relationship ρ is at most exponential in n too, to sum up, the time used for constructing B_φ is at most exponential in n.

Finally, since the emptiness problem for Buchi tree automaton is PTIME [6], the time used for deciding whether the language accepted by B_φ is empty or not is at most exponential in n.

To sum up, the complexity upper-bound for deciding whether an ALC-DL formula is satisfiable or not is EXPTIME. Together with the fact that the satisfiability problem for the propositional dynamic logic is already EXPTIME-complete, we get the following result:

Theorem 2. Satisfiability of ALC-DL formulae is EXPTIME-complete.

5   Related Works

As a combination of the description logic ALC, the propositional dynamic logic PDL and an action theory based on possible models approach, a dynamic description logic named DDL is presented in [8] for describing and reasoning about actions. Compared with DDL, the logic ALC-DL is also capable for describing the preconditions and effects of atomic actions. For example, the formula $< \alpha >true \rightarrow \psi$ states that the formula ψ must be true for the action α to be executed; the formula $[\alpha]\varphi$ states that the formula φ will be true after the execution of the action α. However, in order to set up a complete action theory, some mechanisms out of ALC-DL should be introduced to deal with the frame problem and the ramification problem for action theories. On the other hand, general concept inclusion axioms contained in TBoxes of DDL are restricted to be invariable [9]. However, in ALC-DL, no restrictions are put on these knowledge described by the description logic ALC. To sum up, DDL can be treated as a special case of the logic ALC-DL.

By embracing the description logic ALC into the linear temporal logic LTL, a linear temporal description logic named ALC-LTL was proposed by Baader et al [5]. From the point of view of syntax, ALC-LTL was constructed by replacing all the atomic propositions of LTL with ABox assertions and TBox axioms of ALC. The satisfiability problem of ALC-LTL was investigated by Baader et al and proved to be decidable. In another paper, results on the complexity of this inference problem was also demonstrated with the help of the general Buchi automaton [10]. Compared with ALC-LTL, the logic ALC-DL is constructed with a similar approach. However, since the interpretation structure of *ALC*-DL is more complex than that of *ALC*-LTL, we have to adopt Buchi tree automaton rather than general Buchi automaton to study the satisfiability problem.

Buchi tree automaton was firstly proposed by Vardi et al [6] and proved to have a PTIME complexity on the emptiness problem. The satisfiability problem on formulae of the deterministic propositional dynamic logic DPDL was also reduced to the emptiness problem by Vardi et al [6], and demonstrated to be ExpTime. With a similar approach, Lutz et al [3] transformed the satisfiability problem of the propositional dynamic logic extended with negation of atomic actions into the emptiness problem of Buchi tree automaton, and by which demonstrated that the satisfiability problem was still decidable with the complexity of ExpTime-complete. This paper is inspired by both Vardi et al.'s work and Lutz et al.'s work. In fact, the Buchi tree automaton constructed for *ALC*-DL formula can be treated as a variant of the Buchi tree automaton constructed by Lutz et al [3]. On the one hand, mechanism for dealing with negation of atomic actions is deleted from Lutz et al.'s Buchi tree automaton. On the other hand, some mechanisms for handling the knowledge described by the description logic ALC is introduced here.

6   Conclusion

The representation ability of *ALC*-DL is reflected in two aspects. On the one hand, propositions used in the propositional dynamic logic PDL are upgraded to formulas of the description logic ALC. Therefore, *ALC*-DL offers considerable expressive power going far beyond PDL, while reasoning is still decidable. On the other hand, *ALC*-DL extends the representation ability of the description logic ALC from static domains to dynamic domains. With *ALC*-DL, knowledge based described by ALC can be connected by actions and therefore evolutions of these knowledge based can be investigated and well organized.

One of our future work is to design efficient decision algorithms for *ALC*-DL. Another work is to put *ALC*-DL into application, by studying the description and reasoning of semantic Web services, and the evolution and organization of ontologies.

Acknowledgments. This work is supported by the National Natural Science Foundation of China (Nos. 60903079, 60963010, 61163041), the Natural Science Foundation of Guangxi Province (No.2012GXNSFBA053169), and the Science Foundation of Guangxi Key Laboratory of Trusted Software (No. KX201109).

References

[1] Harel, D., Kozen, D., Tiuryn, J.: Dynamic Logic. MIT Press, Cambridge (2000)
[2] Giacomo, G.D., Massacci, F.: Combining deduction and model checking into tableaux and algorithms for Converse-PDL. Information and Computation 162(1-2), 117–137 (2000)

[3] Lutz, C., Walther, D.: PDL with negation of atomic programs. Journal of Applied Non-Classical Logic 15(2), 189–214 (2005)

[4] Göller, S., Lohrey, M., Lutz, C.: PDL with intersection and converse is 2EXP-complete. In: Proceedings of the 10th International Conference on Foundations of Software Science and Computation Structures, pp. 198–212. Springer, Berlin (2007)

[5] Baader, F., Ghilardi, S., Lutz, C.: LTL over description logic axioms. In: Proceedings of the 17th International Conference on Principles of Knowledge Representation and Reasoning, pp. 684–694. AAAI Press, Cambridge (2008)

[6] Vardi, M.Y., Wolper, P.: Automata-theoretic techniques for modal logics of programs. Journal of Computer and System Sciences 32(2), 183–221 (1986)

[7] Hromkovic, J., Seibert, S., Wilke, T.: Translating regular expressions into small ε-free nondeterministic finite automata. In: Proceedings of the 14th Annual Symposium on Theoretical Aspects of Computer Science, pp. 55–66. Springer, Belin (1997)

[8] Shi, Z.Z., Dong, M.K., Jiang, Y.C., Zhang, H.J.: A logical foundation for the semantic Web. Science in China, Ser. F: Information Sciences 48(2), 161–178 (2005)

[9] Chang, L., Shi, Z.Z., Gu, T.L., Zhao, L.Z.: A family of dynamic description logics for representing and reasoning about actions. Journal of Automated Reasoning 49(1), 19–70 (2012)

[10] Baader, F., Bauer, A., Lippmann, M.: Runtime Verification Using a Temporal Description Logic. In: Ghilardi, S., Sebastiani, R. (eds.) FroCoS 2009. LNCS, vol. 5749, pp. 149–164. Springer, Heidelberg (2009)

On the Support of Ad-Hoc Semantic Web Data Sharing

Jing Zhou[1,2], Kun Yang[1], Lei Shi[1], and Zhongzhi Shi[2]

[1] School of Computer Science,
Communication University of China, Beijing, 100024, China
{zhoujing,yangkun,shilei_cs}@cuc.edu.cn
[2] The Key Laboratory of Intelligent Information Processing,
Institute of Computing Technology,
Chinese Academy of Sciences, Beijing, 100190, China
shizz@ics.ict.ac.cn

Abstract. Sharing Semantic Web datasets provided by different publishers in a decentralized environment calls for efficient support from distributed computing technologies. Moreover, we argue that the highly dynamic ad-hoc settings that would be pervasive for Semantic Web data sharing among personal users in the future pose even more demanding challenges for enabling technologies. We propose an architecture that is based upon the peer-to-peer (P2P) paradigm for ad-hoc Semantic Web data sharing and identify the key technologies that underpin the implementation of the architecture. We anticipate that our (current and future) work will offer powerful support for sharing of Semantic Web data in a decentralized manner and becomes an indispensable and complementary approach to making the Semantic Web a reality.

Keywords: Semantic Web, RDF triples, storage organization, data sharing, decentralized query processing.

1 The Semantic Web, Data, and Data Storage

The Semantic Web was intended to transform heterogeneous and distributed data into the form that machines can directly process and manipulate by providing the data with explicit semantic information, thus facilitating *sharing and reusing of data across applications*. To make the Semantic Web a reality, one important way is publishing a large amount of data encoded in standard formats on the Web. These data and the associations between them are both essential for supporting large scale data integration and the automated or semi-automated querying of data from disparate sources.

In the early days of the Web, documents were connected by means of hyperlinks, or *links*, and data was not provided independently of their representation in the documents. It is obvious that the ultimate objective of the Semantic Web calls for a shift of the associative linking from documents to data [7]—the Semantic Web turns into the Web of Linked Data whereby machines can explore and locate related data.

Z. Shi, D. Leake, and S. Vadera (Eds.): IIP 2012, IFIP AICT 385, pp. 147–156, 2012.

Virtually, in the Semantic Web one can create links between any Web data that is identified by Uniform Resource Identifier (URI) references with the Resource Description Framework (RDF) [13] being utilised to describe the association between the data that the link represents. The RDF is a standard model for exchanging data on the Web as well as a language for describing Web resources or anything that is identifiable on the Web. The basic RDF data model adopts triples in the form of (*subject, predicate, object*) to depict the attributes of Web resources. RDF triples are also referred to as RDF statements and stored in data stores known as *triplestores*.

A triplestore may contain millions of triples and is mainly responsible for not only storage but also efficient reasoning over and retrieving triples in it. Typically, triplestores come in three forms: (1) RDF triples are stored in relational database management systems (DBMSs), (2) an XML database is employed to store RDF data, or (3) a proprietary information repository is developed to accommodate RDF triples. Among others, the practice of building triplestores on top of relational DBMSs has gained wide adoption thanks to their powerful transaction management. The underlying relational DBMSs deal with the storage and organization of RDF triples in one of the following schemes: triple tables, property tables [19], and vertical partitioning [1]. For a more detailed discussion of individual schemes, interested readers may refer to [5].

2   The Development of Triplestores

Triplestores provide storage and management services to Semantic Web data and hence are able to support data sharing in specialized Semantic Web applications including Semantic Web search engines [10], personal information protection products [9], and wikis. In this section, we will present several typical systems and focus on their logical storage, data description, and query processing mechanisms in particular.

2.1   Single Machine Supported Triplestores

Sesame [3] was developed as a generic architecture for storage and querying of large quantities of Semantic Web data and it allowed the persistent storage of RDF data and schema information to be provided by any particular underlying storage devices. The Storage And Inference Layer (SAIL) was a single architectural layer of Sesame that offered RDF-specific methods to its clients and translated these methods to calls to its certain DBMS. A version of RQL (RDF Query Language) [12] was implemented in Sesame that supported querying of RDF and RDFS [2] documents at the semantic level.

Other single machine supported triplestores include 3store [8] that adopted a relational DBMS as its repository for storage and Kowari[1], an entirely Java-based transactional, permanent triplestore.

---

[1] See http://www.xml.com/pub/a/2004/06/23/kowari.html

2.2   Centralised Triplestores

YARS2 [10] was a distributed system for managing large amounts of graph-structured RDF data. RDF triples were stored as quadruple (*subject, predicate, object, context*) with context indicating the URL of the data source for the contained triple. An inverted text index for keyword lookups and the sparse index-based quad indices were established for efficient evaluation of queries. For scalability reason, the quad index placement adopted established distributed hash table substrates that provided directed lookup to machines on which the quadruple of interest could be located. The Index Manager on individual machines provided network access to local indices such as keyword indices and quad indices. The Query Processor was responsible for creating and optimising the logical plan for answering queries. It then executed the plan over the network by sending lookup requests to and receiving response data from the remote Index Managers in a multi-threaded fashion.

4store [9] was implemented on a low-cost networked cluster with 9 servers and its hardware platform was built upon a shared-nothing architecture. RDF triples were represented as quads of (*model, subject, predicate, object*). All the data was divided into a number of non-overlapping segments according to the RIDs (Resource IDentifiers) that were calculated for the subject of any given RDF triple. Each Storage Node maintained one or more data segments. To allow the Processing Node to discover the data segments of interest, each Storage Node in a local network advertised its service type, contained dataset with a unique name, and the total number of data segments. Hence, the Processing Node could locate the Storage Nodes that host the desired data segments by checking the advertised messages and answer queries from the client.

Furthermore, DARQ [15] and [11] utilising Hadoop and MapReduce are among the systems that process RDF data queries in a centralised way.

2.3   Fully Distributed Triplestores

Piazza [6] was one the few unstructured P2P systems that supported data[2] management in Semantic Web applications. Each node provided source data with its schema, or only a schema (or ontology) to which schemas of other nodes could be mapped. Point-to-point mappings (between domain structures and document structures) were supported by Piazza whereby nodes could choose other nodes with which they would like to establish semantic connections. One of the roles that mappings in Piazza played was serving as storage descriptions that specified what data a node actually stored. A flooding-like technique was employed to process queries and the designers claimed that they focused mostly on obtaining semantically correct answers. The prototype system of Piazza comprised 25 nodes.

---

[2] At the time of writing of [6], most Semantic Web data were encoded in the XML format and hence Piazza mainly dealt with the issues arising from XML data management.

RDFPeers [4] was a distributed and scalable RDF repository. Each triple in RDFPeers was stored at three places in a multi-attribute addressable network (MAAN) by applying hash functions to the subject, predicate, and object values of the triple. Hence, data descriptions were given implicitly as identifiers in the one-dimensional modulo-2^m circular identifier space of MAAN. MAAN not only supported exact-match queries on single or multiple attributes but also could efficiently resolve disjunctive and range queries as well as conjunctive multi-attribute queries. RDFPeers was demonstrated to provide good scalability and fault resilience due to its roots in Chord [18].

S-RDF [20] was a fully decentralized P2P RDF repository that explored ways to eliminate resource and performance bottlenecks in large scale triplestores. Each node maintained its local RDF dataset. Triples in the dataset were further grouped into individual RDF data files according to the type of their subject. The Description Generator took an RDF data file as input to generate a description of the file, which could comprise up to several terms, with the help of ontologies. Considering the semantic relationship that might exist between RDF data files hosted by different nodes, S-RDF implemented a semantics-directed search protocol, mediated by topology reorganization, for efficiently locating triples of interest in a fully distributed fashion. Desired scalability was demonstrated through extensive simulations.

2.4 Cloud Computing-Based Triplestores

SemaPlorer [16], which was awarded the first prize at the 2008 Semantic Web Challenge Billion Triples Track, was built upon the cloud infrastructure. Semantic data from sources including DBPedia, GeoNames, WordNet, personal FOAF files, and Flickr, was integrated and leveraged in the SemaPlorer application. SemaPlorer allowed users to enjoy, in real-time, blended browsing of an interesting area in different context views. A set of 25 RDF stores in SemaPlorer was hosted on the virtual machines of Amazon's Elastic Computing Cloud (EC2), and the EC2 virtual machine images and the semantic datasets were stored by Amazon's Simple Storage Service (S3). SemaPlorer employed Networked Graphs [17] to integrate semantically heterogeneous data. In theory, any distributed data sources can be integrated into the data infrastructure of SemaPlorer in an ad-hoc manner on a very important premise, that is, such data sources can be located by querying SPARQL (SPARQL Protocol And RDF Query Language) endpoints [14]. However, the same premise does not hold in an ad-hoc settings since peers have no knowledge about the specific locations of these data sources in the absence of a fully distributed query forwarding and processing mechanism.

2.5 Summary

In brief, triplestores that were intended to support data sharing and reusing have shifted from centralised to distributed architecture over time. The inherent reasons for this are manifold. To begin with, every single machine is unable to accommodate all existing (and coming) Semantic Web data and process them in memory. Then, copying large amounts of distributed data to one machine and processing them in a centralised manner will inevitably result in issues such as increased network traffic, deteriorated query efficiency, and even infrigement of copyrights sometimes. Furthermore, personal users of the future sharing Semantic Web data would seemingly appear the same way that Web users of today share music, video, and image files; the distinguishing features of Semantic Web data, including being structured and given well-defined associations, however, urge us to come up with new distributed computing technologies. An ad-hoc setting in which Semantic Web data sharing should be supported certainly brings about more challenges.

3   Semantic Web Data Sharing – What Is Still Missing?

Hall pointed out in [7] that the great success of the early Web is attributed in part to the power of the network effect, that is, more people will join a network to get the benefits as its value increases and meanwhile people contribute more information to the network, thus further increasing its value. The network effect is expected to exhibit in the Semantic Web and promote its success. Hence, research communities and organizations were encouraged to publish and share data that can be mapped onto URIs and links to other data, thus increasing the value of the data and the Semantic Web. Ever since, triplestores maintained by individual publishers for storing RDF data from particular domains came into existence. These data stores provide generic features such as data access, retrieval, and deletion. Meanhile, system designers manage to increase the scalability and robustness of the data stores through enabling techniques to deal with the ever-increasing volume of Semantic Web data.

For Semantic Web data sharing among different publishers, querying is an indispensable mechanism. Regardless of the paradigm in which RDF data is stored (centralised or distributed), existing query mechanisms are rather simple and most assume that the target data is within two hops away, that is, the data can be simply located by directly interrogating some central directory node[3].

For more complex scenarios, for example a fully distributed ad-hoc environment in which the target data may be more than two hops away, Semantic Web researchers have yet to supply an efficient querying solution. Approaches from the P2P computing community, in the meantime, generally do not take into account the inherent semantic associations between Semantic Web data, and hence the quality of the query results they yield has room for improvement.

Another important factor that makes supporting ad-hoc Semantic Web data sharing a pressing issue, relates to the paradigm in which people will share RDF data on their personal computers sometime in the future. As RDF converters are becoming available for many kinds of application data, we anticipate that

---

[3] Triplestores built upon distributed hash tables are an exception to this but still not applicable to solving our problem because of the data placement issue, that is, any single RDF triple needs to be placed at some location. In our scenario, the tripletore does not have the issue, which excludes the use of distributed hash tables.

large amounts of RDF data will be generated in personal computers. These data might be references to literature (e.g. BibTex), information in spreadsheets (e.g. Excel), or even data from the GPS (Global Positioning System) receiver. They would be carried around and shared with others at will–just like what we do with document files, music files, or video files in our computers. In most cases, Semantic Web data sharing among personal computers will typically occur in an ad-hoc environment[4] where querying becomes much more complicated in the absence of a central directory node.

Though it has long been recognised and followed that publishing and sharing large amounts of RDF data on the Web is an important approach to making the Semantic Web a reality, we argue that providing efficient techniques and technologies to support ad-hoc Semantic Web data sharing, which is currently missing from the overall approach, is a complementary and indispensable solution the importance of which should never be underestimated. In an ad-hoc environment, Semantic Web data will only be shared in a desired way if efficient mechanisms for describing, manipulating, querying, and analysing data are developed.

4   An Architecture for Ad-Hoc Semantic Web Data Sharing

In this section, we will present the main components in an architecture for ad-hoc Semantic Web data sharing based on the P2P paradigm. This is followed by describing the way that all the components in the architecture collaborate to resolve a query for RDF data in an ad-hoc scenario.

4.1   Building an Architecture on Top of Chord

Our decentralized RDF data repository consists of many individual nodes or peers (each representing a user and her client program) and extends Chord [18] with RDF-specific retrieval techniques. We assume that some of the nodes are willing to host indices for other nodes and self-organize into a Chord ring. Nodes that are reluctant to do so will need to be attached to one of the nodes on the ring. In Fig. 1 we show a peer network of five nodes[5] in a 4-bit identifier space. Node identifiers N1, N4, N7, N12, and N15 correspond to actual nodes that are willing to host indices for other peers. In the meantime, node identifiers D1, D2, and D3 represent three actual nodes that are attached to N12 and share information about their RDF data with N12. Hence, N12 becomes associated with all RDF data shared by D1, D2, and D3. For instance, if N4 stores in its finger table an item regarding a pointer to D3 which stores specific RDF triples, the item will only involve node identifier N12 rather than D3.

---

[4] This is very much like the way that Internet users share music and video files in a P2P fashion.

[5] Nodes that are attached to any node on the Chord ring are not taken into account.

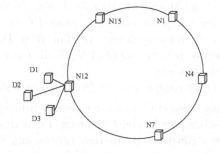

Fig. 1. A peer network of five nodes in a 4-bit identifier space

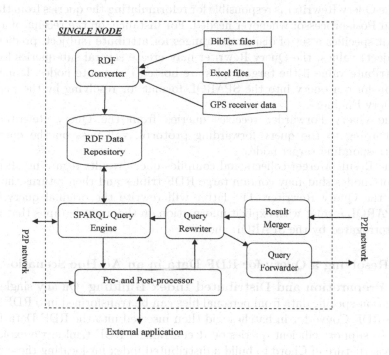

Fig. 2. Single node in ad-hoc Semantic Web data sharing system

Each node has an IP address by which it may be contacted. The overall index will be spread across the nodes on the ring. Unlike RDFPeers [4], our system applies hash functions to the subject (s), predicate (p), object (o), subject and predicate (sp), predicate and object (po), and subject and object (so) types of an RDF triple and stores the information (a pointer to the node, for instance) about the node that shares the triple at six places on a Chord ring.

As shown in Fig. 2, each node on the Chord ring is comprised of seven components: the RDF Converter, SPARQL Query Engine, RDF Data Repository, Pre- and Post-processor, Query Rewriter, Query Forwarder, and Result Merger.

Both the RDF Converter and the SPARQL Query Engine can be obtained by downloading from the Web and some coding is needed for enabling them to interact with other components in the architecture. The RDF Data Repository, or the triplestore, is assumed to be implemented by XML databases, and then is mainly responsible for data storage.

The functionality of the other components is described as follows.

- The Pre- and Post-processor accepts incoming queries from either the external application or any other peer in the P2P network. It directs queries for RDF triples to the Query Rewriter for further processing. When the targe RDF triples are finally obtained, the result will be transmitted to the Pre- and Post-processor for proper presentation to the query originator.
- The Query Rewriter is responsible for reformulating the queries from the Pre- and Post-processor whenever needed. For instance, on the receipt of a query that specifies a set of disjunctive ranges for attribute (subject, predicate, or object) values, the Query Rewriter may create several sub-queries for each attribute value if the targe triples are hosted by remote nodes. It may also transform a query into the SPARQL format for resolving by the SPARQL Query Engine.
- The Query Forwarder receives queries from the Query Rewriter and, according to the query forwarding protocol, will pass on the queries to corresponding target nodes.
- The Result Merger collects and compiles query results regarding all the remote nodes that may contain targe RDF triples and then returns the result to the Query Rewriter. The latter will rewrite the original query in the SPARQL format with explicit information on the named graphs that will be interrogated by the SPARQL query.

4.2   Resolving a Query for RDF Data in an Ad-Hoc Scenario

Data Preparation and Distributed Index Building   On any single node, application-specific data from personal files can be transformed into RDF triples by the RDF Converter in batches and then inserted into the RDF Data Repository. To support efficient queries on decentralized RDF triples, we exploit the overlay structure of Chord to build a distributed index for locating these triples. We store pointers to the node that maintains a specific RDF triple six times, each obtained by applying hash functions to the class type of the subject, predicate, object, subject and predicate, subject and object, and predicate and object of the triple. Each pointer will be stored at the successor node of the hash key of the corresponding attribute type combination.

Decentralized Query Processing   Generally, resolving a query for remote RDF triples in the proposed network is divided into two phases. In the first phase[6], a

---

[6] Note that during the first phase, all queries (except the original one) processed by the Query Rewriter, Query Forwarder, and Result Merger are meant to discover the potential peers that may contain the RDF data of interest. The target RDF data is obtained in the second phase by issuing a SPARQL query.

query from the external application is submitted to a single node which will instruct its Pre- and Post-processor to accept the incoming query. The query is further passed onto the Query Rewriter which will reformulate the query when needed and then send it to the Query Forwarder. The Query Forwarder determines the target node to which the query should be routed. When the Result Merger obtains the query result about the potential target nodes in the network, it will provide the information to the Query Rewriter. The Query Rewriter retrieves the original query, reformulates it in the SPARQL format using the information from the Result Merger, and submits it to the SPARQL Query Engine. In the second phase, the SPARQL Query Engine resolves the SPARQL query by querying against all relevant named graphs. The SPARQL query result containing the RDF data of interest is returned to the Pre- and Post-processor for presentation to the external application.

5   Conclusions and Future Work

We proposed in this position paper a P2P architecture for Semantic Web data sharing in an ad-hoc environment. To fully support the implementation of the architecture, we also need to address the following issues in the future work: (1) decentralized query processing mechanisms that resolve exact-match queries, disjunctive queries, conjunctive queries, and range queries for RDF data and (2) enhancement techniques for RDF data query performance by taking advantage of the semantic relationship between Semantic Web data and by dynamically changing the network topology based on the local knowledge of peer nodes.

Acknowledgments. The first author is grateful for the discussion with Prof. Gregor v. Bochmann from Ottawa University, Canada. This work is funded by China Postdoctoral Science Foundation (No.20100470557). We would also like to acknowledge the input of the National Natural Science Foundation of China (No. 61035003, 60933004, 60970088, 60903141, 61072085, 61103198), National High-tech R&D Program of China (863 Program) (No.2012AA011003), National Science and Technology Support Program2012BA107B02).

References

1. Abadi, D.J., Marcus, A., Madden, S.R., Hollenbach, K.: Scalable Semantic Web Data Management Using Vertical Partitioning. In: VLDB 2007 the 33rd International Conference on Very Large Data Bases, pp. 411–422. VLDB Endowment (2007)
2. Brickley, D., Guha, R.: Rdf vocabulary description language 1.0: Rdf schema. World Wide Web Consortium (2004), http://www.w3.org/TR/rdf-schema/
3. Broekstra, J., Kampman, A., van Harmelen, F.: Sesame: A Generic Architecture for Storing and Querying RDF and RDF Schema. In: Horrocks, I., Hendler, J. (eds.) ISWC 2002. LNCS, vol. 2342, pp. 54–68. Springer, Heidelberg (2002)

4. Cai, M., Frank, M.: Rdfpeers: a scalable distributed repository based on a structured peer-to-peer network. In: The 13th International Conference on World Wide Web, pp. 650–657. ACM, New York (2004)
5. Du, X.Y., Wang, Y., Lv, B.: Research and Development on Semantic Web Data Management. J. Softw. 20(11), 2950–2964 (2009)
6. Halevy, A.Y., Ives, Z.G., Mork, P., Tatarinov, I.: Piazza: Data management infrastructure for Semantic Web applications. In: The 12th International Conference on World Wide Web, pp. 556–567. ACM, New York (2003)
7. Hall, W.: The Ever Evolving Web: The Power of Networks. International Journal of Communication 5, 651–664 (2011)
8. Harris, S., Gibbins, N.: 3store: Efficient bulk RDF storage. In: The 1st International Workshop on Practical and Scalable Semantic Web Systems, pp. 1–15 (2003)
9. Harris, S., Lamb, N., Shadbolt, N.: 4store: the Design and Implementation of a Clustered RDF store. In: The 5th International Workshop on Scalable Semantic Web Knowledge Base Systems, pp. 94–109 (2009)
10. Harth, A., Umbrich, J., Hogan, A., Decker, S.: YARS2: A Federated Repository for Querying Graph Structured Data from the Web. In: Aberer, K., Choi, K.-S., Noy, N., Allemang, D., Lee, K.-I., Nixon, L.J.B., Golbeck, J., Mika, P., Maynard, D., Mizoguchi, R., Schreiber, G., Cudré-Mauroux, P. (eds.) ASWC 2007 and ISWC 2007. LNCS, vol. 4825, pp. 211–224. Springer, Heidelberg (2007)
11. Farhan Husain, M., Doshi, P., Khan, L., Thuraisingham, B.: Storage and Retrieval of Large RDF Graph Using Hadoop and MapReduce. In: Jaatun, M.G., Zhao, G., Rong, C. (eds.) Cloud Computing. LNCS, vol. 5931, pp. 680–686. Springer, Heidelberg (2009)
12. Karvounarakis, G., Alexaki, S., Christophides, V., Plexousakis, D., Scholl, M.: RQL: A Declarative Query Language for RDF. In: The 11th International Conference on World Wide Web (WWW 2002), pp. 592–603. ACM, New York (2002)
13. Klyne, G., Carroll, J.J.: Resource Description Framework (RDF): Concepts and Abstract Syntax. Technical report, W3C Recommendation (2004)
14. Prudhommeaux, E., Seaborne, A.: Sparql query language for rdf. W3C Recommendation (2004), http://www.w3.org/TR/rdf-sparql-query/
15. Quilitz, B., Leser, U.: Querying Distributed RDF Data Sources with SPARQL. In: Bechhofer, S., Hauswirth, M., Hoffmann, J., Koubarakis, M. (eds.) ESWC 2008. LNCS, vol. 5021, pp. 524–538. Springer, Heidelberg (2008)
16. Schenk, S., Saathoff, C., Staab, S., Scherp, A.: SemaPlorer-Interactive Semantic Exploration of Data and Media based on a Federated Cloud Infrastructure. Journal of Web Semantics: Science, Services and Agents on the World Wide Web 7(4), 298–304 (2009)
17. Schenk, S., Staab, S.: Networked Graphs: A Declarative Mechanism for SPARQL Rules, SPARQL Views and RDF Data Integration on the Web. In: The 17th International Conference on World Wide Web, pp. 585–594. ACM, New York (2008)
18. Stoica, I., Morris, R., Karger, D., Kaashoek, M.F., Balakrishnan, H.: Chord: A scalable peer-to-peer lookup service for Internet applications. In: The 2001 Conference on Applications, Technologies, Architectures, and Protocols for Computer Communications, pp. 149–160. ACM, New York (2001)
19. Wilkinson, K., Sayers, C., Kuno, H.A., Reynolds, D.: Efficient RDF Storage and Retrieval in Jena 2. In: The 1st International Workshop on Semantic Web and Databases, pp. 131–150 (2003)
20. Zhou, J., Hall, W., Roure, D.D.: Building a distributed infrastructure for scalable triple stores. J. Comput. Sci. Tech. 24(3), 447–462 (2009)

An Architecture Description Language Based on Dynamic Description Logics

Zhuxiao Wang[1], Hui Peng[2], Jing Guo[3], Ying Zhang[1], Kehe Wu[1], Huan Xu[1], and Xiaofeng Wang[4]

[1] School of Control and Computer Engineering, State Key Laboratory of Alternate Electrical Power System with Renewable Energy Sources, North China Electric Power University, Beijing 102206, China
{wangzx,yingzhang,wkh,xuhuan}@ncepu.edu.cn
[2] Education Technology Center, Beijing International Studies University, Beijing 100024, China
penghui@bisu.edu.cn
[3] National Computer Network Emergency Response Technical Team/Coordination Center of China, Beijing 100029, China
guojing.research@gmail.com
[4] Institute of Computing Technology, Chinese Academy of Sciences, Beijing 100190, China
wangxiaofeng@ict.ac.cn

Abstract. ADML is an architectural description language based on Dynamic Description Logic for defining and simulating the behavior of system architecture. ADML is being developed as a new formal language and/or conceptual model for representing the architectures of concurrent and distributed systems, both hardware and software. ADML embraces dynamic change as a fundamental consideration, supports a broad class of adaptive changes at the architectural level, and offers a uniform way to represent and reason about both static and dynamic aspects of systems. Because the ADML is based on the Dynamic Description Logic DDL($\mathcal{SHON}$ (D)), which can represent both dynamic semantics and static semantics under a unified logical framework, architectural ontology entailment for the ADML languages can be reduced to knowledge base satisfiability in DDL($\mathcal{SHON}$(D)), and dynamic description logic algorithms and implementations can be used to provide reasoning services for ADML. In this article, we present the syntax of ADML, explain its underlying semantics using the Dynamic Description Logic DDL($\mathcal{SHON}$(D)), and describe the core architecture description features of ADML.

Keywords: Architecture Description Languages, Knowledge Representation and Reasoning, Software Architecture, Dynamic Description Logics, Dynamic Adaptation.

1 Introduction

ADML is a promising Architecture Description Language (ADL) towards a full realization of the representing and reasoning about both static and dynamic aspects of

Z. Shi, D. Leake, and S. Vadera (Eds.): IIP 2012, IFIP AICT 385, pp. 157–166, 2012.
© IFIP International Federation for Information Processing 2012

concurrent and distributed systems. Concurrent and distributed systems, both hardware and software, can be understood as a world that changes over time. Entities that act in the world (which can be anything from a monitor to some computer program) can affect how the world is perceived by themselves or other entities at some specific moment. At each point in time, the world is in one particular state that determines how the world is perceived by the entities acting therein. We need to consider some language (like the Architecture Dynamic Modeling Language, ADML) for describing the properties of the world in a state. By means of well-defined change operations named transition rules in ADML, transition rules can affect the world and modify its current state. Such transition rules denote state transitions in all possible states of the world.

In this paper we describe the main features of ADML, its rationale, and technical innovations. ADML is based on the idea of representing an architecture as a dynamic structure and supporting a broad class of adaptive changes at the architectural level. However, simultaneously changing components, connectors, and topology in a reliable manner requires distinctive mechanisms and architectural formalisms. Many architecture description languages[1-4] are dynamic to some limited degree but few embrace dynamic change as a fundamental consideration. ADML is being developed as a way of representing dynamic architectures by expressing the possible change operations in terms of the ADML constructors.

ADML can be viewed as syntactic variants of dynamic description logic. In particular, the formal semantics and reasoning in ADML use the DDL($\mathcal{SHON}$(D)) dynamic description logic, extensions of description logics (DLs) [5] with a dynamic dimension [6-9]. So the main reasoning problem in ADML can be reduced to knowledge base (KB) satisfiability in the DDL($\mathcal{SHON}$(D)). This is a significant result from both a theoretical and a practical perspective: it demonstrates that computing architectural ontology entailment in ADML has the same complexity as computing knowledge base satisfiability in DDL($\mathcal{SHON}$(D)), and that dynamic description logic algorithms and implementations can be used to provide reasoning services for ADML.

In the following sections, we firstly present an overview of the capabilities of ADML in Section 2. It covers the basic language features and includes a few small examples. Furthermore, we demonstrate the descriptions of transition rules can be formalized as actions in the DDL($\mathcal{SHON}$(D)). In Section 3, we summarize basic ADML syntax with an overview of ADML semantics. We show that the main reasoning problem in ADML can be reduced to knowledge base (KB) satisfiability in the DDL($\mathcal{SHON}$(D)) dynamic description logic. Finally, we summarize the paper in Section 4.

2   An Overview of ADML

ADML is intended as a new formal language and/or conceptual model for describing the architecture of a system. ADML is built on a core ontology of six types of entities for architectural representation: components, connectors, systems, ports, roles, and behaviors. These are illustrated in Figure 1 and Table 1. Of the six types, the most basic elements of architectural description are components, connectors, systems, and behaviors. It's important to recognize that ADML is based on the idea of representing an architecture as a dynamic structure. In other words, ADML may also be used as a way of representing reconfigurable architectures by expressing the possible reconfigurations in terms of the ADML structures (like behaviors). For example, an architectural model might include behaviors that describe components that may be added at run-time and how to attach them to the current system.

2.1   ADML Design Element Types

As a simple illustrative example, Figure 1 shows the architectural model of a secure wireless remote-access infrastructure, which is represented as a graph of interacting components. Nodes in the graph are termed components, which represent the primary computational elements and data stores of the system. Typical examples of components include such things as terminals, gateways, filters, objects, blackboards, databases, and user interfaces. Arcs are termed connectors, and represent communication glue that captures the nature of an interaction between components. Examples of connectors include simple forms of interaction, such as data flow channel (e.g., a Pipe), a synchronous procedure call, and a particular protocol (like HTTP). Table 1 contains an ADML description of the architecture of Figure 1. In the software architecture illustrated in Table 1, the secure access gateway (SAG), and the wireless terminal are components. The component exposes its functionality through its ports, which represents a point of contact between the component and its environment. The wireless terminal component is declared to have a single send-request port, and the SAG has a single receive-request port. The connector includes the network connections between the wireless terminal and the SAG. A connector includes a set of interfaces in the form of roles, which may be seen as an interface to a communication channel. The rpc connector has two roles designated caller and callee. The topology of this system is defined by listing a set of attachments, each of which represents an interaction between a port and some role of a connector.

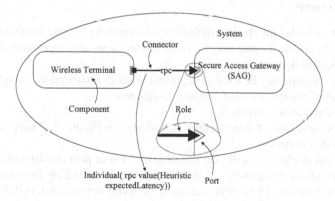

Fig. 1. Elements of an ADML Description

Table 1. The architectural model in ADML

Element Axioms	Property Axioms	Facts
Element(System complete Thing restriction (hasComp minCardinality (1))) ;	ObjectProperty(has Comp domain(unionOf (System Comp)) range(Comp));	Individual(sys type (System)); Individual(term type (Comp)); Individual(SAG type (Comp)); Individual(send-request type (Port)); Individual(receive-request type (Port));
Element(Comp complete Thing unionOf (restriction (hasPort minCardinality (1)) restriction (hasComp minCardinality (1))));	ObjectProperty(hasPort domain(Comp) range(Port)); ObjectProperty(hasRole domain(Connector) range(Role));	Individual(caller type (Role)); Individual(callee type (Role)); Individual(sys value(hasComp term)); Individual(sys value(hasComp SAG)); Individual(sys value(hasConnector rpc));
Element(Connector complete Thing restriction (hasRole minCardinality (2)));	DatatypeProperty(Heuris tic domain(Connector) range(Float)); ObjectProperty(attached domain(Port) range(Role));	Individual(term value(hasPort send-request)); Individual(SAG value(hasPort receive-request)); Individual(rpc value(hasRole caller)); Individual(rpc value(hasRole callee)); Individual(rpc value(Heuristic expectedLatency)); Individual(send-request value(attached caller)); Individual(receive-request value(attached callee));

2.2 Components

A component provides an abstract definition of externally visible behavior—i.e., the behavior that is visible to, and may be observed by, a system containing the component. A component defines

1) the ports which may be used to represent what is traditionally thought of as an interface: a set of operations available on a component.

2) its states and state transitions.

The syntax structure of components is outlined in Figure 2. Many features are omitted from this overview.

A component declares sets of port constituents. Those port constituents are visible to the system. Other components of the system can be wired up (by connectors) to those port constituents. Thus communication between components is defined by the ports.

A component must provide the objects and functions named in the port constituents. Thus, for example, other components can be connected to call its provided

```
type_declaration ::=                          declaration_list : := ADML facts - see Section 3
  type identifier is component_expression ';'
component_expression ::=                       state_transition_rule : :=
  component                                       Transition '(<' trigger ',' transition_body '>)' ';;'
    {component_constituent}
  end [component]                               trigger ::= pattern - see Section 2.3
component_constituent : :=
  { port_declaration}                          pattern ::= ADML axioms- see Section 3
  | behavior
    [declaration_list]                         transition_body : := {state assignment}
    begin                                       [ restricted_pattern ]
      { state_transition_rule }

port_declaration : := Port name '= {'
  {property_declaration;}
  {representation_declaration;}
'}'
```

Fig. 2. Outline of component syntax

functions. Conversely, a component may call its requires functions and assume they are connected to call provided functions of other components. Connectors between required and provided functions define synchronous communication.

A component specifies the types of events it can observe and generate by declaring, respectively, in and out ports. Connectors in the system can call the in ports of a component, thus generating in events which the component can observe; conversely, the component can call its out ports, thereby generating events which the system can observe. Thus connectors define asynchronous communication between components.

A component optionally contains a behavior which consists of a set of objects, functions, and transition rules. The set of objects, functions is described as ADML facts (see Section 3). Facts declared in a component model state. Transition rules model how the component react to patterns of observed events by changing its states and generating events. The behavior of a component in a system is constrained to be consistent with the component's behavior.

A component observes in events from the system. It reacts by executing its transition rules and generating out events which are sent to other components. A component observes calls to its provides functions; the component must declare functions in its behavior that are executed in response to these calls. When a component calls its requires functions it depends upon the system to connect those calls to provides functions in other components.

2.3   Transition Rules

One of the key ingredients of the behaviors in a component is a set of transition rules that model how the component react to patterns of observed events by changing their states and generating events. A state transition rule has two parts, a trigger and a body. A trigger is an event pattern (a finite set of facts). A body is an optional set of state assignments followed by a restricted pattern which describes a finite set of facts. In our description frameworks for transition rules, functional descriptions are essentially the state-based and use at least pre-state and post-state constraints to characterize intended executions of a transition rule. On the basis of the above consideration, we give some formal definitions related to transition rules:

Definition 1 (Atomic transition rules). An atomic transition rule is a tuple Transition(t)= Transition(<trigger, transition_body>)= Transition(<Pre, Effects>), where Pre is a finite set of facts specifying the preconditions for the execution of t; and Effects is a finite set of facts holding in the newly-reached world by the transition rule's execution. A function body is treated as if it was a transition rule that is triggered by a function call; function calls are treated as events.

A close observation on the state transition rule reveals some resemblance between transition rules and DDL($\mathcal{SHON}$(D)) actions. As mentioned in [9-12], the formulas in both Pre and Effects are conferred with well-defined semantics encoded in some TBox, which specifies the domain constraints in consideration.

The execution semantics of transition rules are as follows. If any of the preconditions of the transition rules are satisfied, the process of arbitrarily choosing one of the transition rules, executing its rule body is repeated until none of the preconditions are satisfied.

Executing a rule body consists of changing the state of the behavior part (by calling operations of facts declared there), generating new events defined by the instance of the restricted pattern, and adding the new events to the execution of the system.

Composite transition rules are constructed from atomic transition rules with the help of classic constructors in ADML. Both atomic and composite transition rules are transition rules:

Definition 2 (Transition rules). Transition rules are built up with the following rule:

t_i, t_j ::=   Transition(t) | Transition(t_i test) | choiceOf(t_i t_j) | sequenceOf(t_i t_j) | Transition(t_i iteration), where t is an atomic transition rule; t_i, t_j denote transition rules.

We name sequenceOf(t_i t_j), choiceOf(t_i t_j), Transition(t_i iteration) and Transition(t_i test) as sequence, choice, iteration and test transition rules, respctively.

Remark: Note that test Transition(t_i test) is used to check the executability of the transition rules, which can be reduced to the satifiablity-checking of preconditions of the component transition rules. Test transition rules' execution effects no changes to the world.

ADML is a promising Architecture Description Language (ADL) towards a full realization of the representing and reasoning about both static knowledge and dynamic knowledge in concurrent and distributed systems. In addition to the features inherited from $\mathcal{SHON}$(D), i.e., expressive power in static knowledge representation and decidability in reasoning, DDL($\mathcal{SHON}$(D)) still employs actions to capture functionalities of transition rules. Hence it is intuitive to model transition rules by actions in DDL($\mathcal{SHON}$(D)). As demonstrated in this section, the functionalities of transition rules can be semantically transformed into actions in DDL($\mathcal{SHON}$(D)) by a proper domain ontology (TBox). As a result, all kinds of reasoning tasks concerning the functionalities of transition rules thus can be reduced to the reasoning about actions in DDL($\mathcal{SHON}$(D)).

3   ADML as the DDL($\mathcal{SHON}$(D)) Dynamic Description Logic

ADML is very close to the DDL($\mathcal{SHON}$(D)) Dynamic Description Logic which is itself an extension of the $\mathcal{SHON}$(D) Description Logic [5] (extended with a dynamic dimension[6,9]). ADML can form descriptions of components, connectors, and systems using some constructs. Given the limited space available, in this article I will not delve into the details of the ADML syntax. ADML axioms, facts, and transition rules are summarized in Table 2 below. In this table the first column gives the ADML syntax for the construction, while the second column gives the DDL($\mathcal{SHON}$(D)) Dynamic Description Logic syntax.

Because ADML includes datatypes, the semantics for ADML is very similar to that of Dynamic Description Logics that also incorporate datatypes, in particular DDL($\mathcal{SHON}$(D)).

The specific meaning given to ADML is shown in the third column of Table 2. A DDL(X) model is a tuple $M = (W, T, \Delta, I)$, where,

W is a set of states;

$T : N_A \rightarrow 2^{W \times W}$ is a function mapping action names into binary relations on W;

Δ is a non-empty domain;

I is a function which associates with each state $w \in W$ a description logic interpretation $I(w) = <\Delta, \cdot^{I(w)}>$, where the mapping $\cdot^{I(w)}$ assigns each concept to a subset of Δ, each role to a subset of $\Delta \times \Delta$, and each individual to an element of Δ.

What makes ADML an architecture description language for concurrent and distributed systems, is not only its semantics, which are quite standard for a dynamic description logic, but also the use of transition rules for changes at the architectural level, the use of datatypes for data values, and the ability to use that dynamic description logic algorithms and implementations to provide reasoning services for ADML.

Table 2. ADML axioms, facts, and transition rules

ADML Syntax	DDL Syntax	Semantics
Elements		
Element(A partial $C_1 ... C_n$)	$A \sqsubseteq C_1 \sqcap ... \sqcap C_n$	$A^I \subseteq C_1^I \cap ... \cap C_n^I$
Element(A complete $C_1 ... C_n$)	$A = C_1 \sqcap ... \sqcap C_n$	$A^I = C_1^I \cap ... \cap C_n^I$
SubElementOf ($C_1 \quad C_2$)	$C_1 \sqsubseteq C_2$	$C_1^I \subseteq C_2^I$
EquivalentElements ($C_1 ... C_n$)	$C_1 = ... = C_n$	$C_1^I = ... = C_n^I$
DisjointElements ($C_1 ... C_n$)	$C_i \sqcap C_j = \perp, i \neq j$	$C_i^I \cap C_j^I = \phi, i \neq j$
Datatype(D)		$D^I \subseteq \mathfrak{A}_D$
Datatype Properties (U)		
DatatypeProperty(U super(U_1) ... super(U_n))	$U \sqsubseteq U_i$	$U^I \subseteq U_i^I$
DatatypeProperty(U range(D_1) ... range(D_l))	$\top \sqsubseteq \forall U.D_i$	$U^I \subseteq \mathfrak{A} \times D_i^I$
DatatypeProperty(U Functional)	$\top \sqsubseteq \leq \infty U$	U^I is functional

Table 2. *(continued)*

ADML Syntax	DDL Syntax	Semantics
SubPropertyOf(U_1 U_2)	$U_1 \sqsubseteq U_2$	$U_1^I \subseteq U_2^I$
EquivalentProperties(U_1 ... U_n)	$U_1 = ... = U_n$	$U_1^I = ... = U_n^I$
Object Properties		
ObjectProperty(R super(R_1) ... super(R_n))	$R \sqsubseteq R_i$	$R^I \subseteq R_i^I$
ObjectProperty(R domain(C_1) ... domain(C_m))	$\geq 1\ R \sqsubseteq C_i$	$R^I \subseteq C_i^I \times \varDelta$
ObjectProperty(R range(C_1) ... range(C_l))	$\top \sqsubseteq \forall R.C_i$	$R^I \subseteq \varDelta \times C_i^I$
ObjectProperty(R Functional)	$\top \sqsubseteq \leq \infty R$	R^I is functional
ObjectProperty(R Transitive)	$Tr(R)$	$R^I = (R^I)^\Downarrow$
SubPropertyOf(R_1 R_2)	$R_1 \sqsubseteq R_2$	$R_1^I \subseteq R_2^I$
EquivalentProperties(R_1 ... R_n)	$R_1 = ... = R_n$	$R_1^I = ... = R_n^I$
Annotation		
AnnotationProperty(S)		
Facts		
Individual(o type(C_1) ... type(C_n))	$\iota \in C_i$	$o^I \in C_i^I$
Individual(o value(R_1 o_1) ... value(R_n o_n))	$<o, o_i> \in R_i$	$< o^I, o_i^I > \in R_i^I$
Individual(o value(U_1 v_1) ... value(U_n v_n))	$<o, v_i> \in U_i$	$< o^I, v_i^I > \in U_i^I$
SameIndividual(o_1... o_n)	$o_1 = ... = o_n$	$o_1^I = ... = o_n^I$
DifferentIndividual(o_1... o_n)	$o_i \neq o_j, i \neq j$	$o_i^I \neq o_j^I, i \neq j$
negationOf (φ)	$\neg \varphi$	$(M,w) \not\models \varphi$
disjunctionOf (φ φ' ...)	$\varphi \vee \varphi'$	$(M,w) \models \varphi$ or $(M,w) \models \psi$
diamondAssertion(π φ)	$<\pi>\varphi$	$\exists w' \in W.((w, w') \in T(\pi)$ and $(M,w') \models \varphi)$
Transition rules		
Transition(α)	α	$T(\alpha) = T(P, E) = \{ (w, w') \mid \textcircled{1} (M,w) \models P, \textcircled{2} (M,w) \models \neg \varphi$ for every $\varphi \in E$, $\textcircled{3} (M,w') \models E$, $\textcircled{4} C^{I(w')} = (C^{I(w)} \cup \{u^I \mid C(u) \in E\}) \setminus \{u^I \mid \neg C(u) \in E\}$, and $\textcircled{5} R^{I(w')} = (R^{I(w)} \cup \{(u^I, v^I) \mid R(u, v) \in E \}) \setminus \{(u^I, v^I) \mid \neg R(u, v) \in E\}. \}$
Transition (φ test)	$\varphi?$	$\{(w, w) \mid w \in W$ and $(M,w) \models \varphi\}$
choiceOf (π π' ...)	$\pi \cup \pi'$	$T(\pi) \cup T(\pi')$
sequenceOf (π π' ...)	$\pi ; \pi'$	$\{ (w, w') \mid \exists w''.(w, w'') \in T(\pi)$ and $(w'', w') \in T(\pi') \}$
Transition (π iteration)	π^*	reflexive and transitive closure of $T(\pi)$

4   Summary and Outlook

In this paper we presented ADML a new formal language and/or conceptual model for representing system architectures. We described the main features of ADML, its rationale, and technical innovations. By embracing transition rules into ADML, ADML combine the static knowledge provided by the system requirements with the dynamic descriptions of the computations provided by transition rules, and support the representing and reasoning about both static knowledge and dynamic knowledge in concurrent and distributed systems. ADML can be viewed as syntactic variants of dynamic description logic DDL($\mathcal{SHON}$(D)). The functionalities of the transition rules are abstracted by actions in DDL($\mathcal{SHON}$(D)), while the domain constraints, states, and the overall system objectives are encoded in TBoxes, ABoxes and DL-formulas, respectively. So the main reasoning problem in ADML can be reduced to knowledge base (KB) satisfiability in the DDL($\mathcal{SHON}$(D)) dynamic description logic. Afterwards, dynamic description logic algorithms and implementations can be used to provide reasoning services for ADML. ADML has evolved from several sources:1) Rapide[13] (a concurrent event-based simulation language for defining and simulating the behavior of system architectures.), 2) Acme[14] (a common representation for software architectures), 3) DDL[6][9] (for event patterns and formal constraints on concurrent behavior expressed in terms of description logics). While ADML is still too new to tell whether it will succeed as a community-wide tool for architectural development, we believe it is important to expose its language design and philosophy to the broader software engineering community at this stage for feedback and critical discussion.

In the future, we plan to investigate how we can leverage Distributed Dynamic Description Logics (D3Ls) to support a larger variety of heterogeneous systems with our approach. Another issue for future work is the design of "practical" algorithms for DDL($\mathcal{SHON}$(D)) reasoning.

Acknowledgements. This work is supported by the Fundamental Research Funds for the Central Universities (No.11QG13) and the National Science Foundation of China (No.71101048).

References

1. Dashofy, E.M., Van der Hoek, A., Taylor, R.N.: A comprehensive approach for the development of modular software architecture description languages. ACM Transactions on Software Engineering and Methodology 14(2), 199–245 (2005)
2. Azevedo, R., Rigo, S., Bartholomeu, M.: The ArchC architecture description language and tools. International Journal of Parallel Programming 33(5), 453–484 (2005)
3. Mishra, P., Dutt, N.: Architecture description languages for programmable embedded systems. IEE Proceedings-Computers and Digital Techniques 152(3), 285–297 (2005)
4. Pérez, J., Ali, N., Carsí, J.Á., Ramos, I.: Designing Software Architectures with an Aspect-Oriented Architecture Description Language. In: Gorton, I., Heineman, G.T., Crnković, I., Schmidt, H.W., Stafford, J.A., Ren, X.-M., Wallnau, K. (eds.) CBSE 2006. LNCS, vol. 4063, pp. 123–138. Springer, Heidelberg (2006)

5. Baader, F., Calvanese, D., McGuinness, D., Nardi, D., Patel-Schneider, P.F.: The description logic handbook: theory, implementation, and applications. Cambridge University Press (2003)
6. Shi, Z., Dong, M., Jiang, Y., Zhang, H.: A logical foundation for the semantic web. Science in China, Ser. F 48(2), 161–178 (2005)
7. Artale, A., Franconi, E.: A temporal description logic for reasoning about actions and plans. J. Artif. Intell. Res. 9, 463–506 (1998)
8. Baader, F., Lutz, C., Milicic, M., Sattler, U., Wolter, F.: Integrating description logics and action formalisms: First results. In: Proc. Natl. Conf. Artif. Intell., vol. 2, pp. 572–577 (2005)
9. Chang, L., Shi, Z., Gu, T., Zhao, L.: A Family of Dynamic Description Logics for Representing and Reasoning About Action. J. Autom. Reasoning, 1–52 (2010)
10. Wang, Z., Yang, K., Shi, Z.: Failure Diagnosis of Internetware Systems Using Dynamic Description Logic. J. Softw. China 21, 248–260 (2010)
11. Wang, Z., Guo, J., Wu, K., He, H., Chen, F.: An architecture dynamic modeling language for self-healing systems. Procedia Engineering 29(3), 3909–3913 (2012)
12. Wang, Z., Zhang, D., Shi, Z.: Multi-agent based bioinformatics integration using distributed dynamic description logics. In: Int. Conf. Semant., Knowl., Grid., China, pp. 66–71 (2009)
13. Luckham, D.C., Vera, J.: An Event-Based Architecture Definition Language. IEEE Transactions on Software Engineering 21(9), 717–734 (1995)
14. Garlan, D., Monroe, R., Wile, D.: Acme: an architecture description interchange language, CASCON First Decade High Impact Papers. USA, pp. 159–173 (2010)

Query Expansion Based-on Similarity of Terms
for Improving Arabic Information Retrieval

Khaled Shaalan[1], Sinan Al-Sheikh[2], and Farhad Oroumchian[3]

[1] The British Univ. in Dubai, PO BOX 345015, Dubai, UAE
Khaled.shaalan@buid.ac.ae
[2] IBM, Dubai Internet City, PO BOX 27242, Dubai, UAE
sinan@ae.ibm.com
[3] University of Wollongong in Dubai, PO Box 20183, Dubai, UAE
FarhadOroumchian@uowdubai.ac.ae

Abstract. This research suggests a method for query expansion on Arabic Information Retrieval using Expectation Maximization (EM). We employ the EM algorithm in the process of selecting relevant terms for expanding the query and weeding out the non-related terms. We tested our algorithm on INFILE test collection of CLLEF2009, and the experiments show that query expansion that considers similarity of terms both improves precision and retrieves more relevant documents. The main finding of this research is that we can increase the recall while keeping the precision at the same level by this method.

Keywords: Arabic, Arabic NLP, Arabic Information Retrieval, Query Expansion, EM algorithm.

1   Introduction

Information Retrieval (IR) is the process of finding all relevant documents responding to a query from unstructured textual data. The traditional model for IR assumes that each document is represented by a set of keywords, so-called index terms. An index term is simply a word whose semantics contributes to the document topic. The challenge increases when the number of documents stored grows, the content carries different topics, few words are used in queries, and more clarifications about words in queries are needed.

Arabic language has a very rich set of vocabulary, which with their synonyms introduce a problem to the IR process [15-16]. Many synonyms can contribute to the same meaning of the sentence. An example that shows the challenge in IR using synonyms is the query for "كأس العالم" (the World Cup) which could miss documents represented by the keyword "مونديال", (borrowed from the French "Mondial"). With this set of vocabulary, it is not very hard to write an entire essay in Arabic about, say, how sports benefit human health, and yet do so without ever using the keywords 'Sports', 'Human', or 'Health'.

Z. Shi, D. Leake, and S. Vadera (Eds.): IIP 2012, IFIP AICT 385, pp. 167–176, 2012.
© IFIP International Federation for Information Processing 2012

On top of the previous challenge users tend to input very limited set of words as their intended query. Many researchers such as Stefan Klink [8] have indicated that the average words used in a query is around two to three words. It is a challenge to hit users' real need for information using very limited number of words especially when those few words might carry different meanings, like the case in Arabic. Both the limited set of words in users' query and the potential absence of words from this set in the retrieved documents are the motivation behind this research. Query expansion is a proposed solution to overcome those two problems and successfully retrieved documents that were previously over looked.

Query expansion is considered as a Meta-level process that is used to add more information to clarify the user's query. It is the process of rebuilding new informed queries from an existing one in order to improve the retrieval performance and help in matching additional documents. Many query expansion techniques can be used. They are classified into two categories: automatic expansion based on linguistic knowledge and semi-automatic based on user feedback. In [1], it demonstrates an approach where a query is expanded by adding more synonyms. Whereas, in [2] it shows how a query is expanded by stemming its terms and adding common suffixes and prefixes. Semi-automatic expansion algorithms have been used to add/rebuild input query from user feedback like Probabilistic Relevance Feedback (PRF) expansion algorithm [4-5]. Essentially that algorithm Compares the frequency of occurrences of a term in documents that user marked as relevant with terms in the whole document collection. So if a term occurs in the documents marked as relevant more frequently than in the whole document collection it will be assigned a high weight.

The proposed technique depends on the co-occurrence of words while expanding queries. A paragraph about "كأس العالم" (world cup) uses common words such as "كرة" "القدم"(football), "كرة"(ball), "أهداف"(goals), "حماسة"(excitements), "كأس البطولة"(championship cup) ... etc. Those words are also present in the documents that do not have the exact match of keywords "كأس العالم" (world cup) however, it has the word "مونديال" which also means (world cup).

The proposed technique starts by analyzing documents that have the exact wording of the query in order to identify a list of co-occurring contextual words. This list of words will be used to expand the current query. The expanded query will then be used to pull other documents. New set of documents do not necessarily have the exact words as the original query. This way, it was possible to expand a query based-on similarity of terms for improving Arabic Information Retrieval.

The remaining of this paper is structed as follows. Section 2 presents a background appraisal showing other people work in this area. Section 3 gives a detailed description of the proposed algorithm. Section 4 explains how EM was used to optimize and improve query expansion. Section 5 shows the testing experiments we conducted and points of improvement that the proposed solution provided. Finally section 6 concludes and sums up the main findings.

2   Background

Matthew W. Bilotti [11] discussed "Query Expansion Techniques for Question Answering" in his thesis. He discussed five query expansion techniques, two term expansion methods and three term-dropping strategies. His results show that there are well-performing query expansion algorithms that can be experimentally optimized for specific tasks.

Hayel Khafajeh, and others [12] compare the performance of search engine before and after expanding queries. Their approach to expand queries was based on Interactive Word Sense Disambiguation (WSD). They found that expanding polysemous query terms by adding more specific synonyms will narrow the search into the specific targeted request and thus causes both precision and recall to increase; on the other hand, expanding the query with a more general (polysemous) synonym will broaden the search which would cause the precision to decrease. Their method of expanding queries depends on user feedback for the results.

Hayel Khafajeh and others [9] also worked on automatic Arabic thesauri that can be used in any special field or domain to improve the expansion process. Their efforts concluded that the association thesaurus improved the recall and precision over the similarity thesaurus. However, it has many limitations over the traditional information retrieval system in terms of recall and precision level.

T. Rachidi, and others [10] depended extensively on Arabic root extraction to build expanded queries. They also relied on three concept thesauri while expanding their queries. The first one was built manually, the second was built automatically from crawled XMLs documents and the last one was built automatically from an automatic categorization for the crawled documents. They reported a %75 improvement on their first experiment using query expansion.

3   The Proposed System

Fig. 1 shows the structure of our Arabic information retrieval system. It consists of two fundamental stages: indexing and querying. The indexing stage handles pre-processing, term selection and indexing. In the indexing stage, we also build a relationship database which is useful in runtime processing of top 10 documents and extracting their best terms. The querying stage handles pre-processing of user queries and retrieval of relevant documents using the previously indexed text. The two stages are interdependent. Indexing is designed such that it facilitates querying.

The contribution of this paper lies in the way we are expanding the original query. The indexing stage consists of three steps. It starts by passing Arabic documents into a pre-processing and noise removal phase. The data pre-processing and noise removal step takes care of cleaning up the Arabic text. The tasks performed by this step include: Duplicate white spaces removal, excessive tatweel (or Arabic letter Kashida) removal, HTML tags removal, and Handling special characters (i.e. {!@#$^%&*;+:()_}).

Following the data pre-processing and noise removal step, Arabic documents are passed into baseline indexing and relationship database construction in parallel. For the baseline indexing, the system parses Arabic documents and stems its words before it produces the baseline index. The relationship database construction builds the relationship database which is an SQL database that links each word in a document to all documents that it has occurred in.

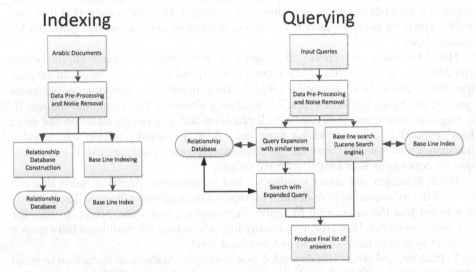

Fig. 1. Arabic Information Retrieval System Architecture

The querying stage consists of five steps. It takes both the baseline index and the relationship database as input. This stage handles the pre- and post-processes performed on queries fed to the system. The first step of the system is pre-processing of the input queries. The pre-processing follows the same steps for both queries and documents. In addition to that it also cleans the query from all stop words. Query processing proceeds with parallel steps: baseline search and query expansion with similar terms. The baseline step applies the baseline retrieval approach and retrieves the baseline documents. The query expansion component expands the query with similar terms and then retrieves another set of documents using expanded query. The details of the query expansion step are described in the next section.

Finally, the last step is to combine both retrieved sets of documents. While merging two lists of documents duplicate answers will be promoted.

4   Query Expansion

Query expansion is the process of adding extra data to the input query in order to provide more clarity. Expanding queries in this work consists of three steps:

- Extracting top 10 documents
- Extracting top 100 keywords out of the top 10 documents, and
- Eliminating irrelevant keywords using EM distance.

The remaining words are then added to the original query to construct the expanded version of the query. Fig. 2 represents the steps followed when expanding a query in this paper:

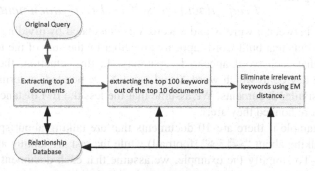

Fig. 2. Steps of Expanding Queries

For every input query, the system selects top 10 documents for each word of the query. Those documents are selected based on the importance of the query word to the entire text of the document. This approach was built on the assumption that the frequency of the word in a single document determines how important the word is to the subject after removing all stop words.

The importance of a word in a document is estimated based on the frequency of the word itself to the entire text of the document. As shown in equation (1).

$$IMP(X) = [1 - \frac{Freq(X)}{word\_Counter(D)}]*100 \tag{1}$$

The importance of word X to document D is calculated as the percentage of appearance of word X in document D. The frequency of word X is divided by the total number of words in document D.

Next, the system selects the top 100 most important keywords used in these top 10 documents. The 100 words are selected using the same principle and equation as that used to select the top 10 documents. In other words, the system calculates the importance of each word in these documents and selects the top 100 words. EM distance is then used to eliminate keywords that are most likely not related to the original query that will be discussed in more details in the following section.

5   EM Algorithm

Expectation Maximization (EM) is a statistical method used for finding the maximum likelihood of parameters. EM is typically used to compute maximum likelihood estimates given incomplete samples. It is guaranteed to find a local optimum of data log likelihood. For the purpose of this research, EM is used to indicate similarity between two words based on their co-occurrence in a set of documents using equation (2).

$$EM(X,Y) = 1 - \frac{Freq\_documents(X,Y)}{Freq\_documents(X) + Freq\_documents(Y)} \qquad (2)$$

EM distance between a word X and a word Y is calculated by dividing the total number of documents that both words appeared together by the sum of the total number of documents that each word appeared separately. In this situation, the EM distance indicates the degree to which word X and word Y are bonded, in terms of their concurrence in similar documents. We assume that the less the EM distance between two terms, the more bonded they are.

For an example if there are 10 documents that are talking about sports. Seven of which are talking about "كرة القدم" (football) while the rest are talking about "كرة السلة" (basketball). To simplify the example, we assume that each document that is talking about "كرة القدم" (football) has used this keywords once in its context and similarly the documents that are talking about "كرة السلة" (basketball). Calculating the EM distance between the word "كرة" (ball) and the word "القدم" (foot) using equation (3) will return:

$$EM (\text{"القدم"}, \text{"كرة"}) = 1 - \frac{7}{7+10} = 1 - 0.411 = 0.598 \qquad (3)$$

Similarly calculating the distance between the word "كرة" (ball) and the word "السلة" (basket) using equation (4) will return:

$$EM (\text{"السلة"}, \text{"كرة"}) = 1 - \frac{3}{3+10} = 1 - 0.23 = 0.77 \qquad (4)$$

You can see from the example above that the word "القدم" (foot) has shorter EM distance to the word "كرة" (ball) than when comparing to the EM distance between "كرة" (ball) and "السلة" (basket). Hence, the word "كرة" (ball) is more likely to come with the word "القدم" (foot).

An iterative approach is used to determine the best EM distance to describe the relevance between Arabic words. After experimentation, a threshold of 0.86 is reached as the optimal EM distance. The EM distances between the top 100 words and the query words are calculated, and all pairs of words with the EM distance more than 0.86 are considered not related to the query words. All other words that have an EM distance less than 0.86 are considered similar to the query words. These groups of words are used in the query expansion.

6   Evaluation

The evaluation used for this work is based on TREC evaluation procedure [6] that consists of a set of documents, a set of test topics and their relevance judgments. The

INFILE corpus from CLEF 2009 initiative is selected as the evaluation corpus. Lucene search engine was used for conducting the baseline search step with the original query. TRECEVAL software was used in calculating the precision, recall and other performance measures.

The INFILE corpus [17] from CLEF 2009 test consists of 50 different queries. Each query has a title, description and few keywords. The query titles consist of 2 to 6 words, while query description varies from few words to few lines. Keywords were not used in evaluating our approach.

We experimented with different parameters in the system in order to find the best approach. Table 1 shows the best runs for the baseline approach running the 50 queries. The baseline approach used only stemmed title words and was able to retrieve 1,007 of 1,195 the relevant documents. The second run used query description in its search and found 17 additional relevant documents. Combining title and description resulted in retrieving 1,078 relevant documents which has 71 extra relevant documents over retrieval with titles only. This shows that there are unique information about users' needs in both query titles and descriptions that complements each other. However, most studies on user behavior suggest that users rarely use long queries while searching on the web. Therefore run 2 and 3 are unlikely scenarios and the title search in run 1 is a better simulation of the user search behavior.

Table 1 shows that the precision decreased in the second run even though the number of query terms increased because of the length of the descriptions. This is due to the type of the terms found in the description itself and how tightly they are related to the meaning of the query. Query description might have the same meaning of the query, but it might not use as relevant keywords as those found in query title. Although the run with description only was able to pull more relevant documents, it was not able to put them in higher ranks. Moreover, when searching for both title and descriptions together, the system performance improved over both runs. The third run found more relevant documents and had higher precision in low recall which indicates better ranking. Unfortunately the queries in real life are closer to title only search rather than the second or third runs. This is because users do not write long descriptions for their searches. Many researchers such as Stefan Klink [8] have concluded that the average words used in a query is around two to three words which is very close to title only search in the first run in Table 1.

Run 1 (search for title only), Run 2 (search for description only), and Run 3 (search for title and description)

Many experiments were conducted to find the best configuration for query expansion. Table 2 lists two of the best runs with the proposed approach for the same queries. Run 4 shows the performance of title only search with query expansion. After expanding user query, the system was able to retrieve 12 more documents than the baseline and it has a better precision also (2.1% on recall 0). Run 4 used an EM distance equal to 0.86. All words that have EM distance less than 0.86 were considered similar to the query words. Run 5 is based on top of run 4. The only difference between run 4 and run 5 is that all terms in run 5 (documents and queries) have been stemmed. All the steps for processing queries and documents have been followed for run 5 on stemmed corpus. Run 5 is there to show that stemming [13] [14] didn't add

Table 1. Precision-Recall and the Precision at document cutoff for three baseline approaches

Recall	Run 1	Run 2	Run3	Doc. cut off	Run 1	Run 2	Run3
0	0.6182	0.6016	0.6374	At 5 docs	0.44	0.3520	0.4600
0.1	0.5481	0.5094	0.5954	At 10 docs	0.404	0.3420	0.4280
0.2	0.4763	0.4016	0.5309	At 15 docs	0.3667	0.3240	0.3960
0.3	0.4325	0.3535	0.5017	At 20 docs	0.331	0.3040	0.3740
0.4	0.384	0.3078	0.4486	At 30 docs	0.2873	0.2613	0.3307
0.5	0.3652	0.2715	0.4141	At 100 docs	0.1554	0.1466	0.1714
0.6	0.3018	0.2195	0.3317	At 200 docs	0.0889	0.0863	0.0971
0.7	0.2734	0.1817	0.2896	At 500 docs	0.0384	0.0381	0.0414
0.8	0.2178	0.1490	0.2222	At 1000 docs	0.0201	0.0205	0.0216
0.9	0.144	0.0930	0.1487				
1	0.079	0.0553	0.0794				

Table 2. Precision-Recall table and the Precision at document cutoff for best 2 trials of the Proposed Approach

Recall	Run 4	Run 5	Doc. cut off	Run 4	Run 5
0	0.6314	0.6381	At 5 docs	0.4560	0.4520
0.1	0.5609	0.5645	At 10 docs	0.4000	0.3900
0.2	0.4743	0.4819	At 15 docs	0.3693	0.3600
0.3	0.4428	0.4430	At 20 docs	0.3290	0.3300
0.4	0.3918	0.3948	At 30 docs	0.2887	0.2893
0.5	0.3677	0.3694	At 100 docs	0.1556	0.1544
0.6	0.3082	0.3063	At 200 docs	0.0886	0.0884
0.7	0.2744	0.2818	At 500 docs	0.0384	0.0385
0.8	0.2212	0.2194	At 1000 docs	0.0202	0.0202
0.9	0.1501	0.1451			
1	0.0776	0.0778			

much value to the overall precision of the system. On the contrary it resulted in finding less relevant documents.

7   Conclusion

This research investigates query expansion using EM algorithm to improve the number of relevant documents retrieved. It also studies the best EM distance for Arabic words that describes the similarity between them. Results are compared with Lucene search results, which are used as a baseline. The test data used is the INFILE test corpus from CLEF 2009.

Moreover, the major contributions of this research are: 1) improving Arabic Information Retrieval through expanding Arabic queries with similar index terms, and 2) the list of suggested query terms and finding the best EM distance to define similarity between Arabic words.

Our experiments prove that expanding queries retrieves more relevant documents as shown in the evaluation section for queries than the baseline. Moreover, it also improves the overall recall precision for the final list of retrieved documents. The runs of the baseline show that there are unique information about users' information needs in both query titles and descriptions that complement each other. The experimental runs show that the proposed system is able to improve Arabic retrieval process while maintaining the same precision.

The EM distance is a major factor in the overall success of this system. It eliminates the unnecessary retrieved answers that the system was retrieving based on dissimilar keyword. It helps in focusing on those keywords that add value to the overall performance and query expansion and prevents the system from expanding the queries to dissimilar keywords. In future work we would like to experiment with EM algorithm focusing on text windows rather than whole documents for calculating the EM distance.

References

1. Staff, C., Muscat, R.: Expanding Query Terms in Context. In: Proceedings of Computer Science Annual Workshop (CSAW 2004), pp. 106–108. University of Malta (2004)
2. Bacchin, M., Melucci, M.: Expanding Queries using Stems and Symbols. In: Proceedings of the 13th Text REtrieval Conference (TREC 2004), Genomics Track, Gaithersburg, MD, USA (November 2004)
3. Martínez-Fernández, J.L., García-Serrano, A.M., Villena Román, J., Martínez, P.: Expanding Queries Through Word Sense Disambiguation. In: Peters, C., Clough, P., Gey, F.C., Karlgren, J., Magnini, B., Oard, D.W., de Rijke, M., Stempfhuber, M. (eds.) CLEF 2006. LNCS, vol. 4730, pp. 613–616. Springer, Heidelberg (2007)
4. Manning, C.D., Raghavan, P., Schtze, H.: Relevance feedback and query expansion. In: Introduction to Information Retrieval. Cambridge University Press, New York (2008)
5. Crestani, F.: Comparing neural and probabilistic relevance feedback in an interactive Information Retrieval system. In: Proceedings of the IEEE International Conference on Neural Networks, Orlando, Florida, USA, pp. 3426–2430 (June 1994)
6. http://trec.nist.gov (visited on September 2010)
7. Magennis, M., van Rijsbcrgen, C.: The potential and actual effectiveness of interactive query expansion. In: Proceedings of ACM Special Interest Group in Information Retrieval Conference (SIGIR 1997), pp. 324–332 (1997)
8. Klink, S., Hust, A., Junker, M., Dengel, A.: Improving Document Retrieval by Automatic Query Expansion Using Collaborative Learning of Term-Based Concepts. Document Analysis Systems, 376–387 (2002)
9. Khafajeh, H., Kanaan, G., Yaseen, M., Al-Sarayreh, B.: Automatic Query Expansion for Arabic Text Retrieval Based on Association and Similarity Thesaurus. In: Proceedings he European, Mediterranean & Middle Eastern Conference on Information Systems (EMCIS), Abu Dhabi, UAE (2010)
10. Rachidi, T., Bouzoubaa, M., ElMortaji, L., Boussouab, B., Bensaid, A.: Arabic user search Query correction and expansion. In: Proceedings of COPSTIC 2003, Rabat, December 11-13 (2003)

176   K. Shaalan, S. Al-Sheikh, and F. Oroumchian

11. Bilotti, M.: Query expansion techniques for question answering. Master's thesis, Massachusetts Institute of Technology (2004)
12. Al-Shalabi, R., Kanaan, G., Yaseen, M., Al-Sarayreh, B., Al-Naji, N.: Arabic Query Expansion Using Interactive Word Sense Disambiguation. In: 2nd International Conference on Arabic Language Resources & Tools, MEDAR, Cairo, Egypt, pp. 156–158 (April 2009)
13. Zitouni, A., Damankesh, A., Barakati, F., Atari, M., Watfa, M., Oroumchian, F.: Corpus-Based Arabic Stemming Using N-Grams. In: Cheng, P.-J., Kan, M.-Y., Lam, W., Nakov, P. (eds.) AIRS 2010. LNCS, vol. 6458, pp. 280–289. Springer, Heidelberg (2010)
14. Attia, M.: Arabic tokenization system. In: Proceedings of the Workshop on Computational Approaches to Semitic Languages: Common Issues and Resources, pp. 65–72 (2007)
15. Farghaly, A., Shaalan, K.: Arabic Natural Language Processing: Challenges and Solutions. ACM Transactions on Asian Language Information Processing (TALIP) 8(4), 1–22 (2009)
16. Habash, N.Y.: Introduction to Arabic Natural Language Processing (Synthesis lectures on human language technologies). Morgan & Claypool (2010)
17. Besançon, R., Chaudiron, S., Mostefa, D., Timimi, I., Choukri, K.: The INFILE Project: a Cross-lingual Filtering Systems Evaluation Campaign. In: Proceedings of the Sixth International Language Resources and Evaluation (LREC 2008), Marrakech, Morocco (2008)

Towards an Author Intention Based Computational Model of Story Generation

Feng Zhu[1,2] and Cungen Cao[1]

[1] Key Laboratory of Intelligent Information Processing, Institute of Computer Technology,
Chinese Academy of Science, Beijing, China
freecafeabc@gmail.com, cgcao@ict.ac.cn
[2] Graduate University of Chinese Academy of Science, Beijing, China

Abstract. This paper addresses the problem of plot controllable story generation model. Since most of the previous works focus on logically flawless plot, the main challenge is the need for more generic method to generate dramatic and interesting plot. Motivated by this background, this paper proposes a computational method for plot controllable story generation. Firstly, we use planning as a model of plot generation. Then we utilize author intentions as plot constraints to force the planner to consider substantially more complex plans. Finally we integrate author intentions into planning and develop a plot controllable Graphplan algorithm. Experimental results demonstrate the effectiveness of our approach.

Keywords: story generation, narrative generation, author intention, graph plan.

1 Introduction

Story as entertainment, in the form of oral, written, or visual storytelling, plays a central role in many forms of entertainment media, including novels, movies, television, and theatre. Since the emergence of the first computational story generation system, TALE-SPIN, developed by Meehan in 1976, automated story generation has been extensively studied, with applications ranging from computer games to education and training [1-3]. While a majority of these studies are on automatic generation of logically flawless plot, the artistic properties of story, such as plot controllability, which is an essential story element for the reader's enjoyment, has received less attention. This paper addresses one of the central problems in the automatic creation of plot controllable story, which keeps the reader engaged in the interesting plot, giving them high entertainment value.

Since story plot can be naturally modeled as a sequence of actions, AI planning has emerged as the technology of choice in the field and a range of planning approaches have been successfully applied to the task of story generation. However the characteristics of a "good" plan, such as optimality, aren't necessarily the same as those of a "good" story, where convoluted or redundant plot sequences may offer more reader interest. For example, if a planner were given an initial state in which a character was rich and an outcome state in which the character was rich, the planner would simply indicate that there was no problem to solve. While a character that begins rich, then

Z. Shi, D. Leake, and S. Vadera (Eds.): IIP 2012, IFIP AICT 385, pp. 177–185, 2012.

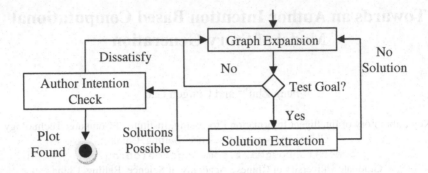

Fig. 1. A Framework of Plot Controllable Story Generation Model

becomes poor, and finally regains the state of being rich may be an interesting plot. We argue that this limits the applicability of off-the-shelf planners as techniques for creating stories.

In this paper, we describe a plot controllable story generation model that cleverly integrates author intentions into Graphplan based planner to generate more dramatic and interesting story plots. As illustrated in Fig. 1, the framework contains two phases: Graphplan phase and author intention check phase. In the Graphplan phase, the algorithm alternates execution of graph expansion and solution extraction until all solutions in current expansion level are found or it is proven that no solution exists. In the intention check phase, all possible solutions will be tested whether they satisfy author intentions or not. We conclude with a discussion of the limitations and future directions of this research. The main contribution of our work is that we use Graphplan as a computational story generation model and integrate author intentions into model to generate plot controllable stories.

2   Related Work

A majority of the works have looked to adapt the plan generation algorithm itself to structure the story along aesthetic lines. For example, to balance logical causal progression of plot and character believability, Riedl and Young describe a refinement search planning algorithm the IPOCL (Intent-based Partial Order Causal Link) planner - that, in addition to creating causally sound plot progression, reasons about character intentionality by identifying possible character goals that explain their actions and creating plan structures that explain why those characters commit to their goals [4]. Whilst Cheong use the Longbow planning algorithm to approximate the reader's planning related reasoning in order to demonstrate how the narrative structure can be post-processed to help promote suspense [5]. To produce conflict plot, Ware presents a CPOCL (Conflict Partial Order Causal Link) planner which allows narrative conflict to arise in a plan without destroying causal soundness [6]. To capture character's psychology, Pizzi uses HSP and its underlying real-time search algorithm to generate interactive affective plot according to the emotion of the story characters [7].

To make the generative plot under the control of the author, Riedl describe a general mechanism, called author goals, which can be used by human authors to assert authorial intent over generative narrative systems [8]. Similarly, Cavazza have developed an approach to planning with trajectory constraints. The approach decomposes the problem into a set of smaller sub problems using the temporal orderings described by the constraints and then solves these sub problems incrementally [9]. Other researchers have proposed the use of search based drama management. The Carnegie Mellon University Oz project uses drama manager to prevent uninteresting and poorly structured stories from emerging [10]. A drama manager oversees and coordinates character agent behavior in order to coerce interesting and well-structured performances out of the autonomous agents.

3   Generating Stories with Controllable Plot

3.1   Planning as a Model of Plot Generation

There are many similarities between plan and story at the level of plot generation. In particular, a plot is a sequence of events that describes how the story world changes over time. In a plot, change is instigated by goal-directed actions of story world characters, although the story world can also be changed through unintentional acts such as accidents and forces of nature. Likewise, a plan is a set of ordered operators that transforms a world from one state to another state. If operators of a plan are events that can happen in a story world, then a plan can be a model of a plot generation.

Planners are implementations of algorithms that solve the planning problem: given a domain theory, an initial state I, and a goal situation G consisting of a set of propositions, and a sound sequence of actions that maps the initial state into a state where G is true. The domain theory is a model of how the world can change. For example, one can use STRIPS-like (Stanford Research Institute Problem Solver) operators that specify what operations can be performed in the world when they are applicable, and how the world is different afterwards. Various algorithms have been developed that solve planning problems including POP (Partial-Order Planners) [11], CSP (Constraint Satisfaction Planners) [12], Graphplan Planners [13] and HSP (Heuristic Search Planners) [14]. Since a plan can be used as a model of plot generation, a planning algorithm can also be used as a model of the dramatic authoring process to create plots. Thus, the creation of a plot can be considered a problem solving activity if one considers the plot of a story to be the sequence of story-world events that achieves some outcome desired by the author.

Graphplan is a planner for STRIPS domains. Planning in Graphplan is based on the concept of a data structure called the planning graph. It is considered a breakthrough in terms of efficiency regarding previous approaches to planning, and has been refined into a series of other, more powerful planners, such as IPP and STAN2, whose efficiency has been empirically verified in several planning algorithm competitions. The Graphplan algorithm performs a procedure close to iterative deepening, discovering a new part of the search space at each iteration. It iteratively expands the planning graph by one level, and then it searches backward from the last level of this graph for a solution. The iterative loop of graph expansion and search is pursued until either a plan is found or a failure termination condition is met.

3.2   Author Intentions as Constraints

In our computational representation of story plot, author intentions are implemented as
a special type of plan steps that should occur in the intermediate world state. Author
intentions can be used to force the planner to consider substantially more complex
plans in which some intermediate steps, such as the character becomes poor or encounters robber, must be integrated into the resultant plans. Author intentions are
provided at the time of planner initialization and describe world events or actions that
must be achieved at some intermediate time during plan execution. If more than one
author intention is given, there can be pre-specified temporal links between them so
that author intentions must occur in a particular order.

To implement the ability for a planner to act on author intentions, we use state trajectory constraints, which is introduced in PDDL3.0 (Planning Domain Definition
Language), as a reference [15]. State trajectory constraints assert conditions that must
be met by the entire sequence of states visited during the execution of a plan. We
recognize that there would be value in allowing propositions asserting the occurrence
of action instances in a plan, rather than simply describing properties of the states
visited during execution of the plan. The language provides a number of modal operators for expressing these constraints. In our computational model we used the basic
modal operators: sometime, sometime-before, sometime-after, at-most-once and
at-end. The sometime operator allows us to specify events that must occur in the story
but not known when these events will happen. The sometime-before operator and
sometime-after operator give a temporal order over these events. The at-most-once
operator indicates that the events will happen not more than once. The at-end operator
specifies the final events in the story. Fig. 2 shows an instance of author intentions.

(sometime (lose Tom Wedding-Ring))

(sometime (rob Jack Tom Wedding-Ring))

(sometime-before (lose Tom Wedding-Ring) (rob Jack Tom Wedding-Ring))

(at-end (marry Tom Mary))

(at-most-once (lose Tom Wedding-Ring))

Fig. 2. An Instance of Author Intentions

In this way, the existence of author intentions serves two important purposes. First,
author intentions constrain the story search space such that it is impossible for a planner
to produce a story that does not meet certain criteria imposed by the human author. That
is, the planner cannot consider any plan in which the events described by an author
intention will not be achieved during plan execution. Second, author intentions control
plot generation to make the story more interesting and dramatic.

3.3   Integrating Author Intentions into Plot Planning

Although the Graphplan method has achieved good results in the field of traditional
planning, it is also confronted with the problem we describe in the introduction section

when applying Graphplan in story generation directly. To remedy this problem, we extend the original Graphplan algorithm which integrates author intentions into planning. This method is called the plot controllable story generation method.

PC-Graphplan(A, s0, g, C)

1   i ←0, ∇←φ, P0 ← S0, G←<P0>

2   until [g⊆ Pi and g2∩μPi =φ] or Fixedpoint(G) do

3   i ← i+1

4   G← Expand(G)

5   if g ⊄ Pi or g2∩μPi ≠ φ then return(failure)

6   Π← Extract (G, g, i)

7   if Fixedpoint(G) then η←|∇(k)| else η←0

8   while Π=failure or not satisfy (Π, C) do

9   i←i+1

10   G← Expand(G)

11   Π← Extract(G, g, i)

12   if Π=failure and Fixedpoint(G) then

13   if η=|∇(k)| then return failure

14   η←|∇(k)|

15   return(Π)

Fig. 3. Plot Controllable Graphplan Algorithm

An outline of the plot controllable Graphplan algorithm is shown in Fig. 3. The input of the algorithm is <A, s0, g, C> where: A is the union of all ground instances of operators in O; s0 and g are an initial state and goal condition of the world; and C, a set of constraints with author intentions. Upon successful termination of the algorithm, the output is a plan P that satisfies the constraints and achieves all final goals. The first step in the algorithm is the initial expansion of the graph until either it reaches a level containing all goal propositions without mutex or it arrives at a fixed-point level in G. If the latter happens first, then the goal is not achieved. Otherwise, a search for all solutions in current expansion level is performed. If no solution is found or all solutions are not satisfied constraint set C at this stage, the algorithm iteratively expands, and then searches the graph G. This iterative deepening is pursued even after a fix-point level has been reached, until success or the termination condition is satisfied. This termination condition requires that the number of no good tuples in ∇(k) at the fixed-point level k, stabilizes after two successive failures.

Evidently, the PC-Graphplan algorithm does not change the intrinsic complexity of planning, which is PSPACE-complete in the set-theoretic representation. Since the expansion of the planning graph and the author intention check are performed in polynomial time, this means that the costly part of the algorithm is in the search of the planning graph. Furthermore, the memory requirement of the planning-graph data structure can be a significant limiting factor.

4   Results

The central hypothesis of our work is that author intentions can be used for controlling plot in the story generation process. In this section we present the results of a qualitative evaluation that support this hypothesis, via analysis of a selection of sample plot. For the evaluation we developed a Graphplan style planner which is an implementation of the algorithm outlined in fig. 2. In these experiments performance was found to be acceptable for story generation purposes. Our results are encouraging and support the hypothesis that these constraints can be used for controlling the plot generation.

(define (problem marry-a-girl)

    :inits (and ((single Tom) (single Mary) (loves Tom Mary) (has Tom Money)))

    :author_goal (married Tom Mary)

    :author_intention (sometime (lose Tom Wedding-Ring)))

Fig. 4. Planning Problem Definition in "marry a girl"

The uses of author intentions in the story marry a girl illustrate their necessity in controlling the story plot. Fig. 4 shows the problem initialization in a PDDL-like language. An initial state defines characters, character traits, and relevant props and features of the world. The outcome is the author goal: Tom married Mary. The planning system is also initialized with an action library that describes ways in which the world can change. For example, characters can buy things if they have money and one character can marry another character if they love each other.

    The author intention defines that a significant feature of generation is that Tom will lose the wedding ring at some point. Note that the initialization parameters do not indicate how the author intention or the outcome is achieved, only that they must be achieved. Author intentions are necessary for controlling the plot. Without author intention, a planning algorithm could naively generate the following: Tom buys wedding ring. Tom proposes Mary. Tom marries with Mary. The author intention prevents this by forcing the planner to consider substantially more complex plans in which Tom loses the wedding ring. Fig. 5 shows one possible plot that can be generated by respecting the author intention. A_i is the set of actions whose preconditions are nodes in P_{i-1}. P_i is defined as the union of P_{i-1} and the sets of positive effects of actions in A_i. Solid arrows are preconditions of actions or positive effects of actions. Dashed arrows are negative effects of actions. The result plot is: Tom buys wedding ring. Tom loses the wedding ring. Tom finds the wedding ring. Tom proposes Marry. Tom marries with Mary.

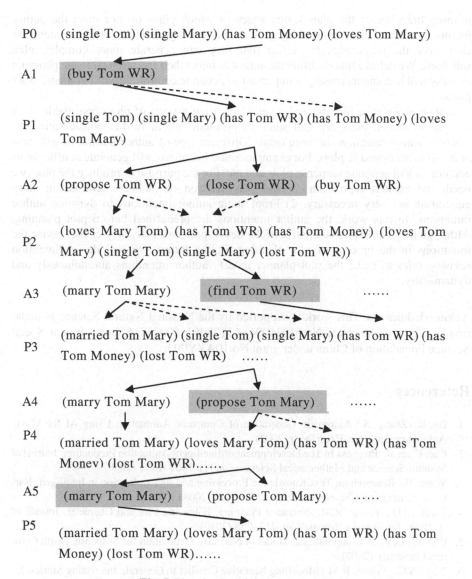

P0   (single Tom) (single Mary) (has Tom Money) (loves Tom Mary)

A1   (buy Tom WR)

P1   (single Tom) (single Mary) (has Tom WR) (has Tom Money) (loves Tom Mary)

A2   (propose Tom WR)   (lose Tom WR)   (buy Tom WR)

P2   (loves Mary Tom) (has Tom WR) (has Tom Money) (loves Tom Mary) (single Tom) (single Mary) (lost Tom WR))

A3   (marry Tom Mary)   (find Tom WR)   ......

P3   (married Tom Mary) (single Tom) (single Mary) (has Tom WR) (has Tom Money) (lost Tom WR) ......

A4   (marry Tom Mary)   (propose Tom Mary)   ......

P4   (married Tom Mary) (loves Mary Tom) (has Tom WR) (has Tom Money) (lost Tom WR)......

A5   (marry Tom Mary)   (propose Tom Mary)   ......

P5   (married Tom Mary) (loves Mary Tom) (has Tom WR) (has Tom Money) (lost Tom WR)......

Fig. 5. Planning Result in "marry a girl"

5   Conclusions and Future Work

The author intentions described in this paper is an attempt to enable human authors to inject their individual preference and requirements into the story generation. In general, author intentions constrain the planner to produce plans with particular structures by pruning branches of the plan search space in which plans do not meet the author intentions. Aside from controlling plot in the story generation, Author intentions also have the pragmatic side effect that they can generate more complex plan solutions. We believe that enabling the author to inject their intentions into the planning process will become increasingly important to generate controllable and dramatic story plots.

Although we have investigated several important aspects of plot controllable story generation model, there are still some of problems worth further consideration. 1) Classify author intentions in more detail. Different type of author intentions will generate different dramatic plots. For example, some intentions will generate conflict plots and others will generate suspense plots and etc. For the purpose of produce the plots we need, the explicit representation and classification of author intentions in story generation are very necessary. 2) From static author intentions to dynamic author intentions. In this work, the author intentions are predefined before plot planning. Although they can effective control of the development of plot, they cannot revise the intentions in the process of plot planning. In the future, we will design intention revision rules to make the plot planner modify author intentions autonomously and dynamically.

Acknowledgments. This work is supported by the National Natural Science Foundation of China under grant No.61035004, 61173063, 30973716 and National Social Science Foundation of China under grant No.10AYY003.

References

1. Lu, R., Zhang, S.: Automatic Generation of Computer Animation: Using AI for Movie Animation. Springer, Heidelberg (2002)
2. Cao, C., et al.: Progress In The Development of Intelligent Animation Production. Journal of Systems Science and Mathematical Sciences 28(11), 1407–1431 (2008)
3. Wang, H.: Research on Text Knowledge Processing and its Application in Intelligent Narrative Generation. Chinese Academy of Sciences (2008)
4. Riedl, M.O., Young, R.M.: Narrative Planning: Balancing Plot and Character. Journal of Artificial Intelligence Research 39, 217–268 (2010)
5. Cheong, G.Y.: A Computational Model of Narrative Generation for Suspense. North Carolina University (2007)
6. Ware, S.G., Young, R.M.: Modelling Narrative Conflict to Generate Interesting Stories. In: Proceedings of Artificial Intelligence in Interactive Digital Entertainment. AIIDE (2010)
7. Pizzi, D.: Emotional Planning for Character-based Interactive Storytelling. Teesside University (2011)
8. Riedl, M.O.: Incorporating authorial intent into generative narrative systems, Intelligent Narrative Technologies II: Papers from the 2009 Spring Symposium (Technical Report SS-09-06), pp. 91–94. AAAI Press, Palo Alto, CA (2009)
9. Porteous, J., Cavazza, M.: Controlling narrative generation with planning trajectories: the role of constraints. In: Proceedings of the 2nd International Conference on Interactive Digital Storytelling, pp. 234–245 (2009)
10. Kelso, M.T., Weyhrauch, P., Bates, J.: Dramatic Presence. The Journal of Tele Operators and Virtual Environments 2(1), 1–15 (1993)

11. Penberthy, J.S., Weld, D.: UCPOP: A Sound, Complete, Partial Order Planner for ADL. In: Proc. KR 1992 (1992)
12. Do, M., Kambhampati, S.: Planning as constraint satisfaction: solving the planning graph by compiling it into CSP. Artificial Intelligence 132(2), 151–182 (2001)
13. Blum, A., Furst, M.: Fast planning through planning graph analysis. In: Proc. IJCAI (1995)
14. Blai, B., Héctor, G.: Planning as heuristic search. Artificial Intelligence, 5–33 (2001)
15. Edelkamp, S., Jabbar, S., Nazih, M.: Large-Scale Optimal PDDL3 Planning with MIPS-XXL. In: Proceedings of Planning Competition. ICAPS (2006)

Adaptive Algorithm for Interactive Question-Based Search

Jacek Rzeniewicz[1], Julian Szymański[1], and Włodzisław Duch[2]

[1] Department of Computer Systems Architecture,
Gdańsk University of Technology, Poland
julian.szymanski@eti.pg.gda.pl, jrzeniewicz@gmail.com
[2] Department of Informatics, Nicolaus Copernicus University, Toruń, Poland
Google: W. Duch

Abstract. Popular web search engines tend to improve the relevance of their result pages, but the search is still keyword–oriented and far from "understanding" the queries' meaning. In the article we propose an interactive question-based search algorithm that might come up helpful for identifying users' intents. We describe the algorithm implemented in a form of a questions game. The stress is put mainly on the most critical aspect of this algorithm – the selection of questions. The solution is compared to broadly used decision trees learning algorithms in terms of knowledge inconsistencies tolerance, efficiency and complexity.

Keywords: information retrieval, semantic search, decision trees.

1 Introduction

Recent years we have observed semantic technology slowly making its way into the web search [1]. Result pages of the most popular searchers like Google, Yahoo or Bing, provide us not only with references to the websites containing desired keywords, but also with additional information on the entities those sites represent. Hints with similar queries are displayed to the users as they type, helping them to express themselves more precisely and receive more relevant results. Finally, synonyms are used to broaden the search.

Moreover, a number of native semantic searchers has been launched last years eg. Duckduckgo, Hakia. However, despite offering categorization of results by context (eg. Yippy!), that aims to lead to better relevance, they generally haven't attracted wide audience yet. Finally, the question answering field has been developed significantly [2]. A number of both general (eg. Yahoo! Answers, Ask.com) and vertical (eg. StackExchange) Q&A engines are available on the Web, offering billions of answers.

On the other hand, broadly used search engines still cannot be considered as semantic in terms of "understanding" users' queries. The same query formulated different ways is usually served with different results [3]. Search is still mostly keyword-oriented. Users need to know proper words related to the subject in order to set the searcher on the right track and obtain relevant results. This is

Z. Shi, D. Leake, and S. Vadera (Eds.): IIP 2012, IFIP AICT 385, pp. 186–195, 2012.

easy when user is searching the Web for a famous artist or a nearby restaurant, but gets tricky if he or she doesn't know any keywords (index phrases) that are distinctive for the desired entity. Let us consider this example: you saw a suricate in a ZOO and after returning home you want to learn more about it on the Web. The problem is, you can't recall its name and any keywords you can think of relate only to its appearance. You will spend a few minutes adjusting those keywords and clicking result links forth and back, or scrolling image results until you come across some information about suricates. When lacking proper keywords, users are required to devote a significant amount of time and effort looking for the information. That work should be performed by the search engine.

When imagining a perfect semantic searcher, it is handy to consider the notion of a Human Search Engine (HSE) [3]. This is a hypothetical searcher acting like a human being in terms of understanding the query. Back to the suricate example – how could the HSE help the user to find desired information? We imagine it being able to direct the search by prompting the user to provide more specific details. This way, the stress would be put less on the keywords selection and more on what the user interacting with the search engine has in mind – meaning of the query and particularly relevant answer in the form of provided search results.

Some of today's searchers limitations might be overcome by a web searcher displaying the ability to take over the initiative and ask questions in the form of simple dialogues. Such an approach requires an effective algorithm: one that asks the minimum number of questions to recognize the appropriate meaning of the user's query.

In this article we address just that: a question-based search algorithm, especially its critical aspect – the selection of questions. The structure of this paper is as follows: in section 2 we introduce a question-based search model in a form of a word game. In section 3 the question-based search algorithm is described and compared to typical decision trees algorithms. We also gave an outline of our future plans in section 4.

2   Questions Game – A Computational Linguistic Challenge

A simple model of dialogue–based search can be approximated in the form of a game of questions. This is a word game where one player thinks of some concept and the other one asks a series of yes/no questions regarding that object. After at most twenty questions the latter guesses what was the concept his partner was thinking of.

The question-based search algorithm described in the following sections was implemented in the form of a questions game that is available on the Internet at http://kask.eti.pg.gda.pl/winston. There users can play the questions game with an avatar called Winston. The avatar took the role of asker and guesser, site visitors need to pick up a concept and answer avatar's questions. The core of the knowledge base was imported from the WordNet dictionary [4].

To test the approach we limit the domain to the information related to the animals branch of WordNet.

Apart from providing the means for the search algorithm evaluation, the questions game has another application. Upon completing each game, some knowledge can be added to the system. This is not only limited to the user answers during the game. After clearing up what the subject of the game was, the player is requested to answer yet another question: *what can you tell me about the [subject]?* User can reply to that by typing in a fact in a form of *object–relation–feature*. For example, to make the fact that suricates are burrowing animals come across, one can type *can* and *burrow* into the text fields. This way both the amount of assertions and the system's vocabulary expand.

3   The Question-Based Search Algorithm

In the first section the need for a question-based search algorithm has been justified. The task of assigning a proper meaning to user's input generally relates to the classification problem. However, before any algorithm is discussed, a few assumptions regarding the problem's conditions ought to be stated:

1. Matching target object may not be present in the system.
2. System's knowledge about given concept may be incomplete or partially incorrect.
3. Knowledge base may contain false assertions or represent a different point of view than user's.

3.1   Decision Trees

One of the machine learning approaches that can be adopted to the search problems are decision trees algorithms. There are many of their variants: ID3 [5], C4.5 [6] or latest successor See5/C5.0 [7]. Those algorithms provide efficient mathematical model, so it requires a little number of steps and computing power to finalize the search: using binary decisions in nodes it is possible to index over one million classes performing only 20 tests. However, there are strong disadvantages to the decision trees approach when confronted with the stated assumptions. The process of searching develops towards a single hypothesis. Usually it is unlikely that there will be a target concept that matches given input very well. Thus we are interested in receiving a set of most matching answers rather than just one. Decision trees don't facilitate that. Neighbor of the top matching leaf is not necessarily the second best match.

Another crunch with decision trees is that the search paths are only optimized globally. According to the assumptions made in the beginning of this section, it can happen that the system's and user's knowledge differs. Occurrence of an answer that does not conform to the system's knowledge makes the decision tree take a wrong path, which must end up fatal for the overall search.

The next issue is that for a construction of a typical decision tree we usually have many examples and only a few classes. In the case of web search the problem

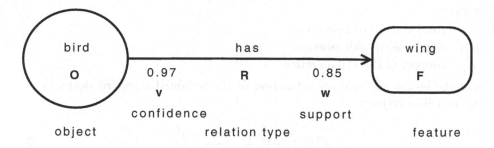

Fig. 1. Example assertion in the vwORF notation

is opposite: there are many potential classes (web pages relevant to user query) and each of them is described with a single example.

3.2 Knowledge Representation for the Question-Based Search Algorithm

The approach presented here is mainly based on fuzzy semantic network as an underlying knowledge model [8]. Specifically, the network consists of concepts interconnected with relations by connections called triples of knowledge, expressed in the vwORF notation:

O – the object described
R – the type of relation, denoting how the object is related to the feature
F – the feature expressing given property of the object
v – confidence: the real number in the range $\langle 0, 1 \rangle$ describing, how this atom of knowledge is reliable. 0 denotes no certainty at all, 1 means the information is sure.
w – support: the real number in range $\langle -1, 1 \rangle$ describing, to what extent the feature is related to the object. Allows to express adjectives such as "always" (1), "frequently" (~0.5), "seldom" (~-0.5) or "never" (-1).

Figure 1 depicts an example assertion expressed in the vwORF notation. Here the fact that birds have wings was associated with parameters support $w = 0.85$ and confidence $v = 0.97$. This means that a bird most usually has a wing and that this assertion is almost sure to be true. The vwORF notation is capable of expressing elementary sentences. More complex assertions, e.g. that birds usually have two wings, cannot be expressed this way, but this could be enabled by allowing a whole triple on the "object" side of another triple.

As mentioned before, the questions game that is running the system is not only used to evaluate the efficiency of the search algorithm, but also to provide new assertions to the knowledge base. Therefore it is necessary to update triple's w and v parameters whenever a new assertion related to this triple appears. The support value is a simple weighted mean of individual assertions:

$$w = \frac{\sum_{i=1}^{N} |\alpha_i| \cdot \alpha_{i_w}}{\sum_{i=1}^{N} |\alpha_i|} \tag{1}$$

where:

N – total number of assertions

$|\alpha_i|$ – weight of the i-th assertion

α_{i_w} – support of the i-th assertion

and the confidence value is a function of the weighted standard deviation of individual assertions:

$$\sigma^2 = \frac{\sum_{i=1}^{N} |\alpha_i| \cdot (w - \alpha_i)^2}{\sum_{i=1}^{N} |\alpha_i|} \tag{2}$$

$$v = \frac{1}{2} \cdot (1 + cos\,(\pi\sigma)) \tag{3}$$

where σ is a non–negative value denoting the weighted standard deviation.

In the Figure 2 the relationship between the standard deviation and confidence is presented. When all the individual assertions indicate equal feature support, the standard deviation is 0 and the confidence value equals 1. The confidence decreases as standard deviation goes up for incoherent assertions; in a boundary case of opposite assertions (such of support values 1 and -1), standard deviation approaches 1 and confidence tends to 0.

The semantic network is an economic (in terms of cognitive economy [9] that allows to reduce redundancy) and powerful model [10] that allows flexible complex model utterances. Such knowledge representation is easily browsable for humans using interactive graphs [11] and convenient to process, i.e. to perform inferences based on relation types interpretation that are related to the edges. On the other hand, only little amount of the overall information stored in the semantic network is stated explicitly and therefore it is hard to do any live computations on it. For this reason, for the sake of making the search computations easier, the knowledge is translated into the form of a semantic space.

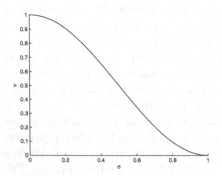

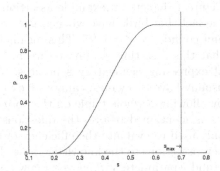

Fig. 2. Relationship between assertions standard deviation and the triple's confidence

Fig. 3. CDV's probability of inclusion into S depending on similarity to ANSV vector

The semantic space is an N-dimensional space where one dimension is associated with a unique pair (*relation*, *feature*). Those pairs are referred to as questions. Each concept in the semantic space is represented by a Concept Description Vector (CDV) [12]. Every element e of each CDV holds two real numbers e_w and e_v denoting respectively support of the connection and the confidence of knowledge. Those values are copied from corresponding triple if exists or inherited from another concept, provided that there is a positive support triple with relation IS_A pointing to that concept.

3.3 The Algorithm

The search algorithm operates on the semantic space constructed from the semantic network. Over the consecutive iterations, a subspace S of candidate vectors is maintained. Any CDV can enter or leave S in between two iterations. The answers that user gives build up a vector ANSV. Each time a new answer appears, similarity s between ANSV and every CDV is computed:

$$\cos\left(\phi\right) = \frac{\sum_{q:ANSV[q]\neq 0} CDV[q] \cdot ANSV[q]}{\sqrt{\sum_{q:ANSV[q]\neq 0} CDV[q]} \cdot \sqrt{\sum_{q:ANSV[q]\neq 0} ANSV[q]}} \tag{4}$$

$$s = \max\left(\cos\left(\phi\right), 0\right) \tag{5}$$

where $CDV[q]$ denotes the confidence and support product of CDV's element for dimension q. Applying the max function in the end of calculations puts similarity into the range $\langle 0, 1\rangle$, which makes computations more convenient.

The main flow of the question-based search algorithm is as follows:

```
1. Include each CDV into S.
2. Ask the most informative question according to S.
3. Obtain an answer to the question and update ANSV.
4. Compute similarity between every CDV and ANSV.
5. If maximum number of questions has been asked:
   return vector with the highest similarity to ANSV.
6. Rearrange S so that it include most similar vectors.
7. Go back to step 2.
```

Note that upon adding a new answer to the ANSV, all vectors' similarities are updated, even those excluded from S. This way no concept can ever be permanently eliminated, even if it doesn't go well with answers in the beginning.

It is also more accurate to always take all possible dimensions into consideration when evaluating a CDV. In practice it often happens that system asks questions that are unrelated to what the user thinks of. But somehow even those seemingly unrelated and unspecific features can help a lot to identify the right object [13,14]. Global evaluation of a vector guarantees the most accurate results.

3.4 Selection of a Question

In each iteration system picks a question that is the most informative across the candidates subspace S. Any CDV is included into S with probability p:

$$p = \begin{cases} 0, & \text{if } s \leq s_l \\ \frac{1}{2} - \frac{1}{2}\cos\left(\pi \frac{s-s_l}{s_u-s_l}\right) & \text{if } s_l < s < s_u \\ 1, & \text{if } s \geq s_u \end{cases} \tag{6}$$

where:

s_u – safety threshold parameter equal to $s_{max} - 0.1$
s_l – rejection threshold parameter equal to $s_{max} - 0.5$
s_{max} – highest similarity between CDV and ANSV in this iteration

Figure 3 presents a sample plot of inclusion probability.

Once the subspace S is defined, information gain IG for each question q can be computed:

$$IG = -\frac{w_\oslash}{w} \cdot \left(\frac{w_\ominus}{w_\oslash} \log_2 \frac{w_\ominus}{w_\oslash} + \frac{w_\oplus}{w_\oslash} \log_2 \frac{w_\oplus}{w_\oslash} \right) \tag{7}$$

where:

w – total weight of all CDV in S
$w_\oslash$ – total weight of all CDV in S such that $CDV[q] \neq 0$
$w_\oplus$ – total weight of all CDV in S such that $CDV[q] > 0$
$w_\ominus$ – total weight of all CDV in S such that $CDV[q] < 0$

and a similarity between CDV and ANSV is considered CDV's weight.

The question-based search algorithm described above to some extent can be considered opposite to the decision trees approach. As pointed out before, decision trees return only a single concept, because they are optimized globally and may very often reject correct branches due to local in-compliances between user's input and knowledge base state. Presented algorithm is more careful: in each step we process all possible concepts and evaluate them according to all possessed information. Any rejections are valid only within a single iteration, which eliminates the risk of decisions. At any time the algorithm can deliver a collection of best matching results, precisely defining how much they relate to the user specifications.

On the other hand, such approach lacks significant features that searching in decision trees has. Every step of the algorithm requires considerable amount of computations. Specifically, for a semantic space of N vectors and M questions iteration cost consist of:

- CDV's similarity update: $O(N)$ (assuming ANSV iteration complexity $O(1)$)
- candidates subspace S re-arrangement: $O(N)$
- IG's update: $O(N \cdot M)$

which results in the overall iteration complexity of $O(N \cdot M)$, a considerable requirement compared to decision trees' $O(1)$.

Second major drawback lies in the efficiency of the algorithm. Maintaining a broad set of candidate objects is safe and guarantees each concept a fair chance to be found in the assumed unstable conditions. Consequently, it can happen that obviously incorrect concepts are considered candidates.

Finally, it is noticeable that the algorithm presented here rejects to process according to common sense hierarchy. Assume there is a knowledge base containing 10 dog concepts and 100 related to cats. Human would most likely ask whether the concept is a cat or a dog. The algorithm promotes high-cardinality branches over those less numerous, and such a behaviour is unjustified.

Described question-based search algorithm overcomes significant drawbacks that searching in decision trees has. On the other hand, it introduces a cost of computation complexity and efficiency fall. In the next part of this section we address those issues.

3.5   Question Asking Strategies

Certain patterns of the system state that may result in algorithm's incorrect behaviour were distinguished in previous section. For each of these patterns, there is a strategy determining how to select a question:

1. A gap between the best– and the second best–matching concept: probability of inclusion of a CDV to the candidates subspace S depends on the difference between its similarity to the ANSV and maximum similarity in current step. Therefore it is unfortunate when incorrect concept remains top-matching for a number of iterations, since it worsens other concepts' odds. Appropriate strategy that should be applied here is to ask user about a feature that relates to that object's specific feature.
2. Numerous group of best-matching concepts: it happens that there is a whole group of very similar objects at the top of the candidates list. The algorithm is then unable to select a question separating this group since it maintains a broad candidates subspace S. In such case, S should be reduced only to the size of those top–matching set.
3. Mutually exclusive questions asked one after another: system happens to ask questions inconsistent with the latest answer. In example, once user gives positive answer to "is it a feline?" question, candidates subspace can still contain vectors describing dogs. Therefore a question "is it a canine?" is likely to be asked next. To avoid such contradictions, the following strategy was implemented: if user answers "yes" to a question q containing relation "IS_A", then the candidates subspace S should only be composed of such CDV's that $CDV[q] > 0$. This way, in the scope of one step, the process resembles searching in a decision tree built on the hypernym relationship. The quality of avoiding incorrect rejections is abandoned, but chances for fast identification of specific features increase instead.

The search algorithm has been rearranged to allow applying the most appropriate strategy in each step. Previous logic of question selection was also encapsulated into a strategy. Algorithm's step 2 has been defined as follows:

    a. Instantiate each strategy.
    b. Rank each strategy according to the state of the system.
    c. Draw one strategy with probability proportional to its rank.
    d. Ask a question returned by the selected strategy.

The rank of a strategy depends on the distribution of similarities between CDV's and ANSV, the index of current iteration and the previously picked strategies. Actual question selection is delegated to the picked strategy.

4   Discussion and Future Plans

We have shown that typical approach used in decision trees are not suitable for applications where uncertainties and inconsistencies happen both on the knowledge base level and on the user's side. Our proposed algorithm proved the ability of being resistant to such inconsistencies. On the other hand, it introduces a demand for additional computations and is less efficient than typical decision tress.

Several behaviour strategies have been implemented into the algorithm, and presented adaptive approach improves algorithm's efficiency significantly. It also demonstrates the flexibility of the algorithm – we showed that it is possible to apply a strategy that temporarily turns system's operation into decision trees alike. It seems that by tweaking this adaptive approach specifics we can achieve a high search efficiency in the end.

Our implementation employs lexical database provided by WordNet. The initial results obtained by the algorithm strongly depend on the knowledge quality stored in this repository. It should be stressed that the system interaction with the humans allows to introduce new, and correct already possessed knowledge. We employ this approach to crowdsourcing lexical knowledge [15] and extending the Wordnet by large amount of interactions with people playing the word game.

During the following months we will attempt to overcome major remaining problems. The algorithm promotes branches of higher cardinality over smaller ones – this characteristic may be removed by filtering out the concepts that are too specific. This way, also the amount of computations will decrease. The scheme of assigning weights to answered questions can also improve the relevance of candidate concepts. Finally, implementation of more accurate behaviour strategies will increase the algorithm's efficiency as well. We believe that high flexibility of the approach we work on allows finding the right balance between limiting incorrect rejections and search efficiency.

Successful application of the presented approach has served us as a test bed of the question-based search algorithm. Implementation in limited domain of animals, employing knowledge from the WordNet dictionary, shows the ability of specifying the search according to interaction with the user. Now we scale up the solution and apply the algorithm for improvement the search in Wikipedia. For now we have created a prototype system for refining the query based search results using the category system in Polish and Simple English Wikipedia. The first results, deployed under http://bettersearch.eti.pg.gda.pl are promising, and they indicate the proposed method can be applied for improvement of information retrieval precession.

Acknowledgments. The work has been supported by the Polish Ministry of Science and Higher Education under research grant N N516 432338.

References

1. Studer, R., Yong, Y.: Editorial-special issue semantic search. Web Semantics: Science, Services and Agents on the World Wide Web 9(4) (2012)
2. Forner, P., Peñas, A., Agirre, E., Alegria, I., Forăscu, C., Moreau, N., Osenova, P., Prokopidis, P., Rocha, P., Sacaleanu, B., Sutcliffe, R., Tjong Kim Sang, E.: Overview of the Clef 2008 Multilingual Question Answering Track. In: Peters, C., Deselaers, T., Ferro, N., Gonzalo, J., Jones, G.J.F., Kurimo, M., Mandl, T., Peñas, A., Petras, V. (eds.) CLEF 2008. LNCS, vol. 5706, pp. 262–295. Springer, Heidelberg (2009)
3. Signorini, A., Imielinski, T.: If you ask nicely, i will answer: Semantic search and today's search engines. In: Proceedings of the 2009 IEEE International Conference on Semantic Computing, pp. 184–191. IEEE Computer Society, Washington, DC (2009)
4. Miller, G.A.: Wordnet: a lexical database for english. Communications of the ACM 38(11), 39–41 (1995)
5. Quinlan, J.R.: Induction of decision trees. Machine Learning 1(1), 81–106 (1986)
6. Mitchell, T.M.: Machine Learning. McGraw Hill (1997)
7. Quinlan, R.: Data mining tools see5 and c5.0 (March 2012),
   http://www.rulequest.com/see5-info.html
8. Szymański, J., Duch, W.: Information retrieval with semantic memory model. Cognitive Systems Research 14, 84–100 (2012)
9. Conrad, C.: Cognitive economy in semantic memory. American Psychological Association (1972)
10. Sowa, J.F.: Principles of semantic networks. Morgan Kaufmann Publishers (1991)
11. Szymański, J.: Cooperative WordNet Editor for Lexical Semantic Acquisition. In: Fred, A., Dietz, J.L.G., Liu, K., Filipe, J. (eds.) IC3K 2009. CCIS, vol. 128, pp. 187–196. Springer, Heidelberg (2011)
12. Duch, W., Szymański, J., Sarnatowicz, T.: Concept description vectors and the 20 question game. In: Intelligent Information Processing and Web Mining, pp. 41–50 (2005)
13. Eckersley, P.: How Unique Is Your Web Browser? In: Atallah, M.J., Hopper, N.J. (eds.) PETS 2010. LNCS, vol. 6205, pp. 1–18. Springer, Heidelberg (2010)
14. Sweeney, L.: Simple demographics often identify people uniquely. Technical report, Carnegie Mellon University, School of Computer Science, Data Privacy Laboratory (2000)
15. Szymański, J., Duch, W.: Context search algorithm for lexical knowledge acquisition. Control & Cybernetics 41(1) (2012)

Research of Media Material Retrieval Scheme Based on XPath

Shuang Feng[1] and Weina Zhang[2]

[1] School of Computer Science, Communication University of China, Beijing, China
fengshuang@cuc.edu.cn
[2] Computer NIC Center, Communication University of China, Beijing, China
zhangweina@cuc.edu.cn

Abstract. With more and more media materials appear on the internet, it comes a sharp problem of how to manage these resources and how to search them efficiently. We construct a management system of media material for the reliable wideband network. According to a detailed analysis of the characteristic of media material queries, we proposed a hierarchical indexing mechanism based on XPath language to discover the resources that match a given query. Our system permits users to locate data iteratively even using scarce information. The description of materials is mapped onto the DHT index. Our indexing scheme has good properties such as space efficient, good scalability and resilient to arbitrary linking.

Keywords: P2P, XML, XPath, media material retrieval.

1   Introduction

With the public's enthusiasm for video creation continues to increase, they tend to create and share their works through internet. Personal media production and distribution not only need production tools, but also need mass media materials. The establishment of shared material library through internet is an effective way to solve this problem. Taking into account of the enormous amount of media materials, we can use a distributed network storage system for storage and services. We also need an efficient indexing technology to allow users retrieve the materials accurately and quickly.

2   Related Work

A P2P (Peer to Peer) network is a distributed system. According to the topology of the network, the P2P system can be divided into two kinds: unstructured P2P networks and structured P2P networks.

Inefficient routing and bad scalability are the main shortcomings of unstructured P2P networks. In contrast, the structured P2P network with DHT (Distributed Hash Table) has good scalability, robustness, and can provide accurate location, the rate of search success is also high. However, the DHT technique can only support accurate

Z. Shi, D. Leake, and S. Vadera (Eds.): IIP 2012, IFIP AICT 385, pp. 196–201, 2012.

retrieval and can't support the complex retrieval such as semantic retrieval. The users usually do not know the exact materials they want, they often tends to retrieve the materials by category and interested in the classical materials with a higher utilization rate in that field. So DHT index can't be used directly in media semantic retrieval .

Currently, the research of semantic retrieval based on DHT includes: [1] gets all the results of each keyword and then find the intersection, but because this method requires to transfer a large number of middle documents, it will consume a lot of network band. [2] supports multi-keyword search by using vector space model. [3] uses latent semantic indexing to overcome the affect of synonyms and noise in VSM . [4] builds a theme overlay network to improve the rate of full search. [5] designs a multi-attribute addressable network to support multi-attribute queries and range queries. Most of the above methods analyze the communication cost and retrieval accuracy, but for the particular application fields such as material retrieval, in most cases, users may only provide partial information, therefore, it is essential to build a relation mapping content-based index onto DHT index.

3   Structure of Large-Scale Material Retrieval System

One of the most important researches in P2P retrieval is structured P2P network retrieval based on DHT. Some of the DHT retrieval algorithms such as Chord[6],CAN[7] are designed for fluctuations in the worst-case network. The problem of this kind of work is long delay time. For the reliable wideband network with server clusters, we use a method called distributed storage and centralized retrieval to improve the efficiency of retrieval. Accordingly, the retrieval and management system of large-scale media material is shown in Figure1. It contains the following parts.

- Catalog server: responsible for labeling the materials according to the international standard of media description.
- Mapping server: responsible for mapping media description onto DHT index.
- Retrieval server: help users find relevant materials through a variety of ways. Sub-retrieval is used to share concurrent search pressure.
- Directory server: responsible for managing user's virtual directory and searching the exact location of block-stored materials. It also provides a directory query in DHT way.
- Storage node: the storage space is composed of many distributed storage nodes. They provide services for materials with high demands such as block storage and other services.
- Client: The client may also constitute a storage space for those materials with relative lower demands. And as an agent, the client is used to access the whole storage space and submit media materials to catalog server.

Before submitting a material, the user needs to fill in relevant information about the material and point out both the share scope and the way of storage (local storage or distributed storage).Then the catalog server will label the material according to international standards. Thus, a standard catalog file is produced and the catalog file is submitted to mapping server. The mapping server will map the description of materials onto DHT index and save this relationship to the retrieval server. The retrieval server accepts the user's search requests and returns relevant query results. Then the user chooses one of the results and the client will tell the storage system which material the user chooses. The storage system searches its own directory server and returns a set of IP address that stores that material. Finally, the client downloads the material by P2P.

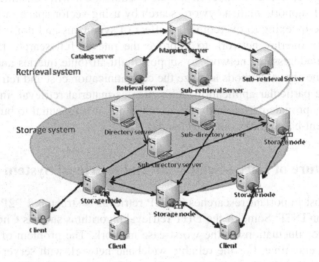

Fig. 1. Structure of large-scale materials retrieval system

4   Services of Large-Scale Material Index

How to build an efficient index to help users find what they want quickly among large-scale materials? Relative works about the management of large-scale distributed materials includes P2P-based file system(OceanStore [8],Tsinghua Granary [9] systems) and distributed indexing(Chord, CAN and Pastry [10]).However, the main problem is that if we use DHT directly on the retrieval of semantic materials, we can't support complex queries such as semantic queries. But in many cases, users may only provide part of information that they want search. Therefore, it is essential to map contend-based index onto DHT index.

4.1   Mapping Description of Materials onto DHT Index

To solve the problem of large-scale material retrieval, our system is organized hierar-chically. As shown in Figure 2. Here are the roles of these three layers. Media description mapping layer maps the description of media onto the DHT-based index. The index of distributed media materials are organized by distributed index layer. We use Pastry, a kind of widely-used distributed hash table technology, to compute indexes and route objects. Storage layer is used to store media materials. The man-agement of distributed content-based media resources is done by this architecture.

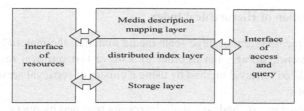

Fig. 2. System architecture of large-scale materials retrieval

In the P2P file sharing systems, file search is based on file names and file names can be seen as a brief description of the file. Further expanding this point of view, we can create media material description with XML. Data query is based on these descriptions. Figure 3 shows the XML description of the media file.

```
<program>
  <director>
    <LastName>Zhang</LastName>
    <FirstName>Yimou</FirstName>
  </director>
  <title> Red Sorghum</title>
  <type>movie</type>
  <date>1987</date>
  <format>general</format>
</program>

        a)    document d1
```

```
<program>
  <director>
    <LastName>Zhang</LastName>
    <FirstName>Yimou</FirstName>
  </director>
  <title> 2008 Olympic opening ceremony</title>
  <type>art</type>
  <date>2008</date>
  <format>wide</format>
</program>

        b)    document d2
```

```
<program>
  <director>
    <LastName>Lang</LastName>
    <FirstName>kun</FirstName>
  </director>
  <title> 2009 Spring Festival Party</title>
  <type>art</type>
  <date>2009</date>
  <format>general</format>
</program>

        c)    document d3
```

Fig. 3. Descriptions of document d1,d2,d3

To query a material based on XML description, we can use XPath language. XPath uses path expressions to select nodes or node sets in XML document. A complete query expression is an expression with all the nodes in XML documents which are not empty. as shown in Figure 4. $q1$ is the complete query expression of document d1 in Figure 3. When the complete query expression is processed with hash functions, we can create DHT index. In this way, not only the original semantic information can be preserved, but also the materials can be stored correctly to the distributed storage system. Thus the description of media materials can be mapped onto DHT index.

q1= / program[director [Last Name/Zhang]][Title/Red Sorghum]
     [Type/Movie] [Date/1987] [Format/general]
q2= / program[director [Last Name/Zhang]] [Type/art]
q3= / program/director [Last Name/Zhang][First Name/Yimou]
q4= / program/Title/Red Sorghum
q5= / program/ Type/art
q6= / program/ Last Name/Zhang

Fig. 4. Examples of queries

4.2   Description of Hierarchical Index

In the management system of large-scale media material, the materials are stored in a distributed network storage, but in order to achieve fast retrieval response time, we adopt a centralized retrieval method by using a cluster of retrieval servers. Now we do some definitions:

Given two queries q and q', if all the results returned by query q are included in the results returned by query q', then we said query q' contains query q, denoted by $q' \supseteq q$. Particularly, if query q is the complete query expression of document d, denoted by $q \equiv d$. Figure 5 is a partially ordered tree of Figure 4.

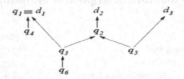

Fig. 5. Partially ordered tree of Figure 4

According to complete query expression, we can build a query expression set for all the queries which are frequently searched. $Q = \{q1, q2, ...qn\}$, where $qi \supseteq q$, we stored $< qi, q >$. For example, according to the document set shown in Figure 3, we can establish a hierarchical index shown in Figure 6. Each rectangle represents an index entry and each index entry stores the corresponding information. Top-level index entry represents a complete material description and other index entries maintain a relation between query conditions. In this way, we can provide an iterative way to help uses find what they want.

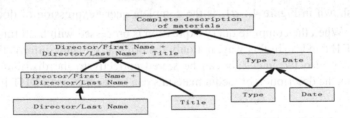

Fig. 6. Example of indexes

4.3   Procedure of Retrieval

If query $q0$ is a complete query expression, we will get the description of the corresponding document directly, or we will get a query sequence $\{q1, q2, ...qn\}$, where $q0 \supseteq qi$ Then the user can choose a query that he is interested as a new query, continue to iterate until he finds the necessary resources. For example, for the query

$q6$ given in Figure 4, query $q3$ is found first. Then we get document d1 and d3 through $q3$. When the given query $q0$ is not included in index entries, but there were some documents that satisfied the query condition. Then we find $qi \supseteq q0$, where qi is the current level of an index entry..

5   Conclusion

The distributed wideband collaborative service environment based on WAN provides technical support for carrying out more and better multimedia production business. This paper builds a retrieval and manage system of large-scale media materials. According to the unique characteristics of materials, we mapped the description of media onto the DHT index and proposed hierarchical indexing mechanism to improve query efficiency. Hierarchical indexing mechanism describes the indexes based on XPath, it helps users find the materials in an iterative way.

Aknowledgements. The paper is supported by the National Key Technology R&D Program of China (2012BAH02F04), and CUC Engineering Planning Project (XNG1125).

References

1. Feng, Z., Li, Z., Zhao, B.Y., et al.: Approximate Object Location and Spam Filtering on Peer-to-peer Systems. In: Endler, M., Schmidt, D.C. (eds.) Middleware 2003. LNCS, vol. 2672. Springer, Heidelberg (2003)
2. Tang, C., Xu, Z., Dwarkadas, S.: On Scaling Latent Semantic Indexing for Large Peer-to-Peer Systems. In: SIGIR 2004 (2004)
3. Tang, C.Q., Yu, Z.H., Mahalingam, M.: pSearch: Information retrieval in structured overlays. ACM SIGCOMM Computer Communication Review (2003)
4. Fu, X.: Research on Construction and Searching Algorithms of Topic Overlay Networks. Computer Science 6 (2007)
5. Cai, M., Frank, M., Chen, J., et al.: MAAN:A multi-attribute addressable network for grid information services. Journal of Grid Computing (2004)
6. Stoica, I., Morris, R., Karger, D., Kaashoek, M., Balakrishnan, H.: Chord: A scalable peer-to-peer lookup service for internet applications. In: Proc. ACM SIGCOMM (2001)
7. Ratnasamy, S., Handley, M., Karp, R., Shenker, S.: A scalable content-addressable network. In: Proc. ACM SIGCOMM (2001)
8. Kubiatowicz, J., Wells, C., Zhao, B., Bindel, D., Chen, Y., Czerwinski, S., Eaton, P., Geels, D., Gummadi, R., Rhea, S.: OceanStore: An architecture for global-scale persistent storage. In: Proc. of the 9th Int'l Conf. on Architectural Support for Programming Languages and Operating Systems, pp. 190–201 (2000)
9. Zheng, W., Hu, J., Li, M.: Granary: architecture of object oriented Internet storage service. In: IEEE International Conference on E-Commerce Technology for Dynamic E-Business, pp. 294–297 (2004)
10. Rowstron, A., Druschel, P.: Pastry: Scalable, Decentralized Object Location, and Routing for Large-Scale Peer-to-Peer Systems. In: Guerraoui, R. (ed.) Middleware 2001. LNCS, vol. 2218, p. 329. Springer, Heidelberg (2001)

Construction of SCI Publications Information System for Statistic

Xie Wu[1], Huimin Zhang[2], and Jingbo Jiang[1]

[1] School of Computer Science and Engineering, Guilin University of Electronic Technology,
Guilin, 541004, China
[2] School of Mathematics and Computing Science, Guilin University of Electronic Technology,
Guilin, 541004, China
iip2012_xiewu@126.com

Abstract. There are over 8000 SCI (Science Citation Index) publications in the
ISI (Institute for Scientific Information) Web of Knowledge database system.
However, the publications are too many and it is difficult for new authors to
choose the most suitable journals or periodicals to submit their research fruits of
high level. So, some valuable information about SCI publications is collected,
and the corresponding database is established. The records from this database
are classified and counted. The statistical results show that the SCI publications
information system is helpful to authors to issue papers.

Keywords: Database system, Science citation index, Publications, Statistics.

1   Introduction

ISI (Institute for Scientific Information) issues or updates periodically the information
of SCI (Science Citation Index) publications and JCR (Journal Citation Reports) every year. The records of SCI journals and periodicals are more than 8000, and it is a
very interest project to feel for some rules about SCI publications. Jacsó and Péter
researched the ISI Web of Science database from three configurations of the h-index,
h-core citation rate and the bibliometric traits [1]. Zhou Ping and Leydesdorff Loet
compared the Chinese scientific and technical papers and citations database and the
SCI data by journal hierarchies and interjournal citation relations [2]. Meho Lokman I
gave a comparison of scopus and web of science about citation counting, citation
ranking, and h-index of human-computer interaction researchers [3]. Li Jinfeng and
his colleagues analyzed the trends on global climate change based on SCIE (Science
Citation Index Expanded) [4]. Ball Rafael made a bibliometric comparison between
two citation databases of SCI and SCOPUS [5]. Some fruits about IF (Impact Factor),
rank, citation and other aspects of SCI journal and periodicals are gained, and this
information may be useful to some new contributors.

However, these fruits are only fit for some special researchers, projects or domains.
Because all the SCI periodicals are generally divided into thirteen classes with many
subdistricts, and the data of publications vary year by year, it is difficult for diverse
authors, especially for some beginners, to find the most appropriate journal to issue

Z. Shi, D. Leake, and S. Vadera (Eds.): IIP 2012, IFIP AICT 385, pp. 202–207, 2012.

high level research fruits when facing over 8000 SCI publications. The role of IF from ISI is also very limited. The SCI IF threshold values from13 JCR academic subjects are different. So, it is not reasonable to assess the level of SCI publications only in terms of IF by four JCR divisions without some comprehensive evaluation indicators. Moreover, some publications belong to the interdisciplinary scopes.

So, motivated by Ref. [1-8], the SCI periodicals and journals information including publication name, abbreviation, ISSN (International Standard Serial Number) number, organization, nationality, location, IF, Chinese name and discipline was collected, and a SCI publications information system is customized and constructed for statistical analysis. The remainder of this paper is organized as follows. Section 2 deals with the system requirement analysis for many new contributors. The relational tables of the SCI publications information system are designed in Sect 3, and the results of the statistical charts of this database system are shows in Sect 4. Finally, the discussion and conclusion are described in Sect 5.

2   Requirement Analysis with Modules

Many new authors are often not familiar with the existing ISI database system or its websites, and journal records are too massive for them, while the useful knowledge about their ideal publications is too little. So, the function requirement of this SCI publications information system mainly involves in the best selection from 8000 SCI records from ISI. This system is primarily divided into several modules as follows.

(1) Records Selection Module
This module introduces the records from the SCI publications information system in detail. The fuzzy selection function is afforded to the new contributors to gain the publication name, abbreviation, ISSN number, organization, nationality, IF, Chinese name, etc.

(2) Data Manipulation Module
It mainly covers the operations including inserting, updating and deleting records information of the SCI publications information system. These manipulations can be carried out through the back platform by DBA (database administration). The system DD (data dictionary) of important tables or attributes can be set up increasingly by the common administrator or super administrator.

(3) Statistical Chart Module
It can implement the functions of producing intuitional statistical charts according to the chosen attribute or attribute sets. From these charts, decision-makers can learn the distribution rules by setting the attribute conditions of organization, nationality, IF and subjects, etc.

(4) Users Management Module
This system can be visited by each super administrator, common administrator, DBA, registered user and guest, and their authority grades associating with this SCI publications information system decrease orderly. The former can grant or revoke the latter through modifying the permission of creating, modifying and dropping users.

3   System Design of Relational Logistic Model

After the detailed demand analysis of publications information system, the logical structure design needs to be followed. The relational logistic model is chosen to store publications information records, and there are several relational tables. These tables are designed and stored in SQL Sever DBMS (database management system). Here, a database named SCIData is created, and all the records of all tables are inserted and updated in SCIData with SQL Sever. These tables are shown as follows.

(1) Publication Basic Information Table

In order to catch some useful information for new authors, these column attributes including publication name, ISSN, publication location, citation rate are collected together to process the raw information of journals or periodicals. The data item of every attributes can be updated easily every year.

Table 1. Publication basic information table

attribute name	Description	data type	constraint
PubNameID	publication ID	Int	primary key
PubFullName	full name	Nvarchar(80)	not null
PubShortName	name for short	Nvarchar(80)	not null
Periodical	full name	Nvarchar(40)	
ISSN	ISSN number	Nvarchar(10)	not null
IF_ID	ID of ISI IF	Int	foreign key
AddressID	address number	Int	foreign key
CountryID	Country number	Int	foreign key
DisciplineID	Discipline ID	Int	foreign key

(2) Country of Publication Table

This table is designed to search the distribution rules of SCI publications with many kinds or numbers. For example, the SCI journals from America or European countries are massive, and few SCI publications belong to those countries from Africa or Latin America. On the other hand, SCI publications in English-speaking countries are far more than that of non-English nations. So, the country is an important factor, and it is a foreign key from the publication basic information table.

Table 2. Country of publication table

attribute name	description	data type	constraint
CountryID	Country ID	Int	primary key
CountryName	Country name	Varchar(80)	check
Language	nation language	Varchar(20)	check

(3) Publication Issue Period Table

The issue periods of 8000 SCI publications can divided into several types, such as annals or yearbook, semiyearly, quarterly, bimonthly, monthly, semimonthly, weekly, tri-annual and daily. This table is helpful to classify the whole SCI journals or periodicals, and it mainly reflects the speed of issuing publications. Authors tend to choose the journals with short periods, yet some important publications with high IF values issue papers for a long time. A new contributor has to make a choice by journal period.

Table 3. Publication issue period table

attribute name	description	data type	constraint
TypeID	TypeID	Int	primary key
Type	Type name	Nvarchar(30)	unique
IssuePeriod	Times in a year	Nvarchar(30)	enumeration
PubNameID	publicationID	Int	foreign key
Irregular	irregular cycle	Nvarchar(30)	
Issueyear	issue year	Datetime	not null

(4) IF Table
This table is one of the most important items to weigh the academic influence of SCI papers and publications. The IF values of every journal or periodical can be gained from ISI JCR every year. It is a very fair and common international rule for all SCI publications.

Table 4. IF table

attribute name	description	data type	constraint
IF_ID	impact factor ID	Int	primary key
IF_Year	IF value by year	Numeric(18, 2)	check
PubNameID	publicationID	Int	foreign kcy

4   System Results with Statistical Charts

The SCI publications information system for statistic is developed with the tools of Microsoft Visual studio 2010 in Asp.net and C# languages, and all the records are stored in Microsoft SQL Sever 2005. All users can select their records by given attribute or attribute sets. Furthermore, one can carry out fuzzy union selections through several fields or attributes.

The statistical charts can be generated from Tab. 1 to Tab. 4 of the SCI publica tions information system. For example, according to the publication issue period table, one of the statistical sub-diagrams is shown as Fig. 1.

In order to find the relationship between the attributions of countries and ISI SCI publications numbers, the above tables can be connected and selected. When the country table is connected to the publication basic information table by CountryID, the histogram is shown as Fig. 2.

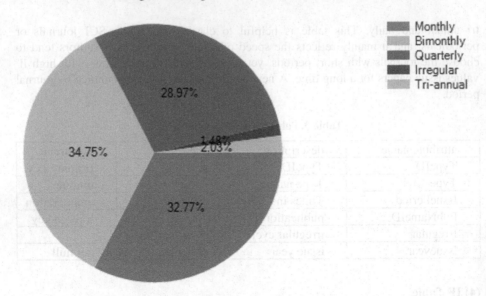

Fig. 1. One of the statistical results by issue period

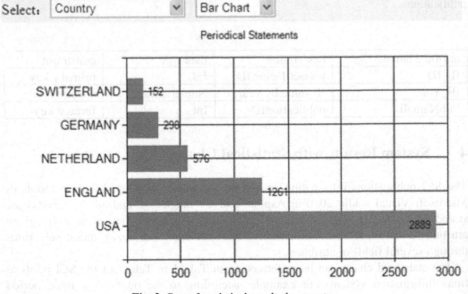

Fig. 2. One of statistical results by country

5   Summary

In this paper, we finish the work of system demand analysis and relational logical model design in detail, and a SCI publications information system is developed with the IDE (Integrated Development Environment) tools of Microsoft Visual Studio 2010 and SQL Server2005. It can be seen that from the results of this system software interfaces and the statistical charts, using this SCI periodical the information system, one can select some useful information by providing several important attributes of publication name, abbreviation, ISSN number, organization, nationality, location, etc. This software is easy to operate with simple and friend interfaces. It is very helpful for many new authors to choose the most appropriate SCI journals or periodicals to submit and issue their research fruits with high efficiency. These statistical charts from the selected data records of relational tables can provide some true and intuitive decision-making information to many new or initial contributors with little subjective assumption.

Acknowledgements. This work is supported by the Guangxi Education Department Foundations (No.201106LX209, 2010C059, 2010JGB035).

References

1. Jacsó, P.: The H-index, H-core Citation Rate and the Bibliometric Profile of the Web of Science Database in Three Configurations. Online Information Review 35(5), 821–833 (2011)
2. Zhou, P., Leydesdorff, L.: A comparison between the China scientific and technical papers and citations database and the science citation index in terms of journal hierarchies and interjournal citation relations. Journal of the American Society for Information Science and Technology 58(2), 223–236 (2007)
3. Meho, L.I., Rogers, Y.: Impact of data sources on citation counts and rankings of LIS faculty: Web of Science versus Scopus and Google Scholar. Journal of the American Society for Information Science and Technology 59(11), 1711–1726 (2008)
4. Li, J.F., Wang, M.H., Ho, Y.S.: Trends in research on global climate change: A Science Citation Index Expanded-based analysis. Global and Planetary Change 77(2), 13–20 (2011)
5. Ball, R., Tunger, D.: Science indicators revisited Science citation index versus SCOPUS: A bibliometric comparison of both citation databases. Information Services and Use 26(4), 293–301 (2006)
6. http://apps.webofknowledge.com
7. Fu, H.Z.: The most frequently cited adsorption research articles in the Science Citation Index (Expanded). Journal of Colloid and Interface Science 379(1), 148–156 (2012)
8. Hung, K.C., Lan, S.J., Liu, J.T·Global trend in articles related to stereotactic published in science citation index-expanded. British Journal of Neurosurgery Colloid and Interface Science 26(2), 258–264 (2012)

Symbolic ZBDD Representations
for Mechanical Assembly Sequences

Fengying Li, Tianlong Gu, Guoyong Cai, and Liang Chang

Guangxi Key Laboratory of Trusted Software, Guilin University of Electronic Technology,
Guilin 541004, China
{lfy,cctlgu,ccgycai,changl}@guet.edu.cn

Abstract. The representations of assembly knowledge and assembly sequences
are crucial in assembly planning, where the size of parts involved is a significant and often prohibitive difficulty. Zero-suppressed binary decision diagram
(ZBDD) is an efficient form to represent and manipulate the sets of combination, and appears to give improved results for large-scale combinatorial optimization problems. In this paper, liaison graphs, translation functions, assembly
states and assembly tasks are represented as sets of combinations, and the symbolic ZBDD representation of assembly sequences is proposed. An example is
given to show the feasibility of the ZBDD-based representation scheme.

Keywords: assembly sequence, assembly knowledge, Zero-suppressed binary
decision diagram.

1   Introduction

Related researches show that 40%~50% of manufacturing cost is spent on assembly,
and 20%~70% of all the manufacturing work is assembly[1, 2]. In order to shorten the
time and reduce the costs required for the development of the product and its manufacturing process, it is desirable to automate and computerize the assembly sequence
planning activity. Typically, a product can have a very large number of feasible assembly sequences even at a small parts count, and this number rises exponentially
with increasing parts count, which renders it staggeringly difficult and even impossible for one to represent all the sequences individually. Thus, the choices of representation for assembly sequences can be crucial in assembly sequence planning, and
there has been a need to develop systematic and efficient methods to represent all the
available alternatives.

In the literature, several representation schemes have been proposed to represent
the assembly sequences. These representations can be classified into two groups:
ordered lists and graphical representations. The ordered list could be a list of tasks,
list of assembly states, or list of subsets of connections. In the ordered lists each
assembly sequence is represented by a set of lists. Although this set of lists might
represent a complete and correct description of all feasible assembly sequences, it is
not necessarily the most compact or most useful representation of sequences. The

Z. Shi, D. Leake, and S. Vadera (Eds.): IIP 2012, IFIP AICT 385, pp. 208–215, 2012.

graphical schemes map the assembly operations and assembly states into specified diagrammatic elements, and share common subsequences and common states graphically in many assembly sequences, which create more compact and useful representations that can encompass all feasible assembly sequences. The most common diagrammatic representation schemes are: precedence diagrams[3], state transition diagrams[4], inverted trees[5], liaison sequence graphs[6], assembly sequence graphs[7] directed graphs[2] and AND/OR graphs[8] etc.

In recent years, implicitly symbolic representation and manipulation technique, called as symbolic graph algorithm or symbolic algorithm[9], has emerged in order to combat or ease combinatorial state explosion. Typically, zero-suppressed binary decision diagram (ZBDD) is used to represent and manage the sets of combinations[10, 11]. Efficient symbolic algorithms have been devised for hardware verification, model checking, testing and optimization of circuits. Symbolic representations appear to be a promising way to improve the computation of large-scale combinatorial computing problems through encoding and searching nodes and edges implicitly.

In this regard, we present the symbolic ZBDD formulation of all the assembly sequences. The subassemblies, assembly states, assembly tasks and assembly sequences are represented by sets of combinations, and the ZBDD representation for them are given. The example shows that the ZBDD formulation is feasible and compact.

2   Zero-Suppressed Binary Decision Diagram

Zero-suppressed binary decision diagram (ZBDD)[10, 11], a variant of ordered binary decision diagram (OBDD), is introduced by Minato for representing and manipulating sets of combinations efficiently. With ZBDDs, the space requirement of the representation of combination sets is reduced and combinatorial problems are solved efficiently.

A combination on n objects can be represented by an n bit binary vector, $(x_n x_{n-1} \cdots x_2 x_1)$, where each bit, $x_k \in \{0, 1\}$, expresses whether the corresponding object is included in the combination or not. A set of combinations can be represented by a set of the n bit binary vectors. We call such sets combination sets. Combination sets can be regarded as subsets of the power set on n objects.

A set of combination can be mapped into Boolean space by using n-input variables for each bit of the combination vector. If we choose any one combination vector, a Boolean function determines whether the combination is included in the set of combinations. Such Boolean functions are called characteristic functions.

By using OBDDs for characteristic functions, we can manipulate sets of combinations efficiently. Because of the effect of node sharing, OBDDs represent combination sets with a huge number of elements compactly. However, there is one inconvenience in that the form of OBDDs depends on the number of input variables. For example, S (abc) and S ($abcd$), shown in Fig. 1(a), represent the same set of combinations $\{a, b\}$, if we ignore the irrelevant input variables c and d. In this case, the OBDDs for S (abc) and S ($abcd$) are not identical. This inconvenience comes from the difference in the model of default variables. In combination sets, default variables are regarded as zero when the characteristic function is true, since the irrelevant object never can be suppressed in the OBDD representation.

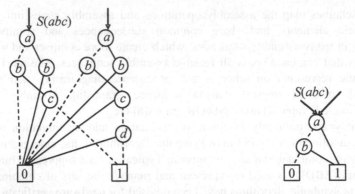

Fig. 1. (a) OBDDs for sets of combinations (b) ZBDDs for sets of combinations

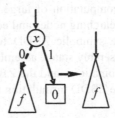

Fig. 2. New reduction rule for ZBDDs

For representing sets of combinations efficiently, ZBDD introduced the following special deletion rules:

Delete all nodes whose 1-edge points to the 0-terminal node, and then connect the edges to the other sub-graph directly as shown in Fig. 2.

This is also called the *pD-deletion* rule. ZBDD does not delete the nodes whose two edges point to the same node, which used to be deleted by OBDD. The zero-suppressed deletion rule is asymmetric for the two edges, as we do not delete the nodes whose 0-edge points to a 0-terminal node.

Fig. 1(b) shows the ZBDD representation for the same sets of combinations shown in Fig. 1(a). The form of ZBDDs is independent of the input domain. The ZBDD node deletion rule automatically suppresses the variables which never appear in any combination. This feature is important when we manipulate sparse combinations.

Another advantage of ZBDDs is that the number of 1-paths in the graph is exactly equal to the number of elements in the combination set. In OBDDs, the node elimination rule breaks this property. Therefore, ZBDDs are more suitable than OBDDs to represent combination sets.

3   Symbolic Representation of Assembly Knowledge

A mechanical assembly is a composition of interconnected parts forming a stable unit. Each part is a solid rigid object, that is, its shape remains unchanged. Parts are interconnected whenever they have one or more compatible surfaces in contact. Surface contacts between parts reduce the degree of freedom for relative motion. These contacts and relative motions are embedded in various logical and physical relations among the parts of the assembly, called as assembly knowledge, and can be extracted directly from the CAD model of assembly.

3.1 Symbolic Representation of Liaison Graph

Liaison graph can explicitly describe various logical and physical contact relations among the parts of the assembly. Liaison graph is a two-tuples $G = (P, L)$, in which P is a set of nodes that represent parts, and L a set of edges that represent any of certain user-defined relations between parts called liaisons. User-accepted definitions of liaisons in a general sense follow the principal literal definition "a close bond or connection" and generally include physical contacts.

Given an assembly and its liaison graph $G = (P, L)$, we can convert the liaison graph to a ZBDD by encoding the parts of the assembly or the elements in P with n binary variables $X=(x_0, x_1, \ldots , x_{n-1})$, where $n = |P|$. Essentially, the liaison set or edge set is a relation on nodes, and each element or a liaison $(a, b) \in L$ is a pair. A liaison $(a, b) \in L$ can be encoded as a combination of binary variables (x_i, x_j), where x_i and x_j are the encoded binary variables corresponding to part a and b respectively. Thus, the liaison graph can be uniquely determined by the following set of combinations:

$C (X) =\{ x_i x_j | x_i \in X, x_j \in X, i \neq j\}$

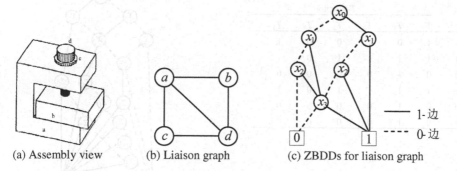

(a) Assembly view   (b) Liaison graph   (c) ZBDDs for liaison graph

Fig. 3. An example of assembly

For example, an assembly shown in Fig. 3(a) includes 4 parts, and its liaison graph is presented in Fig. 3(b), where $L = \{(a, b), (a, c), (a, d), (b, d), (c, d)\}$. We formulate the parts with 4 binary variables by encoding part a, b, c and d as x_0, x_1, x_2 and x_3 respectively. The combination set of relation E is derived as following:

$$C(x_0, x_1, x_2, x_3) =\{x_0x_1, x_0x_2, x_0x_3, x_1x_3, x_2x_3\}$$

Therefore, the liaison graph of the assembly is formulated by a ZBDD corresponding to the $C(x_0, x_1, x_2, x_3)$ as shown in Fig. 3(c).

3.2 Symbolic Representation of Translation Function

The liaison graph provides only the necessary conditions but not sufficient to assembly two components. To be a feasible assembly operation, it is necessary that there is a collision-free path to assembly parts. Gottipolu and Ghosh [6] represented the relative motion between parts of the assembly as a translation function, from which the existence or absence of a collision-free path can be conveniently verified.

The freedom of translation motion between two parts a and b can be represented by $T_{ab} = (T_0, T_1, T_2, T_3, T_4, T_5)$, which is a 1×6 binary function. Hence, it is called the translation function or T-function. It can be defined as:

$$T_{ab} = T_i \rightarrow \{0, 1\}, i = 0, 1, 2, 3, 4, 5$$

Where $T_i=1$ if the part b has the freedom of translation motion with respect to the part a in the direction i, $T_i=0$ if the part b has no freedom of translation motion with respect to the part a in the direction i. Here, direction 1, 2 and 3 indicate the positive sense of X, Y and Z axes ($X+$, $Y+$ and $Z+$) respectively, whereas direction 4, 5 and 6 correspond to the negative sense of X, Y and Z axes ($X-$, $Y-$ and $Z-$) respectively. If the part b has the freedom of translation motion with respect to the part a in the direction i, then the part a has the freedom of translation motion with respect to the part b in the direction $((i+3) \bmod 6)$. Hence, it is enough to give the front half part of the translation function.

Table 1. T-function for the assembly shown in Fig. 3(a)

Pair	T-function					
	T_0	T_1	T_2	T_3	T_4	T_5
(a, b)	1	0	1	0	0	1
(a, c)	1	1	1	1	0	1
(a, d)	0	1	0	0	0	0
(b, c)	1	1	1	1	0	1
(b, d)	0	1	0	0	0	0
(c, d)	0	1	0	0	0	0

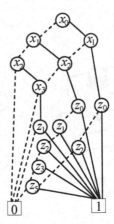

Fig. 4. ZBDDs for translation function

For example, the translation function T of the assembly shown in Fig. 3(a) is shown in table 1.

We can convert the translation function to a ZBDD by encoding the parts of the assembly in P with n binary variables $X=(x_0, x_1, \ldots, x_{n-1})$ and the direction with 6 binary variables $Z=(z_0, z_1, z_2, z_3, z_4, z_5)$, where $n = |P|$. We represent the translation function as set of combinations:

$$T(XZ) = \{ x_i x_j z_k \mid x_i \in X, x_j \in X, z_k \in Z, i \neq j \}$$

The combination set of translation function includes all the pairs (a, b), between which there exists the freedom of translation motion of part b with respect to the part a in the direction i. We can construct the combination set of these translation relations as $T(XZ)$, and thus implicitly formulate the translation functions using ZBDDs.

For example, the translation relations of the assembly in Fig. 3(a) are derived as following:

$$T(x_0, x_1, x_2, x_3, z_0, z_1, z_2, z_3, z_4, z_5) = \{ x_0 x_1 z_0, x_0 x_1 z_2, x_0 x_1 z_5, x_0 x_2 z_0, x_0 x_2 z_1, x_0 x_2 z_2, x_0 x_2 z_3,$$
$$x_0 x_2 z_5, x_0 x_3 z_1, x_1 x_2 z_0, x_1 x_2 z_1, x_1 x_2 z_2, x_1 x_2 z_3, x_1 x_2 z_5, x_1 x_3 z_1, x_2 x_3 z_1 \}$$

Fig. 4 gives the ZBDDs corresponding to the translation relations.

4   Symbolic Representation of Assembly Sequences

A mechanical assembly is a composition of interconnected parts forming a stable unit. Each part is a solid rigid object, that is, its shape remains unchanged. Parts are interconnected whenever they have one or more compatible surfaces in contact. Surface contacts between parts reduce the degree of freedom for relative motion. It is assumed that whenever two parts are put together all contacts between them are established.

A subassembly consists of a unique part or some parts in which every part has at least one surface contact with another part. Although there are cases in which it is possible to join the parts in more than one way, an unique assembly geometry will be assumed for each subassembly. This geometry corresponds to their relative location in the whole assembly. A subassembly is said to be stable if its parts maintain their relative position and do not break contact spontaneously. All one-part subassemblies are stable. To formulate an assembly consisting n parts, n binary variables $(x_1, x_2, \ldots, x_n)$ are demanded, if a part is characterized by a binary variable. Therefore, any subassembly can be characterized by the subset of parts set, and the ith component is presence or absence, respectively, if the nth part is involved in the subassembly or not. Hence, the assembly states can be represented by sets of combinations. For example, the assembly shown in Fig. 3 has four parts, represented by four binary variables x_1, x_2, x_3 and x_4, $\{abc\}$ and $\{cd\}$ are two subassemblies of the assembly, and can be represented by binary variable sets $\{x_1 x_2 x_3\}$ and $\{x_3 x_4\}$ respectively. The initial state and final state are represented as sets of combinations $\{x_1, x_2, x_3, x_4\}$ and $\{x_1 x_2 x_3 x_4\}$ respectively.

The assembly process consists of a succession of tasks, through each of which the subassemblies are joined into a larger subassembly. The process starts with all parts separated, and ends with all parts properly joined together to obtain the whole assembly. It is assumed that exactly two subassemblies are joined by each assembly task, and that after parts have been put together, they remain together until the end of the assembly process. An assembly task is said to be geometrically feasible if there is a collision-free path to bring the two subassemblies into contact from a situation in which they are far apart. And an assembly task is said to be mechanically feasible is it is feasible to establish the attachments that act on the contacts between the two subassemblies. Given two subassemblies characterized by their sets of parts S_1 and S_2, we say that joining S_1 and S_2 is an assembly task if the set $S_3 = S_1 \cup S_2$ characterizes a subassembly. Alternatively, a task can be seen as a decomposition of the output subassembly into the two input subassembly. Therefore, an assembly task τ_i can be characterized by an ordered pair $(\{S_{i1}, S_{i2}\}, S_{i3})$ of its output subassembly and input subassemblies. Since the equation $S_{i3} = S_{i1} * S_{i2}$ holds, an assembly task τ_i can be represented by a set of combinations on $2n$ binary variables (X, Y) in which $X=(x_0, x_1,\ldots, x_{n-1})$ and $Y=(y_0, y_1,\ldots, y_{n-1})$ are the binary variables for subassembly S_{i1} and S_{i3} respectively. For example, for the assembly shown in Fig. 3, if $S_1 = \{ab\}$ and $S_2 = \{c\}$, then joining S_1 and S_2 is an assembly task τ. The assembly task τ is characterized by an ordered pair $\tau = (\{ab, c\}, \{abc\})$.

Given an assembly with n parts, an assembly sequence is an ordered list of n-1 assembly tasks $\sigma = \tau_1 \tau_2 \ldots \tau_{n-1}$, in which the input subassemblies of the first task τ_1 is the separated parts, the output subassembly of the last task τ_{n-1} is the whole assembly, and the input subassemblies to any task τ_i is either a one-part subassembly or the output subassemblies of a task that precedes τ_i. An assembly sequence is said to be feasible if all its assembly tasks are geometrically and mechanically feasible, and the input subassemblies of all tasks are stable. The second input subassembly of task τ_i can be deduced by $S_{i2}= S_{i3}/S_{i1}$, where S_{i1} is the first input subassembly of task τ_i, and S_{i3} is the output subassembly of task τ_i. In this regard, an assembly sequence can be represented by the following set of combinations:

$\varphi_\sigma (X, Y) = \{S_{i1}S_{i3}|\ S_{i1}$ is the first input subassembly of task τ_i, and S_{i3} is the output subassembly of task τ_i, $i = 1, 2,\ldots, n$-1, $\sigma = \tau_1 \tau_2 \ldots \tau_{n-1}\}$

For example, for the assembly shown in Fig. 3, if $\tau_1=(\{a, b\},\{ab\})$, $\tau_2=(\{ab, c\},\{abc\})$, $\tau_3=(\{abc,d\},\{abcd\})$, $\sigma = \tau_1 \tau_2 \tau_3$ is a feasible assembly sequence. The assembly sequence can be represented by the set of combinations:

$$\varphi_\sigma (x_1, x_2,\ldots, x_n, y_1, y_2,\ldots, y_n)=\{\ x_0y_0y_1,\ x_0x_1y_0y_1y_2,\ x_0x_1x_2y_0y_1y_2y_3\}$$

The ZBDD of the assembly sequence is shown in Fig. 5.

An assembly might have a number of different feasible assembly sequences, and many assembly sequences share common subsequences. All the feasible assembly sequences can be represented compactly by the following set of combinations:

$$\psi(X,Y) = \bigcup_\sigma \varphi(X,Y)$$

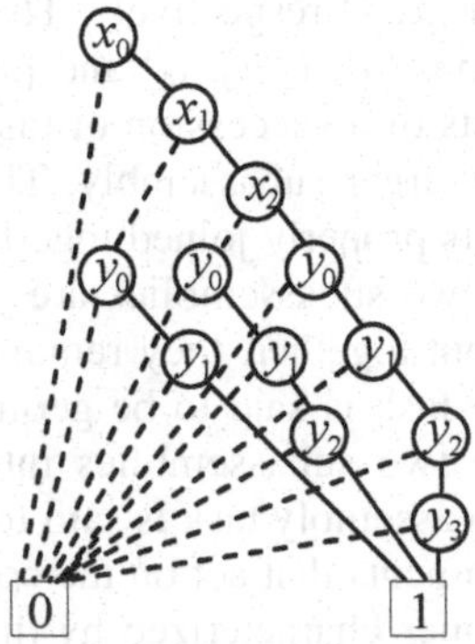

Fig. 5. ZBDD for assembly sequence

5   Conclusion

The choice of representation for assembly sequences has been crucial in assembly sequence planning. However, traditional representations, such as AND/OR graph and Petri nets, face the same challenge that increasing parts count renders it staggeringly difficult and even impossible to represent all the sequences individually. In order to alleviate the state-space explosion problem, a symbolic ZBDD scheme for representing all the feasible assembly sequences is presented in this paper. Validity and efficiency of this symbolic ZBDD-based assembly sequence representation are also demonstrated by the experiment.

Acknowledgments. The authors are very grateful to the anonymous reviewers for their helpful comments. This work has been supported by National Natural Science Foundation of China (No. 60903079, 60963010, 61063002), the Natural Science Foundation of Guangxi Province (No. 2012GXNSFAA053220), and the Project of Guangxi Key Laboratory of Trusted Software (No. PF11044X).

References

1. Gottipolu, R.B., Ghosh, K.: A simplified and efficient representation for evaluation and selection of assembly sequences. Computers in Industry 50(3), 251–264 (2003)
2. Homem de Mello, L.S., Sanderson, A.C.: Representation of mechanical assembly sequences. IEEE Transactions on Robotics and Automation 7(2), 211–227 (1991)
3. Prenting, T., Battaglin, R.: The precedence diagram: A tool for analysis in assembly line balancing. Journal of Industrial Engineering 15(4), 208–213 (1964)
4. Warrats, J.J., Bonschancher, N., Bronsvoort, W.: A semi-automatic sequence planner. In: Proceedings of IEEE International Conference on Robotics and Automation, pp. 2431–2438. IEEE, Piscataway (1992)
5. Bourjault, A.: Contribution to a methodological approach of automated assembly: automatic generation of assembly sequence. Ph. D. Thesis, University of Franch-Comte, Besancon, France (1984)
6. De Fazio, T.L., Whitney, D.E.: Simplified generation of all mechanical assembly sequences. IEEE Journal of Robotics and Automation RA-3(6), 640–658 (1987)
7. Gottipolu, R.B., Ghosh, K.: An integrated approach to the generation of assembly sequence. International Journal of Computer Applications in Technology 8(3-4), 125–138 (1995)
8. Homem de Mello, L.S., Sanderson, A.C.: AND/OR graph representation of assembly plans. IEEE Transactions on Robotics and Automation 6(2), 188–199 (1990)
9. Bryant, R.E.: Symbolic Boolean manipulation with ordered binary decision diagrams. ACM Computing Surveys 24(3), 293–318 (1992)
10. Minato, S.: Zero-suppressed BDDs for set manipulation in combinatorial problems. In: Proceedings of the 30th DAC in Dallas, pp. 272–277. IEEE, Piscataway (1993)
11. Minato, S.: Fast factorization for implicit cube set representation. IEEE Transactions on Computer Aided Design of Integrated Circuits and Systems 15(4), 377–384 (1996)

The Representation of Indiscernibility Relation
Using ZBDDs

Qianjin Wei[1,2], Tianlong Gu[2], Fengying Li[2], and Guoyong Cai[2]

[1] School of Electronic Engineering, XiDian University, Xi'an 710071
[2] Guangxi Key Laboratory of Trusted Software, Guilin University of Electronic Technology,
Guilin 541004
{wei_qj,cctlgu,lfy,ccgycai}@guet.edu.cn

Abstract. The indiscernibility relation is the basic concept in Rough set theory,
a novel representation of indiscernibility relation using Zero-Suppressed BDDs
is proposed in this paper. Through introducing the indiscernibility matrix and
the indiscernibility graph, we put forward the encoding of the variable and give
the characteristic function. Once the characteristic function is constructed, it can
be represented using ZBDDs.And further, combined with an example, we
analyze the effectiveness of this method. It provides a basis for deal with rough
set computing.

Keywords: rough set, Indiscernibility relation, Zero-Suppressed BDDs.

1   Introduction

The problem of imperfect knowledge has been tackled for a long time by philoso-
phers, logicians and mathematicians. Recently it became also a crucial issue for
computer scientists, particularly in the area of artificial intelligence. There are many
approaches to the problem of how to understand and manipulate imperfect
knowledge. The most successful one is, no doubt, the fuzzy set theory proposed by
Zadeh. In this approach sets are defined by partial membership, in contrast to crisp
membership used in classical definition of set. The theory of Rough set, proposed by
Z. Pawlak in 1982, is a new mathematical tool to deal with imprecise, incomplete and
inconsistent data [1].In Rough set theory, expresses vagueness, not by means of
membership, but employing a boundary region of a set. If the boundary region of a set
is empty it means that the set is crisp, otherwise the set is rough. Nonempty boundary
region of a set means that our knowledge about the set is not sufficient to define the
set precisely. The rough set theory has become an attractive field in recent years, and
has already been successful applied in many scientific and engineering fields such as
machine learning and data mining, it is a key issue in artificial intelligence. The indiscernibility relation plays a crucial role in Rough set theory. Due to its importance, the
different representations have been developed. Most existing are indiscernibility
matrix and indiscernibility graph [8-13].

Z. Shi, D. Leake, and S. Vadera (Eds.): IIP 2012, IFIP AICT 385, pp. 216–225, 2012.

Ordered Binary Decision Diagrams, graph-based representations of Boolean functions [18], have attracted much attention because they enable us to manipulate Boolean functions efficiently in terms of time and space. There are many cases in which conventional algorithms can be significantly improved by using BDDs [17][7]. Zero-suppressed BDDs [15] are a new type of BDDs that are adapted for representing sets of combinations. It can manipulate sets of combinations more efficiently than conventional BDDs. especially when dealing with sparse combinations. A Muir, I Düntsch and G Gediga discussed rough set data representation using binary decision diagrams [5], in which, a new information system representation is presented, called BDDIS. Chen Yuming and Miao Duoqian presented searching Algorithm for Attribute Reduction based on Power Graph [6], a new knowledge representation, called power graph, is presented in those paper, therefore, searching algorithms based on power graph are also proposed. Visnja Ognjenovic, etc presents a new form of indiscernibility relation based on graph [16]. In this paper, ZBDD is used to represent the family of all equivalence classes of indiscernibility relation; it has been shown how the indiscernibility relation can be obtained by ZBDD.

2   Indiscernibility Relation in Rough Sets Theory

The basic concepts, notations and results related to the theory of rough set are briefly reviewed in this section, others can be found in [1-4].

2.1   Information Systems for Rough Set

An information system is composed of a 4-tuples as follows:

$$I = < U, Q, V, f >$$

Where
    U is the closed universe, a finite set of N objects $\{x_1, x_2 ... xn\}$
    Q is finite attributes $\{q_1, q_2 ... q_m\}$
    $V = \cup_{q \in Q} V_q$ where V_q is a value of the attribute q, called the domain of attribute q.
    f: $U \times Q \rightarrow V$ is the total decision function called the information function

Table 1. An information table

U	a	b	c
x_1	0	0	0
x_2	0	1	0
x_3	0	1	1
x_4	1	0	0
x_5	0	0	1
x_6	0	1	1
x_7	1	0	1

Such that f(x, q) ∈ Vq for every q∈Q, x∈U .Such that $f(x, q) = v$ means that the object x has the value v on attribute q. An information table is illustrated in Table 1, which has five attributes and seven objects, with rows representing objects and columns representing attributes.

2.2   Indiscernibility Relation and Set Approximation

Let I=<U, Q, V, f> be an information system, then with any non-empty subset P⊆Q there is an associated indiscernibility relation denoted by IND (P), it is defined as the following way: two objects x_i and x_j are indiscernible by the set of attributes P in Q, if $f(x_i, q) = f(x_j, q)$ for every $q∈P$, More formally:

$$IND(P)=\{(x_i,x_j)\in U\times U \mid \forall q\in P, f(x_i,q)=f(x_j,q)\} \tag{1}$$

Where IND (P) is called the P-indiscernibility relation. If $(x_i, x_j) \in$ IND (P), then objects x_i and x_j are indiscernible from each other by attributes from P. Obviously IND (P) is an equivalence relation. The family of all equivalence classes of IND (P) will be denoted by U/ IND (P) , an equivalence class of IND(P) containing x will be denoted by $[x]_{IND (P)}$.

$$U/IND(P)=\{[x]_{IND (P)} \mid x\in U \} \tag{2}$$

$$[x]_{IND (P)}=\{y\in U \mid (x,y) \in IND(P)\} \tag{3}$$

Given any subset of attributes P, any concept X⊆U can be precisely characterized in terms of the two precise sets called the lower and upper approximations. The lower approximation, denoted by $\underline{P}X$, is the set of objects in U, which can be classified with certainty as elements in the concept X using the set of attributes P, and is defined as follows:

$$\underline{P}X=\{ x_i \in U \mid [x_i]_{IND (P)}\subseteq X\} \tag{4}$$

The upper approximation, denoted by $\overline{P}X$, is the set of elements in U that can be possibly classified as elements in X, and is defined as follows:

$$\overline{P}X=\{ x_i \in U \mid [x_i]_{IND (P)}\cap X\neq\varnothing\} \tag{5}$$

For any object x_i of the lower approximation of X, it is certain that it belongs to X. For any object x_i of the upper approximation of X, we can only say that it may belong to X.

According to Table 1, we have the following partitions defined by attribute sets {a}, {b}, and {c}, {d}:

U/IND({a})={{x_1, x_2, x_3, x_5, x_6 },{x_4, x_7}}

U/IND({b})={ {x_2, x_3, x_6}, { x_1, x_4, x_5, x_7}}

U/IND({c})={{x_1, x_2, x_4}, { x_3, x_5, x_6, x_7}}

If we have the set {a, b} we will have:

U/IND({a,b})={{x_1, x_5}, {x_2, x_3, x_6}, {x_4, x_7}}

C. Indiscernibility Matrix

Given an information table I=<U, Q, V, f >, two objects are discernible if their values are different in at least one attribute, the discernibility knowledge of the information system is commonly recorded in a symmetric $|U| \times |U|$ matrix $M_I(c_{ij}$ $(x_i, xj))$, called the indiscernibility matrix of I. Each element c_{ij} (x_i, xj) for an object pair $(x_i, xj) \in U \times U$ is defined as follows:

$$c_{ij}(x_i, x_j) = \begin{cases} \{q \in Q \mid f(x_i, q) \neq f(x_j, q)\} & f(x_i) \neq f(x_j) \\ \varnothing & \text{otherwise} \end{cases} \tag{6}$$

Since $M_I(c_{ij}(x_i, x_j))$ is symmetric and $c_{ii}(x_i, x_i) = \varnothing$, For i=1, 2... m, we represent M_I $(c_{ij}$ $(x_i, x_j))$ only by elements in the lower triangle of M_I $(c_{ij}(x_i, x_j))$, i.e. the $c_{ij}(x_i, x_j)$ is with $1 < i < j < m$.

The physical meaning of the matrix element $c_{ij}(x_i, x_j)$ is that objects x_i and x_j can be distinguished by any attribute in $c_{ij}(x_i, x_j)$, In another words, $c_{ij}(x_i, x_j)$ is defined as the set of all attributes which discern object x_i and x_i. The pair (x_i, x_j) can be discerned if $c_{ij}(x_i, x_j) \neq \varnothing$. The indiscernibility matrix of Table 1 is shown Table 2, for the underlined object pair (x_1, x_2), the entry {b, c, d, e} indicates that attribute b, c, d or e discerns the two objects.

Table 2. Indiscernibility matrix for the information system in Table 1

	x_1	x_2	x_3	x_4	x_5	x_6	x_7
x_1							
x_2	{b}						
x_3	{b,c}	{c}					
x_4	{a}	{a,b}	{a,b,c}				
x_5	{c}	{b,c}	{b}	{a,c}			
x_6	{b,c}	{c}	$\varnothing$	{a,b,c}	{b}		
x_7	{a,c}	{a,b,c}	{a,b}	{c}	{a}	{a,b}	

3   Zero-Suppressed BDDS

3.1   Combination Sets

Sets of combinations often appear in solving combinatorial problems. The representation and manipulation of sets of combinations are important techniques for many applications. A combination on n objects can be represented by an n bit binary vector, $(x_1 x_2 ... x_n)$, where each bit, $x_k \in \{0, 1\}$, expresses whether the corresponding object is included in the combination or not. A set of combinations can be represented by a set of the n bit binary vectors. In other words, combination sets is a subset of the power set of n objects. We can represent a combination set with a Boolean function by using n-input variables for each bit of the vector, If we choose any one combination vector, a Boolean function determines whether the combination is included in the set of combinations. Such Boolean functions are called characteristic functions. For example, given three elements {a, b, c}, consider the set of subsets {{a, b}, {a, c}, {c}}.If we associate each element with a binary variable having the same name, the characteristic function of the set of subsets is $f=abc'+ab'c+a'b'c$ .The first minterm corresponds to the subset {a, b}, and so on. The operations of sets, such as union, intersection and difference, can be executed by logic operations on characteristic functions.

Once the characteristic function is constructed, it can be represented using a OBDD or ZBDD. The two representations of the set of subsets {{a, b}, {a, c}, {c}} are given in Fig.1 (a).In both diagrams, there are three paths from the root node to the1- terminal node, which correspond to the subsets {a, b}, {a, c}, and {c}. By using OBDD for characteristic functions, we can manipulate sets of combinations efficiently. Due to the effect of node sharing, OBDD compactly represent sets of huge numbers of combinations. Despite the efficiency of OBDD, there is one inconvenience that the forms of OBDD depend on the input domains when representing sets of combinations. This inconvenience comes from the difference in the model on default variables. In combination sets, default variables are regarded as zero when the characteristic function is true, since the irrelevant objects never appear in any combination.Unfortunately, such variables cannot be suppressed in the OBDD representation, and many useless nodes are generated when we manipulate sparse combinations. This is the reason why we need another type of OBDD for manipulating sets of combinations.

3.2   Zero-Suppressed BDDs

Zero-suppressed BDDs are a new type of BDD adapted for representing sets of combinations. This data structure is more efficient and simpler than usual BDDs when we manipulate sets in combinatorial problems. They are based on the following reduction rules:

> pD-deletion rule: Delete all nodes whose 1-edge points to the 0-terminal node, and then connect the edges to the other subgraph directly, as shown in Fig.2
>
> Merging rule: Share all equivalent subgraphs in the same manner as with conventional BDDs

Notice that we do not delete the nodes whose two edges point to the same node. The zero-suppressed deletion rule is asymmetric for the two edges, as we do not delete the nodes whose 0-edge points to a 0-terminal node.

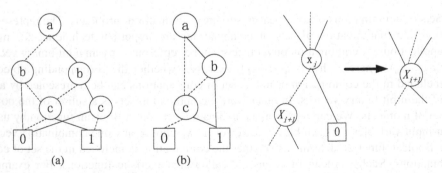

Fig. 1. OBDD and ZBDD for the set of subsets {{a, b}, {a, c}, {c}}

Fig. 2. pD-deletion rule

Fig.1 (b) shows the ZBDDs for the same sets of combinations shown in Fig.1 (a). The form of ZBDDs is independent of the input domain. The ZBDDs node deletion rule automatically suppresses the variables which never appear in any combination. This feature is important when we manipulate sparse combination.

4   ZBDDs Representation of Indiscernibility Relation

4.1   Indiscernibility Graph

It has been shown how the indiscernibility relations can be obtained by a graph. The application of the indiscernibility graph enables the partitioning or the universe of objects represented by their attributes [16].   Let I=<U, Q, V, f> be an information system, where U={x_1, x_2... x_n}, the indiscernibility graph is a tree structure to represent the family of all equivalence classes over U, we call this graph IR-tree. P⊆Q is any subsets of Q, nodes equally distant from root node represent all equivalence classes over U for P. To be able to observe entire indiscernibility classes by a certain attribute, each node needs to be associated with an attribute value label<v_{q1}, v_{q2},...,v_{qm}>,where v_{qm} is a value of the attribute $q_m \in$ P.

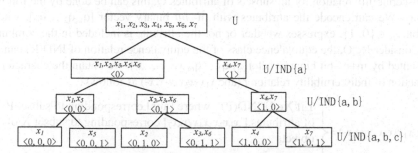

Fig. 3. The indiscernibility graph in Table 1

For example, the indiscernibility graph for the information system in table I is shown in Fig.3. The root node of IR-tree is U= {x_1, x_2, x_3, x_4, x_5, x_6, x_7}, the children are nodes that represent the family of all equivalence class over U for attribute {a}.For the node which represents the first indiscernibility class for the attribute {a}(x_1, x_2, x_3, x_5, x_6,,<0>),its child nodes are equivalence classes for the attribute{a,b},and so on .The leaf nodes represent indiscernibility relation classes for Q={a,b,c}.

4.2   Zero-Suppressed BDDs of Indiscernibility Relation

Let I=<U, Q, V, f> be an information system, where U={x_1, x_2... x_n}, Q= {q_1, q_2... q_m}. By formula (1), for any attribute q∈Q, there is an indiscernibility relation IND({q}).The equivalence class[x_i] $_{IND(\{q\})}$ of any object $x_i \in$ U consists of all objects $x_j \in$ U such that (x_i, x_j) ∈ IND({q}). In other words, an equivalence relation induces a portioning of the universe U, the partitions can be used to build new subsets of the universe, a subset of *k* objects can be represented by an *k-bit* binary vector,

222   Q. Wei et al.

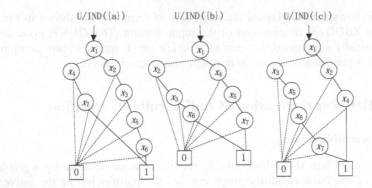

Fig. 4. ZBDDs of indiscernibility relation for attribute {a}, {b}, {c} in Table 1

$[x_1, x_2...x_k]$, in which 1-compent (x_i) means that the object x_i is included in the subset and 0-compent (x_i') means that the object x_i is not included in the subset. For the information system in table 1, the ZBDDs of indiscernibility relation for attribute {a}, {b}, {c} is given in Fig.4.

On the foundation of research above, the crucial problem is to create unique ZBDDs of indiscernibility relation for all subsets of attributes Q, this can be done by the following way. We can encode the attributes with m- *bit* binary vector $[q_1 q_2... q_m]$, where each bit, $q_k \in \{0, 1\}$, expresses whether or not the attribute is included in the combination. Consider R⊆Q, the equivalence class of the equivalence relation of IND(R) can be represented by m+n –bit binary vector$[q_1 q_2... q_m x_1, x_2...x_k]$. we obtain the characteristic function of indiscernibility relation f (abc $x_1x_2x_3x_4x_5x_6x_7$) as follows:

$$f(abc\ x_1x_2x_3x_4x_5x_6x_7)=\begin{cases} 1 \text{ if } X\in U/IND(P) \text{ where [abc] corresponding to subset P} \\ \text{of attributes,}[\ x_1x_2x_3x_4x_5x_6x_7] \text{ corresponding to subset X of} \\ \text{objects} \\ 0 \ otherwise \end{cases}$$

Table 3. characteristic function for the information system in Table 1

abc	$x_1x_2x_3x_4x_5x_6x_7$	f	abc	$x_1x_2x_3x_4x_5x_6x_7$	f
001	0 010111	1	101	0001000	1
001	1101000	1	101	0000001	1
010	0110010	1	110	1000100	1
010	1001101	1	110	0110010	1
011	1001000	1	110	0001001	1
011	0100000	1	111	1000000	1
011	0010010	1	111	0100000	1
011	0000101	1	111	0010010	1
100	1110110	1	111	0001000	1
100	0001001	1	111	0000100	1
101	1100000	1	111	0000001	1
101	0010110	1			

For example, Based on the information table in Table 1, to subset{a,b}, U/IND({a,b})={{x_1, x_5}, {x_2, x_3, x_6}, {x_4, x_7}},we can represent it used three vector:[1101000100],[1100110010]and [1100001001],we obtained the characteristic function as Table 3.

Once the characteristic function is constructed, it can be represented using a ZBDDs,the ZBDDs of f is shown in Fig.5.

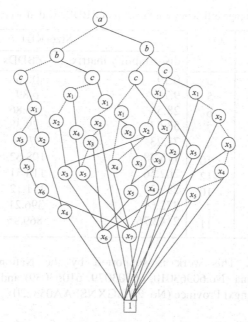

Fig. 5. ZBDDs of indiscernibility relation for information table in Table 1

5   ZBDDs of Indiscernibility Relation and State Space

For evaluating the ZBDD node deletion, it can be proved that the upper bound on the size of the ZBDDs is the total number of elements appearing in all subsets of a set.Meanwhile, the upper bound on the size of the ZBDDs is given by the number of subsets multiplied by the number of all elements that can appear in them. This shows that ZBDDs should be much more compact when representing sets of subsets. The above theoretical upper bound on the ZBDDs size is rarely reached, this means that ZBDDs are particularly effective for representing sets of sparse combinations.

6   Conclusions

The indiscernibility relation is the mathematical basis of the Rough set theory; indiscernibility matrix is a widely accepted representation for it. In this paper, we gave a new representation of indiscernibility relation based on ZBDDs. Through introducing the method by a graph, we put forward the new method by the data structure of Zero-Suppressed BDDs.Then, the encoding variable and the characteristic functions have been introduced, and further ,the implementation steps which construct ZBDDs of indiscernibility relation is presented. At the same time, combined with ZBDDS and concrete example we analyze the effective of this method. the results is shown in Table 4.

Table 4. Storage efficiency comparison of ZBDDs with discernibility matrix

	$|U|$	$|Q|$	Size(KB)	
			discernibility matrix	ZBDDs
1	7	5	0.19	0.16
2	50	5	9.76	6.89
3	80	6	28.58	16.86
4	100	7	53.67	29.38
5	100	9	71.23	37.59
6	150	10	175.78	108.62
7	150	12	202.56	118.51
8	200	10	299.34	151.12
9	400	7	872.69	396.21
10	500	11	2142.76	869.95

Acknowledgment. This work is supported by the National Natural Science Foundation of China (No.60963010, 60903079, 61063039) and the Natural Science Foundation of Guangxi Province (No. 2012GXNSFAA053220)

References

1. Pawlak, Z.: Rough sets. International Journal of Information and Computer Science 11(5), 341–356 (1982)
2. Skowron, A., Rauszer: The discernibility matrices and functions in information systems. In: Slowinski, R. (ed.) Intelligent Decision Support–handbook of Applications and Advances of the Rough Sets Theory, pp. 331–362. Kluwer Academic Publishers, Dordrecht (1992)
3. Pawlak, Z.: Rough set approach to multi-attribute decision analysis. European Journal of Operational Research 72(3), 443–459 (1994)
4. Pawlak, Z., Skowron, A.: Rudiments of rough set. Information Sciences 177(1), 3–27 (2007)
5. Muir, A., Düntsch, I., Gediga, G.: Rough set data representation using binary decision diagrams. Computational Sciences 98(1), 197–211 (2004)
6. Chen, Y., Miao, D.: Searching Algorithm for Attribute Reduction based on Power Graph. Chinese Journal of Computers 32(8), 1486–1492 (2009)
7. Gu, T., Xu, Z., Yang, Z.: Symbolic OBDD representations for mechanical assembly sequences. Computer-Aided Design 40, 411–421 (2008)

8. Yao, Y., Zhao, Y., Wang, J.: On Reduct Construction Algorithms. In: Proceedings of the First International Conference on Rough Sets and Knowledge Technologies, Chongqing, China, pp. 297–304 (2006)
9. Yao, Y., Zhao, Y.: Discernibility matr Simplification for constructing attribute ruducts. Information Sciences 179(7), 867–882 (2009)
10. Thangavel, K., Pethalakshmi, A.: Dimensionality reduction based on rough set theory: A review. Applied Soft Computing (9), 1–12 (2009)
11. Hu, X.H., Cercone, N.: Learning in relational databases: A rough set approach. Computational Intelligence 11(2), 323–337 (1995)
12. Ye, D., Chen, Z.: A new discernibility matrix and the computation of a core. Acta Electronica Sinica 30(7), 1086–1088 (2002)
13. Wang, G.: Calculation methods for core attributes of decision table. Chinese Journal of Computer 26(5), 611–615 (2003)
14. Gu, T., Xu, Z.: Ordered binary decision diagram and it s application, pp. 22–57. Science Press, Beijing (2009)
15. Minato, S.: Zero-suppressed BDDS for set manipulation in combinatorial problems. In: Proceeding of 30th ACM/IEEE Design Automation Conference, pp. 272–277
16. Ognjenovic, V., Brtka, V., Jovanovic, M., Brtka, E., Berkovic, I.: The Representation of Indiscerniblity Relation by Graph. In: Proceeding of 9th IEEE International Symposium on Intelligent Systems and Informatics, pp. 91–94
17. Matsunaga, Y., Fujita, M.: Multi-level Logic Optimization Using Binary Decision Diagrams. In: IEEE International Conference on Computer Aided Design 1989, Santa Clara (November 1989)
18. Bryant, R.E.: Graph-based algorithms for Boolean function manipulation. IEEE Transactions on Computers 35(8), 677–691 (1986)

Symbolic OBDD Assembly Sequence Planning Algorithm Based on Unordered Partition with 2 Parts of a Positive Integer

Zhoubo Xu, Tianlong Gu, and Rongsheng Dong

Guangxi Key Laboratory of Trusted Software, Guilin University of Electronic Technology, Guilin 541004, China
xzbli_11@guet.edu.cn, cctlgu@guet.edu.cn, ccrsdong@guet.edu.cn

Abstract. To improve solution efficiency and automation of assembly sequence planning, a symbolic ordered binary decision diagram (OBDD) technique for assembly sequence planning problem based on unordered partition with 2 parts of a positive integer is proposed. To convert the decomposition of assembly liaison graph into solving unordered partition with 2 parts of positive integer N, a transformation method from subassembly of the assembly to positive integer N is proposed, the judgment methods for the connectivity of a graph and geometrical feasibility of each decomposition combined with symbolic OBDD technique is proposed too, and all geometrically feasible assembly sequences are represented as OBDD-based AND/OR graph. Some applicable experiments show that the symbolic OBDD based algorithm can generate feasible assembly sequences correctly and completely.

Keywords: assembly sequence planning, automatic assembly planning, unordered partition, ordered binary decision diagram.

1   Introduction

Assembly sequence planning is to find the feasible or optimal sequence that put the initially separated parts of an assembly together to form the assembled product, and always plays a key role in determining important characteristics of the tasks of assembly and of the finished assembly. Since 1984, a number of assembly sequence planning systems have been developed, such as precedence relations method [1], cut-set decomposition method [2] and soft computing technique [3], etc. Precedence relations method based on a set of questions whose answers lead to establishment conditions, from which one can reason the geometrically feasible assembly sequence. However, it will become a difficult and error-prone process for all but simplest assemblies. Cut-set decomposition method generates the assembly sequences by application of cut-set algorithm based on an assumption that the sequence of assembly is the reverse of disassembly sequence. This method suffers from the fact that there may be an exponential number of candidate cut-sets, and this may lead to the so-called combinatorial explosion, but it is easy to be programmed, so that it has been the most

Z. Shi, D. Leake, and S. Vadera (Eds.): IIP 2012, IFIP AICT 385, pp. 226–233, 2012.

commonly used method for generating assembly sequence. Soft computing technique can find good solutions quickly for complex products, but cannot generate feasible assembly sequences completely and cannot find optimal solutions.

Essentially, generating assembly sequences is one of typical combinatorial computing problem. It is well known that the number of feasible assembly sequences increases exponentially with the number of parts or components composing the whole products. In recent years, efficient symbolic algorithm based on ordered binary decision diagram (OBDD) or variant have been devised for hardware verification, model checking, testing and optimization of circuit [4,5], and have been demonstrated that the symbolic algorithms can deal with large scale problems which traditional algorithm can't solve. Recently, there has emerged a class of OBDD-based approaches in assembly sequence planning. Gu, Xu and Yang proposed symbolic OBDD representations for mechanical assembly sequences [6], the experimental results show that the storage space of OBDD based representation of all the feasible assembly sequences is less than that of AND/OR graph do. Gu and Liu developed an symbolic OBDD algorithm for assembly sequence planning which is limited to monotone linear assembly [7]. Gu and Xu presented a novel scheme to integrate constraint satisfaction problem model with OBDD for the assembly sequence planning, but the procedure consumes a lot of time with the problem of backtracking [8].

In this regard, this paper references the idea of cut-set decomposition method, and integrates unordered partition with 2 parts of a positive integer which is used to generate all feasible true cut-set with OBDD symbolic technique to generate assembly sequence. Based on symbolic OBDD assembly model of liaison graph and translation function, the one-to-one correspondence between the parts set of assembly $V=\{v_0,v_1,...,v_{n-1}\}$ and sets $W=\{2^0,2^1,...,2^{n-1}\}$ is established firstly, so that any non-empty subset V' of V can be represented as a unique positive integer N ($1 \leq N \leq 2^n-1$). After this, the decompositions of sub-graph of liaison graph with parts set V' can be enumerated by unordered partition with 2 parts of a positive integer, and the connectivity of each segment and geometrical feasibility corresponding to each of decompositions are checked by symbolic OBDD technique, and all geometrically feasible assembly sequences are represented by OBDD-based AND/OR graph. Finally, some applicable experiments show that the novel algorithm can generate feasible assembly sequences correctly and completely.

2   Formulating Assembly Knowledge via OBDD

Liaison graph is one of the role model for generating assembly sequences. Liaison graph is a undirected connected graph $G=<V,E>$, where V is a set of vertices that represent parts, and E is a set of edges that represent any of certain user-defined relations between parts called liaisons.

Given an assembly and its liaison graph $G=<V,E>$, we can convert the liaison graph to an OBDD by encoding the parts of the assembly with a length-l binary number, where $l=\lceil \log_2|V| \rceil$, each parts in V can be encoded as a vector of binary variables $x=[x_0x_1...x_{l-1}]$. Essentially, the liaisons are a relation on parts, so a liaison $(a,b) \in E$ can be encoded as a vector of binary variables $[x_0x_1...x_{l-1}y_0y_1...y_{l-1}]$, where $[x_0x_1...x_{l-1}]$ and $[y_0y_1...y_{l-1}]$ are the encoded binary vectors corresponding to part a and b respectively. Hence, the liaison (a,b) can be represented as a Boolean characteristic function:

$$\xi(x_0, x_1, ..., x_{l-1}, y_0, y_1, ..., y_{l-1}) = \alpha_0 \cdot \alpha_1 \cdot ... \cdot \alpha_i \cdot ... \cdot \alpha_{l-1} \cdot \beta_0 \cdot \beta_1 \cdot ... \cdot \beta_i \cdot ... \beta_{l-1}$$

where α_i is the appearance of variable x_i corresponding to $i+1$ bit code of part a, and β_i is the appearance of variable y_i corresponding to $i+1$ bit code of part b. If $x_i=1$, then $\alpha_i = x_i$, else $\alpha_i = x_i'$. If $y_i=1$, then $\beta_i = y_i$, else $\beta_i = y_i'$. Hence, the liaison graph G can be represented by the following characteristic function:

$$\phi_G(x, y) = \sum \xi(x_0, x_1, ..., x_{l-1}, y_0, y_1, ..., y_{l-1})$$

where $x=[x_0x_1...x_{l-1}]$, $y=[y_0y_1...y_{l-1}]$.

The liaison graph provides only the necessary conditions but not sufficient to assembly two components. To be a feasible assembly operation, it is necessary that there is a collision-free path to assembly parts. Gottipolu and Ghosh [9] defined translational function to represent the relative motion between parts of assembly.

Translational function T: $P \times P \rightarrow \{0,1\}^6$, where P is a set of parts, $\{0,1\}^6$ is a 0-1 vector space with six dimension, each dimension correspond to the one of six directions of triorthogonal Cartesian coordinate system. Here, directions 1, 2, 3, 4, 5 and 6 indicate the six directions $+X$, $+Y$, $+Z$, $-X$, $-Y$, and $-Z$ of X, Y and Z axes respectively.

Let $(a, b) \in P \times P$, the correspond value of T function is 0-1 vector space with six dimension $(T_1(a,b), T_2(a,b), T_3(a,b), T_4(a,b), T_5(a,b), T_6(a,b))$, where $T_i(a,b)=1$ $(i=1,2,...,6)$ if the part b has the freedom of translational motion with respect to the part a in the direction i, $T_i(a,b)=0$ if the part b has no freedom of translational motion with respect to the part a in the direction i.

For any part a and b, if $T_i(a,b)=1$, then it can also be represented by the Boolean function $\xi(x_0, x_1, ..., x_{l-1}, y_0, y_1, ..., y_{l-1})$. Hence, the ith component of translational function T can be represented by the following characteristic function:

$$\Phi_{T_i}(x_0, x_1, ..., x_{l-1}, y_0, y_1, ..., y_{l-1}) = \sum \xi(x_0, x_1, ..., x_{l-1}, y_0, y_1, ..., y_{l-1})$$

3   Formulating Subassembly via Integer

For an assembly with n parts, a subassembly of assembly can be characterized by its set of parts, and can be represented by an n-dimensional binary vector $[x_0x_1...x_{n-1}]$ in which the ith component is 1 or 0, respectively, if the ith part is involved in the subassembly or not. So the assembly can be represented by full one binary vector.

Theorem 1. Let parts set $V=\{v_0,v_1,...,v_{n-1}\}$, and a function $f(v_i)=2^i$ $(i=0,1,...,n-1)$. If a function $g(x_0,x_1,...,x_{n-1})= \sum_{x_i=1} f(v_i)$, where $x_i=0$ or 1, then any integer k $(1 \leq k \leq 2^n-1)$ is correspond to one and only one subassembly.

Proof. For any k, $1 \leq k \leq 2^n - 1$, there must exist an n-dimensional binary vector $[x_0 x_1 \ldots x_{n-1}]$ such that $\sum_{x_i=1} f(v_i) = k$. If $x_i = 1$, it means that part v_i is involved in the sub-assembly, otherwise part v_i is not involved in the subassembly.

Assume such an n-dimensional binary vector $[x_0 x_1 \ldots x_{n-1}]$ is not unique, then there exist at least two n-dimensional binary vectors for the value k. Suppose that $[x_0 x_1 \ldots x_{n-1}]$ and $[y_0 y_1 \ldots y_{n-1}]$ are two n-dimensional binary vectors for the value k. if there are only $x_{i_1}, x_{i_2}, \ldots, x_{i_p}$ which have the value 1, where $0 \leq i_1 \leq i_2 \leq \ldots \leq i_p \leq n-1$, then $g(x_0, x_1, \ldots x_{n-1}) = 2^{i_1} + 2^{i_2} + \ldots + 2^{i_p}$, namely, $k = 2^{i_1} + 2^{i_2} + \ldots + 2^{i_p}$. If there are only y_{j_1}, $y_{j_2}, \ldots, y_{j_q}$ which have the value 1, where $0 \leq j_1 \leq j_2 \leq \ldots \leq j_q \leq n-1$, then $g(y_0, y_1, \ldots y_{n-1}) = 2^{j_1} + 2^{j_2} + \ldots + 2^{j_q}$, that is to say $k = 2^{j_1} + 2^{j_2} + \ldots + 2^{j_q}$. Thus $2^{i_1} + 2^{i_2} + \ldots + 2^{i_p} = 2^{j_1} + 2^{j_2} + \ldots + 2^{j_q}$.

① If k is odd, then it must have $i_1 = 0$, $j_1 = 0$, namely, $j_1 = i_1$. Hence

$$2^{i_2} + 2^{i_3} + \ldots + 2^{i_p} = 2^{j_2} + 2^{j_3} + \ldots + 2^{j_q} \tag{1}$$

Let $l = \min\{i_2, i_3, \ldots, i_p, j_2, j_3, \ldots, j_q\}$, here might as well assume that $l = i_2$, extract lowest common multiple 2^l in both ends of equation (1), we have

$$2^{i_2}(1 + 2^{i_3 - i_2} + \cdots + 2^{i_p - i_2}) = 2^{i_2}(2^{j_2 - i_2} + 2^{j_3 - i_2} \cdots + 2^{j_q - i_2})$$

Hence

$$1 + 2^{i_3 - i_2} + \cdots + 2^{i_p - i_2} = 2^{j_2 - i_2} + 2^{j_3 - i_2} + \cdots + 2^{j_q - i_2} \tag{2}$$

It is evident that the left of the equation (2) is odd, so there must have $2^{j_2 - i_2} = 1$, that is $j_2 = i_2$. Therefore, the following equation holds:

$$2^{i_3 - i_2} + \cdots + 2^{i_p - i_2} = 2^{j_3 - i_2} + \cdots + 2^{j_q - i_2} \tag{3}$$

And it has $j_3 = i_3$ through the process of the proof similar to that of equation (1). For the same reason, there have $j_4 = i_4, \ldots, j_q = i_p$.
② If k is even, it can also proof that $j_1 = i_1$, $j_2 = i_2, \ldots, j_q = i_p$ through applying the similar process of the proof as equation (1). □

For example, the assembly shown in Fig.1 includes 4 parts. We assume that $f(u) = 2^0$, $f(b) = 2^1$, $f(c) = 2^2$ and $f(d) = 2^3$, then according to theorem 1, integer 7 can be represent the parts set $\{a, b, c\}$.

4   Generation of Assembly Sequences

Given an assembly and its liaison graph $G = \langle V, E \rangle$, if a decomposition partition the graph G into two connected subgraph $G_1 = \langle V_1, E_1 \rangle$ and $G_2 = \langle V_2, E_2 \rangle$, where $V_1 \cap V_2 = \varnothing$,

$V_1 \cup V_2 = V$, $E_1 \subseteq V_1 \times V_1$, $E_2 \subseteq V_2 \times V_2$, then this decomposition will be correspond to a possible disassembly. In fact, the decomposition of a graph is to divide a set V with n parts into two non-intersect subset V_1 and V_2, such that $V_1 \cap V_2 = \varnothing$, $V_1 \cup V_2 = V$, and each of induced subgraph of subset V_1 and V_2 are connected.

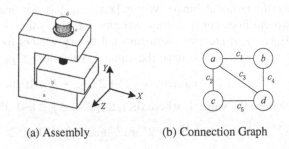

(a) Assembly   (b) Connection Graph

Fig. 1. A Simple Assembly

Unordered partition with 2 parts of positive integer N is corresponding to a possible disassembly operation. For the assembly shown in Fig.1, one of the unordered partition with 2 parts of integer 15 is 3 and 12, the corresponding disassemble operation is to disassemble assembly $\{a,b,c,d\}$ into subassembly $\{a,b\}$ and $\{c,d\}$. So, to find all possible disassembly operation is equal to find all unordered partition with 2 parts of integer N. But it need to check the feasibility of disassembly operation, which include two step, one step is check connectivity of each sub-graph after disassemble operation, the other step is check geometrical feasibility of two subassembly.

For an assembly has n parts, the assembly is correspond to integer $2^n - 1$. We assume unordered partition with 2 parts of positive integer $2^n - 1$ is p and q, p is correspond to the parts set V_1, and q is correspond to the parts set V_2. Judging the connectivity of induced sub-graph of pars set V_1 is as follows:

$$H_1(x,y) = V_1(x) \wedge \phi_C(x,y) \wedge V_1(y) \qquad (4)$$

$$New(x) = \exists y (New(y) \wedge (H_1(x,y) \vee H_1(y,x))) \wedge \overline{Reached(x)} \qquad (5)$$

$$Reached(x) = Reached(x) \vee New(x) \qquad (6)$$

$New(x)$ and $Reached(x)$ are initialized as any part in $V_1(x)$. Equation (5) and equation (6) are iterative until $Reached(x)$ is not changed or equation $Reached(x) = V_1(x)$ is hold. If the equation $Reached(x) = V_1(x)$ is hold, then the induced sub-graph of pars set V_1 is connected, otherwise, it is not connected.

Judging the connectivity of induced sub-graph of parts set V_2 is same to V_1.

If both the induced sub-graph of parts set V_1 and V_2 are connected, checking the geometrical feasibility of subassembly V_1 and V_2 is to check whether the following equation is hold:

$$V_1(x) \wedge V_2(y) \wedge \Phi_{T_i}(x,y) = V_1(x) \wedge V_2(y) \qquad (i=1,2,...,6) \qquad (7)$$

If for all i, equation (7) is not hold, then the assembly of subassembly V_1 to subassembly V_2 is not geometrically feasible, otherwise it is geometrically feasible, and insert the assemble operation into OBDD-based AND/OR graph $\Phi_S(x,y)$.

For example, the assembly shown in Fig.1 includes 4 parts. The pre-state and post-state of a disassembled operation can be characterized by $[x_0x_1x_2x_3]$ and $[y_0y_1y_2y_3]$ respectively, where x_0 and y_0, x_1 and y_1, x_2 and y_2, x_3 and y_3 are encoded binary number corresponding to part a, b, c and d, respectively. If $x_i=1$ then the corresponding part is involved in the subassembly, otherwise it is not involved. We still assume that $f(a)=2^0$, $f(b)=2^1$, $f(c)=2^2$ and $f(d)=2^3$, then the disassembled processes of the assembly are shown in table1.

Table 1. The disasssembled process of assembly shown in Fig.2

N	p	q	V_1	V_2	Disassembled Operation
15	3	12	$\{a, b\}$	$\{c, d\}$	$x_0'x_1'x_2x_3y_0y_1y_2y_3 + x_0x_1x_2'x_3'y_0y_1y_2y_3$
15	7	8	$\{a, b, c\}$	$\{d\}$	$x_0x_1x_2x_3'y_0y_1y_2y_3$
3	1	2	$\{a\}$	$\{b\}$	$x_0'x_1'x_2'x_3'\ y_0y_1y_2'y_3'$
12	4	8	$\{c\}$	$\{d\}$	$x_0'x_1'x_2'x_3'y_0'y_1'y_2y_3$
7	2	5	$\{b\}$	$\{a, c\}$	$x_0x_1'x_2x_3'y_0y_1y_2y_3'$
7	3	4	$\{a, b\}$	$\{c\}$	$x_0x_1x_2'x_3'y_0y_1y_2y_3'$
5	1	4	$\{a\}$	$\{c\}$	$x_0'x_1'x_2'x_3'y_0y_1'y_2'y_3'$

As shown in Table 1, the Boolean characteristic function of all feasible disassembled sequences of the assembly shown in Fig.1 can be formulated as:

$$\Phi_S(x,y)= x_0'x_1'x_2x_3y_0y_1y_2y_3 + x_0x_1x_2'x_3'y_0y_1y_2y_3 + x_0x_1x_2x_3'y_0y_1y_2y_3 + x_0'x_1'x_2'x_3'y_0y_1y_2'y_3' +$$
$$x_0'x_1'x_2'x_3'y_0'y_1'y_2y_3 + x_0x_1'x_2x_3'y_0y_1y_2y_3' + x_0x_1x_2'x_3'y_0y_1y_2y_3' + x_0'x_1'x_2'x_3'y_0y_1'y_2'y_3'$$

Based on an assumption that the sequence of assembly is the reverse of that of disassembly, the OBDD representation of all feasible assembled sequences of the assembly are shown in Fig.2 (all the edges point to the terminal 0 which is omitted for clarity).

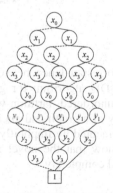

Fig. 2. OBDD Representation of Assembled Sequence of the Assembly Shown in Fig.1

5   Experiments

The symbolic algorithm proposed in this paper has been implemented in windows XP and the software package CUDD[10]. We have tested the performance of our algorithm on some practical assemblies, such as assembly shown in Fig.1, electrical controller(Fig.3) and Centrifugal Pump (Fig.4). Geometrically feasible assembly sequences of the assemblies were generated by the algorithm using Microsoft Visual C++. Experiments were performed on a P4 3GHZ with 512MB of memory. For the assembly shown in Fig.1, the OBDD for the product contains 28 OBDD node and 7 hyperarcs, and the CPU time is 0.015 seconds for generating all assembly sequences. In electrical controll case, the OBDD for the product contains 21442 OBDD node, and the CPU time is 996.437 seconds for generating all assembly sequences. In centrifugal Pump case, the OBDD for the product contains 3850 OBDD node, and the CPU time is 6280.437 seconds for generating all assembly sequences. For the centrifugal pump, it can not be solved in 10 hours by tranditional cut-set decomposition in the same enviorment,   it shows that the symbolic algorithm is better than the tranditional algorithm which is attribute to the symbolic OBDD implicit operation.

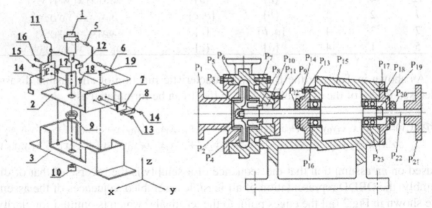

Fig. 3. An Electrical Controller   **Fig. 4.** A Centrifugal Pump

6   Conclusions

In this paper, the one-to-one correspondence between subassembly and integer is established, and a symbolic OBDD algorithm for generating mechanical assembly sequence is presented based on unordered partition with 2 parts of a positive integer method. Judging the connectivity of a graph and checking the geometrical feasibility of a subassembly are both implemented by implicitly symbolic OBDD manipulation. Some applicable experiments show that the novel algorithm can generate feasible assembly sequences correctly and completely.

Acknowledgements. This work hav been supported by National Natural Science Foundation of China (61100025,60963010,61063002) and Science Foundation of Guangxi Key Laboratory of Trusted Software (No. kx201119).

References

1. De Fazio, T.L., Whitney, D.E.: Simplified generation of all mechanical assembly sequences. IEEE Journal of Robotics and Automation 3(6), 640–658 (1987)
2. Homen de Mello, L.S., Sanderson, A.C.: A correct and complete algorithm for the generation of mechanical assembly sequences. IEEE Transactions on Robotics and Automaition 7(2), 228–240 (1991)
3. Wang, J.F., Liu, J.H., Zhong, Y.F.: A novel ant colony algorithm for assembly sequence planning. International Journal of Advanced Manufacturing Technology 25(11), 1137–1143 (2005)
4. Bryant, R.E.: Symbolic boolean manipulation with ordered binary decision diagrams. ACM Computing Surverys 24(3), 293–318 (1992)
5. Burch, J., Clarke, E., Mcmillan, K., et al.: Symbolic model checking:10^{20} states and beyond. Information and Computation 98(2), 142–170 (1998)
6. Gu, T.L., Xu, Z.B., Yang, Z.F.: Symbolic OBDD representations for mechanical assembly sequences. Computer-Aided Design 40(4), 411–421 (2008)
7. Gu, T.L., Liu, H.D.: The symbolic OBDD scheme for generating mechanical assembly sequences. Formal Methods in System Design 33(1/3), 29–44 (2008)
8. Xu, Z.B., Gu, T.L.: A constraint satisfaction problem model and its symbolic OBDD solving for assembly sequence planning problem. Journal of Computer-Aided Design & Computer Graphics 22(5), 803–810 (2010)
9. Gottipolu, R.B., Ghosh, K.: A simplified and efficient representation for evaluation and selection of assembly sequences. Computer in Industry 50(3), 251–264 (2003)
10. Cudd, S.F.: CU decision diagram package release 2.3.1, http://vlsi.Colorado.edu/fabio/CUDD/cuddIntro.html

A Representation Model of Geometrical Tolerances Based on First Order Logic

Yuchu Qin[1], Yanru Zhong[1], Liang Chang[1], and Meifa Huang[2]

[1] School of Computer Science and Engineering,
Guilin University of Electronic Technology, Guilin, China
[2] School of Mechanical and Electrical Engineering,
Guilin University of Electronic Technology, Guilin, China
qinyuchu@163.com

Abstract. Tolerance representation models are used to specify tolerance types and explain semantics of tolerances for nominal geometry parts. To well explain semantics of geometrical tolerances, a representation model of geometrical tolerances based on First Order Logic (FOL) is presented in this paper. We first investigate the classifications of feature variations and give the FOL representations of them based on these classifications. Next, based on the above representations, we present a FOL representation model of geometrical tolerances. Furthermore, we demonstrate the effectiveness of the representation model by specifying geometrical tolerance types in an example.

Keywords: Feature variations, Representation model, Geometrical tolerances.

1 Introduction

Tolerance representation model is a kind of data structure which makes tolerance information be well represented in computers [1]. The existing mainstream representation models can be classified into the following five categories: surface graph model, variational geometric model, tolerance zone model, degree of freedom model, and mathematical definition model [2]. An excellent tolerance representation model is able to represent tolerance information accurately and explain semantics of tolerances reasonably. Most of the existing models can meet the first requirement, but they cannot meet the second one completely. Thus it is an immediate concern to study a tolerance representation model which can well explain semantics of tolerances.

FOL is a formal system for representing and reasoning about knowledge of applications. It can well represent knowledge on the semantic layer. This paper presents a representation model of geometrical tolerances based on FOL. It is organized as follows. First of all, the classifications of feature variations are investigated and the FOL representations of feature variations are given. Next, a representation model of geometrical tolerances is presented. Finally, the effectiveness of the representation model is demonstrated by specifying geometrical tolerance types in an example.

Z. Shi, D. Leake, and S. Vadera (Eds.): IIP 2012, IFIP AICT 385, pp. 234–239, 2012.

2   FOL Representations of Feature Variations

In feature-based CAD systems, tolerances are essentially the variations of geometrical features. Feature variations are the geometrical motions of the real feature compared with the ideal feature. To link closely Geometrical Product Specifications (GPS) to geometric features, Srinivasan [3] classified the ideal features in GPS into seven classes of symmetry based on symmetry group theory (see Table 1).

Table 1. Seven classes for feature variations. $Aut_0(S)$: automorphism of S under small variation. DOFs: degrees of freedom. $T(m)$: m independent translations. $R(n)$: n independent rotations. I: identity variation.

Real integral feature	Associated derived feature	$Aut_0(S)$	DOFs
Spherical	Point	$R(3)$	$T(3)$
Cylindrical	Line	$T(1) \times R(1)$	$T(2), R(2)$
Planar	Plane	$T(2) \times R(1)$	$T(1), R(2)$
Helical	(Point, Line)	$T(1) \times R(1)$	$T(2), R(2)$
Revolute	(Point, Line)	$R(1)$	$T(3), R(2)$
Prismatic	(Line, Plane)	$T(1)$	$T(2), R(3)$
Complex	(Point, Line, Plane)	I	$T(3), R(3)$

For convenience, we use a triple group (M, N, f) to denote the variations of a feature, where M is a constraint feature, N is a constrained feature, and f is a geometrical variation from M to N. For the seven classes for feature variations, if we let M be one of the associated derived features, N be its corresponding real integral feature, we will obtain seven combinations called self-referenced feature variations shown in Table 2.

Table 2. Self-referenced feature variations. DOFs: degrees of freedom. $T(m)$: m independent translations. $R(n)$: n independent rotations.

Name	Constraint feature	Constrained feature	Spatial relation	DOFs
S1	Point	Spherical	Constrain	$T(3)$
S2	Line	Cylindrical	Constrain	$T(2), R(2)$
S3	Plane	Planar	Constrain	$T(1), R(2)$
S4	(Point, Line)	Helical	Constrain	$T(2), R(2)$
S5	(Point, Line)	Revolute	Constrain	$T(3), R(2)$
S6	(Line, Plane)	Prismatic	Constrain	$T(2), R(3)$
S7	(Point, Line, Plane)	Complex	Constrain	$T(3), R(3)$

Let predicate "$CON(x, y)$" denote associated derived feature x constrains its real integral feature y, predicate "$TRA(x, m)$" denote m independent translations of real integral feature x, predicate "$ROT(x, n)$" denote n independent rotations of real integral feature x, the self-referenced feature variations can be represented in FOL as:

$$(\forall rif) \, (\exists adf) \, (CON(adf, rif) \land TRA(rif, m) \land ROT(rif, n)) \qquad (1)$$

where adf $\in$ {Point, Line, Plane}, rif $\in$ {Spherical, Cylindrical, Planar, Helical, Revolute, Prismatic, Complex}, and m, n $\in$ {1, 2, 3}.

Table 3. Twenty-seven basic cross-referenced feature variations. DOFs: degrees of freedom. $T(m)$: m independent translations. $R(n)$: n independent rotations. COI: Coincide. DIS: Disjoint. INC: Include. PAR: Parallel. PER: Perpendicular. INT: Intersect. NON: Nonuniplanar.

	Point2	Line2	Plane2
Point1	C1: *COI*, $T(3)$ C2: *DIS*, $T(3)$	C3: *INC*, $T(2)$ C4: *DIS*, $T(2)$	C5: *INC*, $T(1)$ C6: *DIS*, $T(1)$
Line1	C7: *INC*, $T(2)$ C8: *DIS*, $T(2)$ — — — — — — — — —	C9: *COI*, $T(2)$, $R(2)$ C10: *PAR*, $T(2)$, $R(2)$ C11: *PER*, $T(1)$, $R(1)$ C12: *INT*, $T(1)$, $R(1)$ C13: *NON*, $T(1)$, $R(1)$	C14: *INC*, $T(1)$, $R(1)$ C15: *PAR*, $T(1)$, $R(1)$ C16: *PER*, $R(2)$ C17: *INT*, $R(2)$ — — —
Plane1	C18: *INC*, $T(1)$ C19: *DIS*, $T(1)$ — — — — — —	C20: *INC*, $T(1)$, $R(1)$ C21: *PAR*, $T(1)$, $R(1)$ C22: *PER*, $R(2)$ C23: *INT*, $R(2)$	C24: *COI*, $T(1)$, $R(2)$ C25: *PAR*, $T(1)$, $R(2)$ C26: *PER*, $R(1)$ C27: *INT*, $R(1)$

Similarly, if we let M be one of the associated derived features of a part, N be another associated derived feature of the part, the numbers of the combinations called cross-referenced feature variations are forty-nine. To simplify tolerance design, let M, $N \in \{Point, Line, Plane\}$. Other complex situations can be decomposed into simple situations. Through decomposition, there remain twenty-seven basic cross-referenced feature variations which satisfy M, $N \in \{Point, Line, Plane\}$. Table 3 shows these twenty-seven basic cross-referenced feature variations.

Let predicate "$SR(x, y)$" denote associated derived feature x and associated derived feature y in the same part have spatial relation of SR, where $SR \in \{COI, DIS, INC, PAR, PER, INT, NON\}$, predicate "$TRA(x, m)$" denote m independent translations of associated derived feature x, predicate "$ROT(x, n)$" denote n independent rotations of associated derived feature x, then the basic cross-referenced feature variations can be represented in FOL as:

$$(\forall adf1)\,(\forall adf2)\,(SR(adf1, adf2) \land TRA(adf2, m) \land ROT(adf2, n)) \qquad (2)$$

3   Representation Model of Geometrical Tolerances

Based on variational geometric constraints theory, Jie et al. [4] classified the geometrical tolerances into self-referenced tolerances and cross-referenced tolerances, and gave the geometric feature variations which are specified by each geometric tolerance. From Reference [4], Expression (1) and Expression (2), the FOL representation model of geometric tolerances can be constructed as Table 4 shows.

4   Case Study

The following example [5, 6] shows how the representation model explains semantics of geometrical tolerances and specifies geometrical tolerance types. The parts drawing of the gear pump are shown in Fig. 1. The gear pump consists of three parts: the pump body Part A, the driving gear shaft Part B, and the driven gear shaft Part C. There are three pairs of matting relations: a mate between gear pairs $a1$ and $c1$, a mate between surfaces $a2$ and $b3$, and a mate between surfaces $b2$ and $c2$.

From Figure 1, we can obtain the feature pairs (x, y) (where x is a constraint feature and y is the constrained feature of x) as follows: $(b1\_adf, b2\_adf)$, $(b1\_adf, b3\_adf)$, $(b2\_adf, b3\_adf)$, $(a2\_adf, a1\_adf)$, $(c2\_adf, c1\_adf)$, $(a1\_adf, a1\_rif)$, $(c1\_adf, c1\_rif)$, $(a2\_adf, a2\_rif)$, $(b3\_adf, b3\_rif)$, $(b2\_adf, b2\_rif)$, and $(c2\_adf, c2\_rif)$, where "rif" is real integral feature, and "adf" is associated derived feature.

Table 4. FOL representation model of geometric tolerances. Profile: Profile-any-line, Profile-any-surface. Runout: Circular-run-out, Total-run-out.

	First Order Logic representation model
Self-referenced tolerances	
—	$((\forall planar)\,(\exists plane)\,((CON(plane, planar) \land TRA(planar, 1) \land ROT(planar, 2)) \to Straightness(plane, planar)))\ \Box\ ((\forall cylindrical)\,(\exists line)\,((CON(line, cylindrical) \land TRA(cylindrical, 2) \land ROT(cylindrical, 2)) \to Straightness(line, cylindrical))\ \Box\ ((\forall revolute)\,(\exists line)\,((CON(point, revolute) \land CON(line, revolute) \land TRA(revolute, 3) \land ROT(revolute, 2)) \to (Straightness(point, revolute) \land Straightness(line, revolute)))\ \Box\ ((\forall prismatic)\,(\exists line)\,((CON(line, prismatic) \land CON(plane, prismatic) \land TRA(prismatic, 2) \land ROT(prismatic, 3)) \to Straightness(line, prismatic) \land Straightness(plane, prismatic)))$
□	$(\forall planar)\,(\exists plane)\,((CON(plane, planar) \land TRA(planar, 1) \land ROT(planar, 2)) \to Flatness(plane, planar))$
○	$((\forall spherical)\,(\exists point)\,((CON(point, spherical) \land TRA(spherical, 3) \to Roundness(point, spherical)))\ \Box\ ((\forall cylindrical)\,(\exists line)\,((CON(line, cylindrical) \land TRA(cylindrical, 2)) \to Roundness(line, cylindrical))\ \Box\ ((\forall revolute)\,(\exists point)\,((CON(point, revolute) \land TRA(revolute, 3) \land ROT(revolute, 2)) \to (Roundness(point, revolute) \land Roundness(line, revolute)))$
⌔	$(\forall cylindrical)\,(\exists line)\,((CON(line, cylindrical) \land TRA(cylindrical, 2) \land ROT(cylindrical, 2)) \to Cylindricity(line, cylindrical))$
⌒	$((\forall revolute)\,(\exists point)\,(\exists line)\,((CON(point, revolute) \land TRA(revolute, 3) \land ROT(revolute, 2)) \to (Profile(point, revolute) \land Profile(line, revolute)))\ \Box\ ((\forall prismatic)\,(\exists plane)\,((CON(line, prismatic) \land CON(plane, prismatic) \land TRA(prismatic, 2) \land ROT(prismatic, 3)) \to (Profile(line, prismatic) \land Profile(plane, prismatic))\ \Box\ ((\forall complex)\,(\exists point)\,(\exists line)\,((CON(point, complex) \land CON(line, complex) \land TRA(complex, 3) \to ROT(complex, 3)) \to Profile(point, complex) \land Profile(line, complex)))$
Cross-referenced tolerances	
∥	$((\forall line)\,(\forall line2)\,(PAR(line1, line2) \land TRA(line2, 2) \to ROT(line2, 2) \to Parallelism(line1, line2)))\ \Box\ ((\forall line1)\,(\forall line2)\,(PAR(line1, line2)\ (\forall line2)\,(PAR(plane1, plane2), 1) \land TRA(line2, 1) \land ROT(line2, 1) \to Parallelism(line1, line2)))\ \Box\ ROT(line2, 1) \to {}^{2}Parallelism(plane1, line2))\ \Box\ ((\forall line1)\,(\forall plane2)\,((\forall plane2), 1) \land ROT(plane2, 1) \to Parallelism(line1, plane2)))\ \Box\ ((\forall plane1)\,(\forall plane2)\,(PAR(plane1, plane2) \to TRA(plane2, 2), 1) \to Parallelism(plane1, plane2))$
⊥	$((\forall line1)\,(\forall line2)\,(PER(plane1, line2) \land TRA(line2, 2) \to Perpendicularity(line1, line2)))\ \Box\ ((\forall line1)\,(\forall plane2)\,(PER(line1, plane2) \land ROT(plane2, 2) \to Perpendicularity(line1, plane2)))\ \Box\ ((\forall plane1)\,(\forall plane2)\,(PER(line1, plane2) \to Perpendicularity(plane1, plane2)))\ \Box\ ((\forall line1)\,(\forall line2)\,(PER(line1, line2), 2) \land ROT(line2, 2), 1) \to {}^{p}Perpendicularity(line1, line2)))$
∠	$((\forall plane1)\,(\forall line2)\,(INT(plane1, line2) \land ROT(line2, 2) \to Angularity(plane1, line2))\ \Box\ ((\forall line1)\,(\forall plane2)\,(INT(line1, plane2) \to Angularity(line1, plane2)))\ \Box\ ((\forall plane1)\,(\forall plane2)\,(INT(plane1, plane2) \to Angularity(plane1, plane2)))\ \Box\ ((\forall line1)\,(\forall line2)\,(INT(line1, line2) \land ROT(line2, 2))\ (\forall plane1, plane2), 1) \to Angularity(plane1, plane2), 1) \to Angularity(line1, line2))\ TRA(line2, 1) \land RO?(line2, 1) \to Angularity(line1, line2))$
⊕	$((\forall point1)\,(\forall line2)\,(INC(point1, line2) \land TRA(line2, 2) \to Position(point1, line2))\ \Box\ (\forall point1)\,(\forall line2)\,(DIS(point1, line2) \to Position(point1, line2))\ \Box\ ((\forall line1)\,(\forall point2)\,(INC(line1, point2) \land TRA(point2, 2) \to Position(point1, point2))\ \Box\ ((\forall line1)\,(\forall point2)\,(DIS(line1, point2) \land TRA(point2, 2) \to Position(line1, point2))\ \Box\ ((\forall point1)\,(\forall plane2)\,(INC(point1, plane2) \to Position(point1, point2))\ \Box\ ((\forall point1)\,(\forall plane2)\,(DIS(point1, plane2), 2) \to Position(point1, point2))\ \Box\ ((\forall plane1)\,(\forall point2)\,(INC(plane1, point2), 1) \to Position(plane1, point2))\ \Box\ ((\forall plane1)\,(\forall point2)\,(COL(plane1, point2) \to Position(plane1, point2))\ \Box\ ((\forall plane1)\,(\forall line2)\,(NON(line1, line2) \to Position(line1, line2))\ \Box\ (\forall line1)\,(\forall line2)\,(INC(line1, line2))\ \Box\ (\forall plane1)\,(\forall line2)\,(Position(plane1, line2))\ \Box\ ((\forall plane1)\,(\forall plane2)\,(PAR(plane1, line2) \to TRA(line2, 1) \land ROT(line2, 2) \to Position(line1, line2))\ \Box\ ((\forall line1)\,(\forall plane2)\,(PAR(line1, plane2) \land TRA(plane2, 1) \land ROT(plane2, 2) \to Position(line1, plane2))\ \Box\ ((\forall plane1)\,(\forall plane2)\,(PAR(plane1, plane2) \land TRA(plane2, 1) \to Position(plane1, plane2))$
◎	$(\forall point1)\,(\forall point2)\,(COI(point1, point2) \land TRA(point2, 3) \to Concentricity(point1, point2));\ (\forall line1)\,(\forall line2)\,(COI(line1, line2) \land TRA(line2, 2) \to ROT(line2, 2) \to Coaxiality(line1, line2)$
⫼	$((\forall plane1)\,(\forall plane2)\,COI(plane1, plane2) \land TRA(plane2, 1) \to Symmetry(plane1, plane2))\ \Box\ ((\forall plane1)\,(\forall plane2)\,(INC(plane1, plane2), 1) \land ROT(line2, 2) \to TRA(line2, 1) \land ROT(line2, 1) \to Symmetry(line1, plane2))$
↗	$((\forall line1)\,(\forall plane2)\,(PER(line1, plane2) \land ROT(plane2, 2) \to Runout(line1, plane2))\ \Box\ ((\forall line1)\,(\forall line2)\,(COI(line1, line2) \land TRA(line2, 2) \land ROT(line2, 2) \to Runout(line1, line2)))$

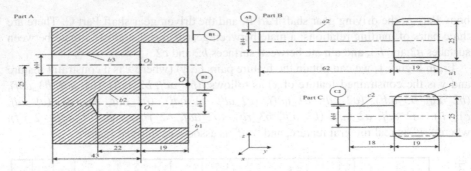

Fig. 1. Parts drawing of the gear pump

For ($b1\_adf$, $b2\_adf$), the associated derived feature of $b1$ is a line (denoted as $line1$), and the associated derived feature of $b2$ is also a line (denoted as $line2$). The spatial relation of $line1$ and $line2$ is **PER**, and $line1$ constrains the rotations about y-axis and z-axis of $line2$. Let predicate "$T(f, p)$" denote the translations along p-axis of associated derived feature f, predicate "$R(f, p)$" denote the rotations about p-axis of associated derived feature f, the above facts can be represented in FOL as "($\exists line1$) ($\exists line2$) (**PER**($line1$, $line2$) $\wedge$ R($line2$, y) $\wedge$ R($line2$, z))". According to the expression "(($\forall line1$) ($\forall line2$) (**PER**($line1$, $line2$) $\wedge$ **TRA**($line2$, 2) $\wedge$ **ROT**($line2$, 2) $\rightarrow$ $Perpendicularity$($line1$, $line2$)))" and the resolution principle in FOL, we have:

($\exists line1$) ($\exists line2$) (**PER**($line1$, $line2$) $\wedge$ R($line2$, y) $\wedge$ R($line2$, z)) $\rightarrow$ $Perpendicularity$($line1$, $line2$)) is satisfiable.

Thus the geometrical tolerance type specified by feature pair "($b1\_adf$, $b2\_adf$)" is perpendicularity tolerance. For the remaining feature pairs, similarly, we have:

($\exists b1\_adf$) ($\exists b3\_adf$) (**PER**($b1\_adf$, $b3\_adf$) $\wedge$ R($b3\_adf$, y) $\wedge$ R($b3\_adf$, z) $\rightarrow$ $Perpendicularity$($b1\_adf$, $b3\_adf$)) is satisfiable.

($\exists b2\_adf$) ($\exists b3\_adf$) (**PAR**($b2\_adf$, $b3\_adf$) $\wedge$ T($b3\_adf$, y) $\wedge$ T($b3\_adf$, z) $\rightarrow$ $Position$($b2\_adf$, $b3\_adf$)) is satisfiable.

($\exists a2\_adf$) ($\exists a1\_adf$) (**COI**($a2\_adf$, $a1\_adf$) $\wedge$ T($a1\_adf$, y) $\wedge$ T($a1\_adf$, z) $\rightarrow$ $Circular$-run-out($a2\_adf$, $a1\_adf$)) is satisfiable.

($\exists c2\_adf$) ($\exists c1\_adf$) (**COI**($c2\_adf$, $c1\_adf$) $\wedge$ T($c1\_adf$, y) $\wedge$ T($c1\_adf$, z) $\rightarrow$ $Circular$-run-out ($c2\_adf$, $c1\_adf$)) is satisfiable.

Table 5. Tolerance specifications of the gear pump

Feature pair	Tolerance type	Tolerance value
($b1\_adf$, $b2\_adf$)	$\perp$ (Perpendicularity)	Tol_1
($b1\_adf$, $b3\_adf$)	$\perp$ (Perpendicularity)	Tol_2
($b2\_adf$, $b3\_adf$)	$\oplus$ (Position)	Tol_3
($a2\_adf$, $a1\_adf$)	$\nearrow$ (Circular-run-out)	Tol_4
($c2\_adf$, $c1\_adf$)	$\nearrow$ (Circular-run-out)	Tol_5
($a2\_adf$, $a2\_rif$)	$\pm$ (Dimensional-tolerance)	Tol_6
($b2\_adf$, $b2\_rif$)	$\pm$ (Dimensional-tolerance)	Tol_7
($b3\_adf$, $b3\_rif$)	$\pm$ (Dimensional-tolerance)	Tol_8
($c2\_adf$, $c2\_rif$)	$\pm$ (Dimensional-tolerance)	Tol_9

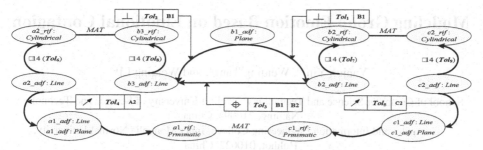

Fig. 2. Tolerance network of the gear pump

Through the above steps, the tolerance specifications are obtained (see Table 5). According to Table 4, the tolerance network of the gear pump is built (see Fig. 2).

5   Summary

This paper presents a representation model of geometrical tolerances based on FOL. With this model, the generation of geometrical tolerance types can be well implemented, and the semantics of geometrical tolerances can be well explained. One future work will focus on constructing the mathematical model of geometrical tolerances. Another work is to research tolerance analysis and tolerance synthesis based on representation model and mathematical model. Moreover, it is also a valuable work to develop a prototype system of computer aided tolerancing based on the above works.

Acknowledgements. The work is funded by the National Science Foundation of China (No. 61163041) and the foundation of Guangxi Key Lab of Trusted Software (No. kx201120).

References

1. Roy, U., Li, B.: Representation and interpretation of geometric tolerances for polyhedral objects–I. Form tolerances. Computer-Aided Design 30(2), 151–161 (1998)
2. Hong, Y.S., Chang, T.C.: A comprehensive review of tolerancing research. International Journal of Production Research 40(11), 2425–2459 (2002)
3. Srinivasan, V.: A geometrical product specification language based on a classification of symmetry groups. Computer-Aided Design 31(11), 659–668 (1999)
4. Hu, J., Xiong, G., Wu, Z.: A variational geometric constraints network for a tolerance types specification. The International Journal of Advanced Manufacturing Technology 24, 214–222 (2004)
5. Salomons, O.W., Haalboom, F.J., Jonge Poerink, H.J., Van Slooten, F., Van Houten, F.J.A.M., Kals, H.J.J.: A computer aided tolerancing tool II: tolerance analysis. Computers in Industry 31(2), 175–186 (1996)
6. Desrochers, A., Riviere, A.: A matrix approach to the representation of tolerance zones and clearances. The International Journal of Advanced Manufacturing Technology 13, 630–636 (1997)

Modeling Group Emotion Based on Emotional Contagion

Yanjun Yin[1,2], Weiqing Tang[3], and Weiqing Li[1]

[1] School of Computer Science and Technology, Nanjing University of Science and Technology,
Nanjing, 210094, China
[2] Computer and Information Engineering College, Inner Mongolia Normal University,
Hohhot, 010022, China
[3] Institute of Computing Technology, Chinese Academy of Sciences, Beijing, 100190, China
ciecyyj@163.com, tang@fulong.com.cn, li_weiqing@139.com

Abstract. Using computer to generate crowd animation to understand the behavior choice and making decision of individuals in crowd has become a trend in several fields. And group emotions have a great impact on group behaviors and group outcomes. Based on the researches of group emotions described by Hatifield, we propose a quantitative method to building the group emotion model which is focus on a group or a crowd, not on an individual. Our aim is to reflect more believable emotion experiences of individuals in social situation; the individuals' emotion is coming not only from the external stimulus, but from others group members through the conscious and unconscious induction of emotion states as well. For the emotional contagion plays a significant role in the development of group emotion, we select personality, emotional expressivity and susceptibility as mainly factors which influence the intensity of group emotions. Simulation is done by using Netlogo software, and the results show that the model is available and embody the fundamental characteristics of group emotion, and virtual individuals in crowd can generate credible emotional experience and response.

Keywords: Group Emotion, Emotional Contagion, Emotion Model, Personality.

1   Introduction

Emotion is an essential part of human life; it not only influences how we think, adapt, learn, behave and how communicate with others, but impacts other people in same situation through emotional contagion as well. Psychologists pointed out that emotional influence on others is completed by emotional contagion. In social interaction, an individual will unconsciously imitate other people's emotional expressions, experience those emotions and then affect his emotions. The process is called as emotional contagion [1]. The essence of emotional contagion is exchange and transfer of emotion which influences others' emotion and further impact their cognition and affective attitude toward the environment which they are facing.

In order to better embody the emotional experience of individuals in social situation, the paper adopts the emotional contagion as the main mechanism of individuals'

Z. Shi, D. Leake, and S. Vadera (Eds.): IIP 2012, IFIP AICT 385, pp. 240–247, 2012.

emotion interaction, and proposes a new method to model the group emotion based on the 'bottom-up' approach which is discussed by Barsade and Gibson [5].

2   Relative Works

Group emotion can be dated back to the study of mass movements in 19th century. Gustave Le Bon[2] pointed out that crowd emotion was induced by collective mind in his book 'The Crowd: A Study of the Popular Mind', and he explained further that collective mind would emerge in a crowd of people and influence group behavior which would not predicted by simply studying an individual. Modern views proposed that crowd emotion is the functional reaction to situations or events related to individuals. Sociological researches showed the relative deprivation is the most important root causing the negative emotion of the individuals in mass events; it believed that the individuals in crowd will felt discontent when they compare their positions to others and realize that they are lower than others. Smith et al[3] described that members of the group assessed the sense of unfairness as a group events rather than an individual event, and the group emotion would be formed in the evaluation process of individuals who should responsible for the unfair and unequal situation. Yingxin[4] viewed the crowd emotion as 'gas field' in study of mass event; he pointed out that the 'gas field' is a kind of special emotional atmosphere formed by mass who vented their discontents. Although the concept of 'gas field' has its limitations, from the perspective of Chinese traditional culture, it interpreted the procession of forming and development of group emotion. Sigal G. Barsade and Donald E.Gibson [5] offered a brief summary of prominent of research on group emotion and suggested that, from a top-down perspective, group emotion has been characterized as powerful forces which dramatically shape individual emotional response, as social norms prescribing feeling and expression, as the interpersonal glue that keeps groups together and as a window to viewing a group's maturity and development. However, from a bottom-up perspective, they proposed that group emotion can be viewed as the sum of its part of individuals' emotion. There are many computational emotion models recently. However, most of these models focus on the individual, no on the group. Rob Duell et al [6] provided firstly the computational group emotion based on the emotional contagion and the work of Barsade and Gibson. In their model [6,7], the emotion level of a special group was defined as the sum of the product of the emotion intensity of group members and an relevance factor respectively, but the relevance factor did not give any explanation on how to calculate it. In the research on the virtual spectators in game, Xiejun[8] provided the level of neighboring emotion of a special individual as the mean emotion of all of his neighbors. Zhangxue and cheng'an[9] provided a way to calculate the group angry based on network review of news. In the way, many factors: index of angry, group correlation and the level of event influence, have been involved, but in fact the group emotion is defined as the weight sum of individual s' emotion.

Based on above theories and technologies, we provided a new way to calculate the group emotion level using emotional contagion. In our method, we define individual's personality, emotional expressivity and susceptibility as main factors to influence the intensity of group emotion.

3   Main Factors of Group Emotion

3.1   Personality

Emotional contagion, proposed by Hatfield et al [1], is defined as the tendency to automatically mimic and synchronize facial expressions, vocalizations, postures, and movements with those of another person and, consequently, to converge emotionally. In social situation, individuals are easier to be infected by others' emotion. However, there are obviously different in the degree of which individuals transmit their emotions or catch others' emotions [10]. From emotional contagion perspective, Verbeke [11] developed a classification method of personality. Based on the ability to infect and the capability to be infected, individuals are divided into four different classifications: charismatic, empathetic, expansive and bland. In general, most descriptions of personality emphasize the distinctive quality of individual. Most widely accepted models of personality have three-dimensional personality model (PEN) and Five Factor model (FFM). In the paper, the definition of personality is based on the work of Verbeke and described as two-dimensional vector, where each dimension is represented by a personality factor. The distribution of the personality factors is modeled by a normal distribution function N with mean μ_i and standard σ_i :

$$personality =< \lambda_{Ex}, \lambda_{Es} > \tag{1}$$

$$\lambda_i = N(\mu_i, \sigma_i^2), \quad \mu_i \in [0,1] \quad for \quad i \in \{Ex, Es\} \tag{2}$$

Where Ex (Emotional expressivity) represents the ability to infect others' emotion through emotional expression, and Es (Emotional susceptibility) denotes the capability to be infected by others through catching others' emotion clue and understanding others' inner feeling. As an example, an expansive individual that has the high ability to infect and low capability to be infected is represented as $personality=<0.8, 0.3>$. However, an empathetic individual that has the low ability to infect and high capability to be infected is represented as $personality=<0.1, 0.9>$, an charismatic individual as $personality=<0.7,0.9>$, a bland individual as $personality=<0.1, 0.2>$.

4   Group Emotion Model

4.1   Emotion Intensity of Individual in Crowd

In psychology, emotion is often defined as a complex state of feeling that results in physical and psychological changes that influence thoughts and behaviors. However, there is a clear difference in the extent. When an individual is alone, his emotion is elicited by external stimulus. The emotional experience of the individual is only response to his cognition and the attitude toward what happened to him, and the emotional expression is the nature revelation of his inner feelings. So in most of computational emotion model, emotion is defines as a function of the individual's personality and external stimulus. However, sociological researches showed that the emotion of an individual in crowd will be influenced by other people. From the emotional contagion perspective, the process of emotional contagion, in which a group member influences the emotion of anther group member and vice versa, through the conscious and unconscious induction of emotion states [12], is primary mechanism through which individual emotions create a group emotion. And the social comparison theory also gave the interpretations about the difference. Individuals tend to compare their behavior and attitude with others that are most like them. In the process of comparison, individuals coordinate his behaviors, attitudes and emotions with those of others people.

Based on these analyses and our previous work [13], we defined the emotion of individual in crowd as a time variation function in which the effect of emotional contagion is considered. For a given individual A, at time step t+1, the intensity of his/her emotion is $I_a(t+1)$, which is

$$I_a(t+1) = I_a(t) + \beta * IC_a(t) + IS_a(t)$$ (3)

Where $I_a(t)$ is the emotion intensity of the individual A in the last time step.$\beta$is an adaptable factor and is set in line with emotional types and environments. $IS_a(t)$ is the level of individual emotion elicited by external stimulus at the time t. if there is no external stimulus at the time t, $IS_a(t)$ is zero. $IC_a(t)$ denotes the intensity emotion caused by others' emotional contagion. It is defined as follow.

$$IC_a(t+1) = \lambda_{Es}(a) * [\omega * I_{special}(t) + (1-\omega) * I_{goupimpact}(t)]$$ (4)

$$I_{special}(t) = \max\{ abs[\lambda_{Ex}(s) * (I_a - I_s(t))] * \tau \mid s \in G \setminus \{a\}\}$$ (5)

Where $\lambda_{Ex}(a)$ represents the emotional susceptibility of individual A, ω the degree of attainment for special individual. Ispecial (t) is the emotional influence of special individual on the individual A, and it is defined based on the absorption model described by Bosse et al [7]. $\lambda_{Es}(s)$ represents the emotional expressivity of individual S. I_a and I_s are the level of emotion of individual A and S. τ is a moderation factor representing the strength of the channel from individual A to S. $I_{groupimpact}(t)$ is the emotion intensity of individual caused by group emotion at the time t, it will be described in detail in next section.

Emotion does not immediately disappear with external stimulus' end, but it is bound to undergo decay, which is a natural depression or a gradually decline because of mental satisfaction or emotional catharsis. The differences of the decay degree of emotion are showed on individuals' personality. In general, the emotional decay of individual with the high ability to express his emotions is faster than that of whom with low ability. And an individual tends to forget the positive emotions more quickly than the negative emotions. In social situation, the emotional decay is influenced by group size. Researcher pointed out that the emotional delay of individuals in the group with large size is lower than one with smaller size. Based on the researching in psychology, we propose a decay function as follow:

$$I_a(t+1) = I_a(t) * [1 - \exp(\frac{-\mu * groupSize}{\lambda_{Ex}(a) * timeStep})]$$ (6)

Where μ is a moderation factor representing different decay speeds for different emotions. For example, its value is smaller when individual experience positive emotion than negative emotion. $\lambda_{Ex}(a)$ is the emotional expressivity of individual A. *timeStep* represents the number of emotional contagion times. *groupSize* is the number of individual in a group.

4.2   Group Emotion

Emotion is an inherent part of mass events. Individuals bring their unique emotional tendencies to the group. The group emotion has been viewed as moving upward from the compositional effects of individual group member emotions. The dynamic of group emotion will be in line with the individuals' emotion. That is, the intensity of group emotion should be related with the 'expected value' of individuals. In the paper, we defined group emotion as follow.

$$I_{group}(t) = \sum_{a \in G} \frac{\lambda_{Ex}(a)}{\sum_{s \in G} \lambda_{Ex}(s)} * I_a(t) \tag{7}$$

Where λ_{Ex} is the emotional expressivity of individuals. $I_{group}(t)$ is represent the emotion intensity of group at time step t. The emotion intensity of individual caused by group emotion is defined as follow:

$$I_{groupimpac\ t}(t) = \frac{\sum_{a \in G} \lambda_{Ex}(a)}{groupSize} * I_{group}(t) - \rho * I_a(t) \tag{8}$$

Where ρ is the weight coefficient representing different influence on individual from different group characteristics, for example group size and group type.

5   Experiences and Result Analysis

A group emotion model based on emotional contagion is proposed in the paper. Our aim is to reflect more believable behavior and emotion of individual in social situation. A large number of simulations have been performed to test the model, using simulation software. In the section, some of simulation results are discussed.

Firstly, we have fixed the emotion levels and personalities of individuals in crowd to explore the effect of group size. The value of group size is from 3 to 100. The emotion intensities are random numbers following a normal distribution with mean 0.3 and standard deviation 0.1, and group members are charismatic with *personality*=<0.9,0.9>. As expected, group size plays a significant role in emotional contagion, and it influences the level of group emotion. The group emotion is easier to outburst in large group than in small group. Although the decaying speed of positive emotion is faster than that of negative emotion, emotional explosion of group is likely to occur if the group is large enough (Fig.1). And in small group, it is difficult to increase the emotion intensity of individuals, especially when the individuals have positive emotions (Fig 2). The results show that the emotion dynamic of group based on the model provided by the paper is in line with the emotional response of human crowd.

Secondly, we have fixed group size at value 20, and studied the effect of the personalities of individuals in group. We set four groups representing four group types: charismatic, empathetic, expansive and bland respectively. The emotion intensities of individuals in these groups were generated randomly following a normal distribute ($\mu=0.5, \theta=0.1$). From the result (Fig.3), we can see the Charismatic group is easier to form emotion climate than others group, especially for negative emotion. And the bland group has low emotion expressivity and susceptibility, so the void of emotional communication and understanding make the bland group difficult to form collective mind.

At last, this model was implemented in the NetLogo[14] environment to simulate the panic crowd. We show an example of the fleeing behavior triggered when a danger is encountered. A fire is treated as a danger to individuals' health. When individuals saw the fire, they would sense the danger and fear/panic is elicited. In our model, the emotion intensity of these individuals is generated random following the normal distribution. The fear/panic will be propagated in the crowd very quickly. All infected individuals (in blue) take the group mind of escaping from the source of danger and performing the flee behavior. There are many individuals who did not saw the danger, their flee behaviors were made based on the information from others emotional expression, for example facial expressions, vocalizations, postures and behavior (Fig.4, Fig.5).

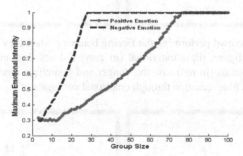

Fig. 1. The maximum emotional intensity of different group sizes. The maximum emotional intensity is the maximum value of group emotion intensity in the whole variable process.

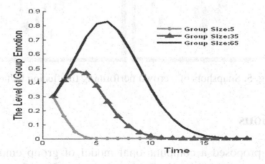

Fig. 2. The change of emotion intensity of groups with different participants

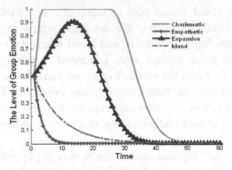

Fig. 3. The change of emotional intensity of groups which have different emotional expressivity and emotional susceptibility

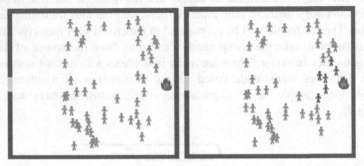

Fig. 4. Snapshots of a crowd performing the fleeing behavior when they feel the existence of a danger. In the left subfigure, these individual (in grey) did not find any danger. In the right subfigure, some individuals (in red) saw the danger and panic/fear emotion was elicited, and they infected others (in blue) emotion through emotional contagion.

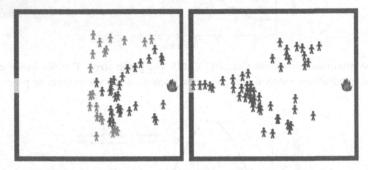

Fig. 5. Snapshots of a crowd performing the fleeing behavior

6   Conclusions

In the paper, we proposed a computational model of group emotion, based on the emotional contagion provided by Hatfield. A large number of experiences have been done. And the simulation results show that the group emotion model is reasonable and it can make the virtual individuals in crowd more believable. For the crowd simulation, the model has some referential value. In a future work, we are interesting to investigate the relationship between emotion and behavior choice of individuals in crowd.

References

1. Hatfield, E., Cacioppo, J.T., Rapson, R.L.: Emotional contagion. Current Directions in Psychological Science 2, 96–99 (1993)
2. Bon, G.L.: The Crowd: A Study of the Popular Mind. Goungxi Normal University Press (2007) (in Chinese)
3. Smith, E.R.: Social identity and social emotions: toward new conceptualizations of prejudice. Affect, Cognition and Stereotyping, 297–315 (1993)
4. Ying, X.: 'Gas Field' and the Occurring Mechanism of Mass Events—the comparison of two cases. Sociological Research 6, 1–9 (2009)
5. Barsade, S.G., Gibson, D.E.: Group Emotion: A View from Top and Bottom. Research on Managing Groups and Teams 1, 81–102 (1998)
6. Duell, R., Memon, Z.A., Treur, J., van der Wal, C.N.: An ambient agent model for group emotion support. In: Affective Computing and Intelligent Interaction and Workshops. ACII 2009, pp. 1–8 (2009)
7. Bosse, T., Duell, R., Memon, Z.A., Treur, J., van der Wal, C.N.: A Multi-agent Model for Emotion Contagion Spirals Integrated within a Supporting Ambient Agent Model. In: Yang, J.-J., Yokoo, M., Ito, T., Jin, Z., Scerri, P. (eds.) PRIMA 2009. LNCS, vol. 5925, pp. 48–67. Springer, Heidelberg (2009)
8. Jun, X.: Research of Effect Model of Spectators in Virtual Game. Nanjing University of Science and Technology. Master dissertation (2011) (in Chinese)
9. Xue, Z., An, C.: The Monitoring and Measurement of Mood of Anger Based on Network Review of News. Science & Technology for Development 09, 44–49 (2010) (in Chinese)
10. Barsade, S.G.: The Ripple Effect: Emotional Contagion and Its Influence on Group Behaior. Administrative Science Quarterly 47, 644–675 (2002)
11. Verbeke, W.: Individual differences in emotional contagion of salespersons: Its effect on performance and burnout. Psychology and Marketing 14(6), 617–636 (1997)
12. Schoenewolf, G.: Emotional Contagion: Behavioral induction in individuals and groups. Modern Psycho-analysis 15, 49–61 (1990)
13. Yin, Y., Li, W., Tang, W.: Spectator's Emotion Modeling Based on Emotional Contagion. In: The International Conference on Automatic Control and Artificial Intelligence (ACAI 2012), vol. 3, pp. 2256–2260 (2012)
14. NetLogo homepage. Center for Connected Learning and Computer-Based Modeling, Northwestern University, http://ccl.northwestern.edu/netlogo/

Hierarchical Overlapping Community Discovery Algorithm Based on Node Purity

Guoyong Cai, Ruili Wang, and Guobin Liu

Guilin University of Electronic Technology, Guilin, Guangxi, China
ccgycai@guet.edu.cn, wangruili1207.1@163.com

Abstract. A hierarchical overlapping community discovery algorithm based on node purity (OCFN-PN) is proposed in the paper. This algorithm chooses the maximal relative centrality as the initial community, which solves the problem of inconsistent results of the community discovery algorithm based on fitness resulting from randomly choosing nodes. Before optimizing and merging communities, the community overlapping degree and the joint-union should be calculated so that the problems of twice merging can be solved. Research results show that this algorithm has lower time complexity and the communities obtained by this algorithm are more suitable for real world networks.

Keywords: hierarchical overlapping community discovery algorithm, node purity, relative centrality, overlapping degree, the joint-union.

1 Introduction

Community structure is a key characteristic of many networks, that is, nodes tend to aggregate into communities and nodes within communities are tightly connected while nodes between communities are loosely connected. Recently most of community discovery algorithms transform community discovery problem into hierarchical segmentation problem[1-5,7,9,10-16], which assumes that individuals of networks only belong to one community and communities of a network are isolated groups, such as the classical GN algorithm[1], Kernighan-Lin algorithm[16], and Fast Newman algorithm[2]. These isolated groups become the subgroups of larger groups until all groups become subgroups of a group, thus, hierarchical structure of the whole network is formed. However, this assumption is only suitable for some networks, for instance, the organization system network or taxonomy networks, but not suitable for most of real world networks. Many researches show that social networks not only have hierarchical structure but also have overlapping communities, in other word, a community is not the subcommunity of another community and its individuals usually belong to many different communities.

Take Facebook, one of online social network sites, as an example, every Facebook user has average 130 friends and these friends may belong to different social groups, such as high school, university, and family[6]. Marlow et al. find that groups appearing in the ego network are corresponding to acquaintances circles of different life stages[8]. An overlapping community discovery algorithm is firstly proposed by Palla et al. in

Z. Shi, D. Leake, and S. Vadera (Eds.): IIP 2012, IFIP AICT 385, pp. 248–257, 2012.

2005, and after that there are many other overlapping community discovery algorithms be proposed[10-15]. However, all of these algorithms have certain limitations, for instance, the inconsistent results problem caused by choosing node randomly and the twice merging problem. In order to deal with these problems, a new hierarchical overlapping community discovery algorithm is presented in this paper, which not only can discover hierarchical structure and high overlapping communities but also has lower time complexity.

2   Problems Description

- Inconsistent results problem

Recently many methods about how to select the initial community are proposed. Assuming that every node belongs to at least one community, Lancichinetti et al. use the way of randomly choosing nodes which have assigned to none of communities as the initial community in the fitness algorithm and the ending condition is that each node belongs to at least one community[14]. Baumes et al. randomly chooses edge as the initial community[15]. Because of the difference of the initial community, the final obtained communities may be different.

- Twice merging problem

The merging process of overlapping community discovery algorithms has the repeating merging problem, for example, there are three communities C1, C2, and C3, in which C1 , C3 and C2, C3 cannot be merged, while community A is merged by C1 and C2, and then community A can be also merged with C3.

- One-sideness community discovery

Complex network researchers find that most social networks not only have hierarchical structure but also may be overlaps between communities in the same level. Many of the existing community discovery algorithms define a community as a k-clique, and use the clique filter algorithm to discover the communities, but, this method cannot discover the hierarchical structure of networks. In order to solve this problem, a fitness algorithm is proposed, which can both find the hierarchical structure of networks and the overlapping communities. However, the fitness algorithm may lead to inconsistent results problem.

3   OCFN-PN Algorithm

3.1   Basic Concepts

Node relative centrality equals to node absolute centrality divides the maximum possible degree of network nodes. Besides, the node having the largest relative centrality is called a core node.

$$f_C = \frac{k_{in}^C}{(k_{in}^C + k_{out}^C)^\partial}$$

Community fitness is defined by equation   , in which k_{in}^C and k_{out}^C are inter-degree and outer-degree separately, k_{in}^C equals to the double of the number of edges whose two endpoints are both in community C, k_{out}^{c} equals to the number of edges that only one endpoint in community C, ∂ is a adjustable parameter and its value directly controls the size of communities and the hierarchical structure of networks, moreover, the bigger the value of ∂, the less the size of the community.

Node purity is defined by equation $P = F_2 / F_1$ in which F1 and F2 are the fitness of community C1 and C1*(C1* is the new community formed after a new node A join into community C1), and $F_1 = f_{C_1}$, $F_2 = f_{C_1^*}$. If P>1, the new node A will be available, and vice versa. Thus, the value of node purity can be used to determine whether put the new node into community or not.

The existing overlapping degree calculating methods only consider the number of overlapping nodes but ignore the degree of overlapping nodes which is an important factor to judge how much the degree of overlapping communities is. For instance, in Fig.1.a and Fig.1.b the number of overlapping nodes of the two communities is both 3, but the degree of overlapping nodes of Fig.1.b is obviously bigger than that of Fig.1.a. In the existing overlapping degree calculating methods Fig.1.a is regarded as Fig.1.b, which would lead to the same overlapping degree even Fig.1.a is quite different with Fig.1.b.

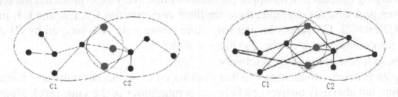

Fig. 1. a (left) and 1.b (right) two cases of overlapping communities

In order to settle this problem, a new overlapping degree calculating method is proposed in this paper. This formula contains the direct overlapping nodes, the indirect overlapping nodes, and another affecting the overlapping degree factor, which is the overlapping node degree.

Let the common nodes of community C1 and C2 be the direct overlapping and the common neighborhood nodes of community C1 and C2 be the indirect overlapping, so the overlapping degree formula can be described as

$$Doverlap(C_i, C_j) = \beta * \frac{|C_i \cap C_j| + \sum_k d(v_k^{ovlp})}{|C_i \cup C_j| + \sum_k d(v_k^{ovlp})} + (1 - \beta) * \frac{|N(C_i) \cap N(C_j)|}{|N(C_i) \cup N(C_j)|} \quad (1)$$

in which $|C_i \cap C_j|$ is the number of overlapping nodes, $\sum_k d(v_k^{ovlap})$ is the total degree of overlapping nodes, $|C_i \cup C_j|$ is the total number of nodes of community Ci and Cj, $|N(C_i) \cap N(C_j)|$ is the common neighborhood nodes of community Ci and Cj,

$|N(C_i) \cup N(C_j)|$ is the total number of neighborhood nodes of community Ci and Cj, β is an adjustable parameter whose value can control the ratio of direct overlapping. The bigger the value of β, the smaller the proportion of direct overlapping, and vice versa. Usually let the value of β be bigger than 0.5 and smaller than 0.8.

$$Q = \frac{|C_i \cap C_j|}{\min(C_i, C_j)}$$

Joint-union is defined by equation $\frac{|C_i \cap C_j|}{\min(C_i, C_j)}$, in which $\min(C_i, C_j)$ means the number of nodes of the community whose nodes are smaller. In this paper if Q of community Ci and Cj is bigger than K (after many test find that it is better when K is 0.7358), the twice merging problem can be solved.

3.2   Basic Idea

Choosing the node with biggest relative centrality as the initial community, using node purity to determine whether put the new node into the initial community or not, and using the fitness algorithm to find the communities of networks. Because the parameter ∂ of the fitness calculating formula and one node is traveled more than one time, the hierarchical structure and the overlapping communities of the given network can be discovered. Besides there is overlap between communities so the discovered communities need to be merged. Before merging communities, the overlapping degree and the joint-union should be calculated. If the overlapping degree is bigger than K and the joint-union is bigger than γ, the two communities should be merged.

OCFN-PN Algorithm

```
Input: network graph G(V, E), in which V is the set of
nodes, E is the set of edges.

Output: the set of communities{ C1,…,Cm }

1       For i=1 To n-1

2            Calculating the degree of every node in network;

3            Using the relative centrality formula calcu-
lates the relative centrality of every node;

4       Using bubble sort algorithm orders the relative
centrality from the biggest to the smallest and stores
the result in list DL;

5       Choosing the first data of list DL-the core node
as the initial community C1 ;
```

```
6      For i=1 To the number of neighborhood nodes of C1
subtracting 1

7         Using fitness calculating formula calculates
all of the fitness of neighborhood nodes of C1 and stores
the results in list FL;

8         Choosing the first data of list FL as a candidate
node;

9         Calculating the purity P of this candidate node;

10     If P> 1

11        Let this candidate node join into C1 and forms
a new community C1*;

12        The number of community C1* pulses 1;

13    else

14        Regarding this candidate node as an una-
vailable node;

15    Calculating the relative centrality of the re-
maining nodes and repeating step 4 to step 13;

16    Until all nodes in network belong to at least one
community;

17    Return {C1,…,Cm} and the number of every community
```

Optimizing and Merging Algorithm

```
Input: the communities obtained by OCFN-PN algorithm

Output: the set of communities {C1,…,Cl}, (1≤m)

1    For i=1 To m

2         Using bubble sort algorithm orders the relative
centrality from the biggest to the smallest and stores
the result in list NL;

3         Choosing the first data in NL;

4         Using overlapping degree calculating formula
calculates the overlapping degree OD between this
community and the other communities;
```

```
5          If OD> 0

6              Calculating the joint-union Q between the
two communities;

7              If Q >K And OD>Υ

8                  Merging these two communities;

9        Choosing the second community in list NL and
repeat step 4 to 8;

10     Repeat 2 to 9 ;

11     Until no community can merge anymore;

12     Return {C1,…,Cl}
```

3.3 Time Complexity

The time complexity of OCFN-PN algorithm is O(n2) and the time complexity of optimizing and merging algorithm is O(m2) separately, so the time complexity of the hierarchical overlapping community discovery algorithm is O(n2).

However, the time complexity of this algorithm is a little bigger than that of the algorithm based on fitness, the reasons are as follow:

1. In order to avoid the problem of inconsistent results, we choose the node with the biggest relative centrality as the initial community.
2. In order to obtain better nature communities, we calculate every node's purity to determine whether the candidate node is available or not.
3. In order to obtain more reasonable communities, we consider the number of overlapping nodes and their degree during the optimization and merge algorithm.

4   The Experimental Result and Its Analysis

The experimental data in this paper comes from Zachary's karate club, the classical data set of social network. This network is described by sociologist Zachary, who use two years to observe the members and social friendships of this club. This network is an undirected graph containing 34 nodes and 78 edges, see Fig.2.

Because Zachary's karate club itself is an overlapping community, the algorithm proposed in this paper can be used. Tab.1 describes the different community structure of Zachary's karate club obtained by this algorithm when the value of ∂ is different and Fig.3 and Fig.4 show the hierarchical structure of Zachary's karate club network when ∂ =1.7.

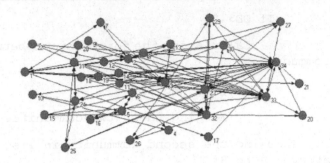

Fig. 2. Zachary's karate club network

Table 1. The communities of Zachary's karate club obtain by OCFN-PN algorithm

∂	β	γ	the communities of Zachary's karate club
0.3	0.7	0.6	{34,9,31,33,15,16,19,21,23,30,24,27,28,10,3,29,32,25,26 ,14,9,13,11,22,18,8,1,20,5,6,7,12,2,17,4}
0.7	0.7	0.6	{34,9,31,33,15,16,19,21,23,30,24,27,28,10,3,29,32,25,26 ,14},{1,2,18,22,20,8,4,14,3,13,12,10,9,31,5,6,12,7,11,17}
1	0.7	0.6	{34,9,31,33,15,16,19,21,23,9,3,30,24,27,28,10,14}, {1,2,18,22,20.31,8,4,14,3,13,12,10,9,31,5,6 }, {12,7,11,17,6,3,29,9,20,32,25,26,14}
1.3	0.7	0.6	{34,9,31,33,15,16,19,21, 9,23,30,27,24,28,10}, {1,2,18,22,20,8,4,14,3,13,12,10}{7,1,5,11,6,17,12,13,18, 22}, {26,24,25,28,32,29)
1.5	0.7	0.6	{34,9,31,33,15,16,19,21, 9,23,30,27,24,28,3,10}, {1,2,18,22,20,8,4,14,3,13,9,12,10},{7,5,11,6,17}, {26,24,25,28,32,29)
1.7	0.7	0.6	{34,23,15,19,31,30,21,31, 9,29,10,16,27,33,10}, {26,24,25,28,32,29),{1,2,13,3,4,18,8,20,9,14,13,22}, {3,14,9,20,10,8,31,29},{7, 5,11,17,6 },{12}

From Tab.1, when ∂ =0.3 all nodes of the network form a community, which makes no sense to communities discovery. And when ∂ =1.7 this algorithm generates six communities, one of which only contains one node. So the value of ∂ is usually bigger than 0.6 but smaller than 1.5.

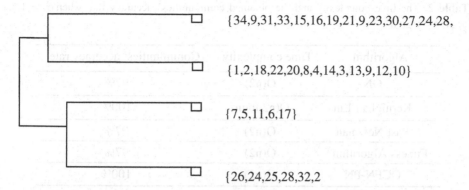

{34,9,31,33,15,16,19,21,9,23,30,27,24,28,

{1,2,18,22,20,8,4,14,3,13,9,12,10}

{7,5,11,6,17}

{26,24,25,28,32,2

Fig. 3. The hierarchical structure of Zachary's karate club network when ∂ is 1.5

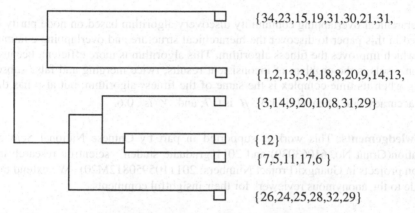

{34,23,15,19,31,30,21,31,

{1,2,13,3,4,18,8,20,9,14,13,

{3,14,9,20,10,8,31,29}

{12}

{7,5,11,17,6 }

{26,24,25,28,32,29}

Fig. 4. The hierarchical structure of Zachary's karate club network when ∂ is 1.7

From Tab.1, Fig.3, and Fig.4, we can conclude that different ∂, β and γ lead to different communities and different hierarchical structure. The adjustable parameter ∂ and β have be introduced above, so here we mainly introduce parameter γ. The function of parameter γ is the same as that of parameter β which is used to determine whether merge communities or not. γ =0 means the nodes of two communities are the same, which makes no sense to the optimizing and merging algorithm. γ =1 means the communities do not need to be merged.

From comparing the time complexity and the obtained communities' accuracy rate of several classical algorithms(see Tab.2), the communities obtained by OCFN-PN algorithm is better for Zachary's karate club.

Table 2. The time complexity and the obtained communities' accuracy rate when ∂ = 1.5, β =0.7, γ =0.6

Algorithm	Time complexity	Communities' accuracy rate
GN	O(n2)	97%
Kernighan-Lin	O(n2 logn)	100%
Fast Newman	O(n2)	97%
Fitness Algorithm	O(n2)	97%
OCFN-PN	O(n2)	100%

5   Conclusions

The hierarchical overlapping community discovery algorithm based on node purity is proposed in this paper to discover the hierarchical structure and overlapping communities, which improves the fitness algorithm. This algorithm is more efficient, because it not only solves the problem of inconsistent results, twice merging and false subset merging when its time complex is the same of the fitness algorithm, but also has the higher accuracy rate when ∂ is 1.5, β is 0.7, and γ is 0.6.

Acknowledgements. This work is supported in part by Chinese National Science Foundation(Grant No. 61063039) and 2011 graduate student   scientific research innovation projects in Guangxi(Project Numbers: 2011105950812M20) . We extend our gratitude to the anonymous reviewers for their insightful comments.

References

1. Girvan, M., Newman, M.E.J.: Community structure in social and biological networks. Proc. Natl. Acad. Sci. USA 99, 7821–7826 (2002)
2. Newman, M.E.J.: Fast algorithm for detecting community structure in networks. Phys. Rev. E 69, 066133 (2004)
3. Palla, G., Derényi, I., Farkas, I., Vicsek, T.: Uncovering the overlapping community structure of complex networks in nature and society. Nature 435, 814–818 (2005)
4. Hofman, J.M., Wiggins, C.H.: Bayesian approach to network modularity. Phys. Rev. Lett. 100, 25870, 409–418 (2008)
5. Clauset, A., Newman, M.E.J., Moore, C.: Finding community structure in very large networks. Physical Review E 70, 066111 (2004)
6. Facebook Press Room, Facebook statistics (February 2009),
   http://www.facebook.com/press/info.php.statistics
7. Duch, J., Arenas, A.: Community detection in complex networks using extremal optimization. Physical Review E 72, 027104 (2005)

8. Marlow, C., Byron, L., Lento, T., Rosenn, I.: Maintained relationships on facebook (March 2009), http://overstated.net/2009/03/09/maintained-relationships-on-facebook

9. Girvan, M., Newman, M.E.J.: Community structure in social and biological networks. PNAS 99(12), 7821–7826 (2002)

10. Clauset, A.: Finding local community structure in networks. Phys. Rev. E 72(2), 26132–26137 (2005)

11. Gregory, S.: An Algorithm to Find Overlapping Community Structure in Networks. In: Kok, J.N., Koronacki, J., Lopez de Mantaras, R., Matwin, S., Mladenič, D., Skowron, A. (eds.) PKDD 2007. LNCS (LNAI), vol. 4702, pp. 91–102. Springer, Heidelberg (2007)

12. Gregory, S.: Finding Overlapping Communities Using Disjoint Community Detection Algorithms. In: Menezes, R., et al. (eds.) Complex Networks: CompleNet 2009. SCI, vol. 207, pp. 47–61. Springer, Heidelberg (2009)

13. Mishra, N., Schreiber, R., Stanton, I., Tarjan, R.E.: Clustering Social Networks. In: Bonato, A., Chung, F.R.K. (eds.) WAW 2007. LNCS, vol. 4863, pp. 56–67. Springer, Heidelberg (2007)

14. Lancichinetti, A., Fortunato, S., Kertesz, J.: Detecting the overlapping and hierarchical community structure of complex networks. New Journal of Physics 11, 033015 (2009)

15. Baumes, J., Goldberg, M., Krishnamoorthy, M., MagdonIsmail, M., Preston, N.: Finding communities by clustering a graph into overlapping subgraphs. In: International Conference on Applied Computing, IADIS 2005 (2005)

16. Radicchi, F., Castellano, C.: Defining and Identifying Communities in Networks. PNAS 101(9), 2658–2663 (2004)

Finding Topic-Related Tweets Using Conversational Thread

Peng Cao[1,2], Shenghua Liu[1], Jinhua Gao[1,2], Huawei Shen[1],
Jingyuan Li[1], Yue Liu[1], and Xueqi Cheng[1]

[1] Institute of Computing Technology, Chinese Academy of Sciences, Beijing, 100190
[2] Graduate School of Chinese Academy of Sciences, Beijing, 100190

Abstract. Microblog has gained more and more users around the world, the popularity of which makes information spreading in microblog the most important and influential activities on the Internet. Therefore, search in microblog is of the most significant issue for both academic and industrial world. Search in webpages has been studied for several decades, but as for microblog it is still an open and brand new question for everyone. Search in microblog is more difficult than that in traditional webpages because of the sparseness of the messages. Search functions in current microblogging services simply match microblog messages with query words, which cannot guarantee the correlation between the retrieving messages and the users' intention. We introduce the concept of conversational thread to gain more information and improve the search result in microblog. We also use SVMRank to train a model to determine the rank of relevance of the queries and messages. Through a series of experiments, we proved that our method is easy to implement, and can improve the precision up to 29% in average.

Keywords: Microblog, Search, Conversational Thread, SVMRank.

1 Introduction

In the age of social web, the microblogging service as a media is prevalent, in which the posts carry massive information and spread aggressively through the Internet. Twitter, as one of the most popular microblogging services generates up to 230 million tweets[1] per day by September, 2011, according to Michael Abbott, Twitter's ex-VP of engineering. A tweet[2] differs from a post in traditional blogs in that its content is typically restricted to a very short length, typically 140 bytes or characters [1]. A tweet is "real time", the value of which drops dramatically after it has been posted for more than a day or even a few hours [2]. Microblog users read tweets in their timelines, which are posted or forwarded by their friends. Furthermore, trending topics are another important feed to help users access interesting tweets. Both reading tweets from friends and trending topics are passive ways to get information. Nevertheless,

---

[1] The posts named under Twitter.
[2] We borrow the name "tweet" to refer the posts in a general microblogging service.

Z. Shi, D. Leake, and S. Vadera (Eds.): IIP 2012, IFIP AICT 385, pp. 258–267, 2012.
© IFIP International Federation for Information Processing 2012

there are requirements to find information with users' intension, such as web search engines. In this paper, we study on the ad-hoc search of tweets, to find topic-related tweets according to user's descriptions, which has become one of the most popular academic and industrial interests. The Text REtrieval Conference (TREC) 2011 proposed this problem as the microblog task for the first time.

Web search known as a traditional search problem has been studied for several decades. Although a collection of classical models and algorithms on crawling, indexing, ranking and evaluation, etc. have been proposed, there are still many challenging issues, especially in searching tweets:

1. Extremely short. The length of a tweet is too short to contain enough topical features. However, those topical tweets are not independently posted in the microblogging services. Thus, considering the context of tweets to enrich its topical features is a key to find the topic-related tweets.
2. Realtime. The tweets in a microblog are posted rapidly. Furthermore the microblog users may have different intensions from the traditional web search engine. They may want to know statuses of particular topics such as some news about famous stars, the newest progress of events, etc. On the contrary, a webpage search engine mostly helps user with the navigational (give me the url of the site I want to reach) or transactional (show me sites where I can perform a certain transaction, e.g. shop, download a file, or find a map) need[3]. The freshness is more important than the relevance in microblog search. Thus the results of ad-hoc tweet searching must be organized in time-reverse order. Thus the ad-hoc searching has a definite intention to find the status of a happening event or topic, so that tweets are called statuses sometimes. Because of the importance of the freshness the ad-hoc searching, therefore, should present the results in a time-reverse order.
3. User authority. In such a social media, users in the microblogging services play a significant role in the value of tweets' content. Thus the importance of a tweet is not only decided by its content, but also the authority of its author.

This paper focuses on searching new and interesting tweets relevant to a given topic description. A query time is also given for each query, indicating the exact time of the query issuance. The search is supposed to be conducted onsite, and those tweets later than the query time should not be returned. The conversation thread is a set of tweets and their replies. It is supposed that tweets in the same conversational thread are more relevant to the same topic than other tweets outside. We introduce the concept of conversational thread as a very useful tweets context to expand the extreme short content. We propose an enhanced BM25 to determine the relevance of the tweet and the query. Furthermore we bring the tweets' freshness and activities as features. We then use SVM rank to learn the importance of those features, and our ranking results are truncated and filtered by relevance orderly. Finally those truncated results are organized in reverse-time order. In the experiments, we use precision@30 for evaluation, and our approach achieves 29% more relevant and fresh results totally.

The rest of the paper is organized as follows. Section 2 describes the related work. Section 3 describes our main method to search and rank the search results. Section 4 describes our experiment results. We conclude in section 5.

2   Related Work

The existing related work about microblog mainly focuses on users [4-8], information flow [9,10], and tweets' content [11-14]. [2-7] studied Twitter from the users' point of view, such as users' intentions, followers and friends, users' network, and the social features of Twitter, etc. [15] proposed a new ranking method RE(Reader Effective). [9] proposed 2 new concepts stickiness and persistence to detect the information diffusion on twitter. [10] judged the information flow among twitter users. [16] proposed some important features of tweets via PCA(primary content analyze). [11] evaluated the credibility of tweets' content via training a classifier based on message-based features, user-based features, topic-based features and propagation-based features. [12] used the instant content from Twitter to detect earthquakes timely, and collected the geographic and damage information from tweets' descriptions and users' locations as soon as possible. There were also many works focus on thread. [17] focused on the discovery of reply relationship in twitter and used simple features to train model to recover the conversational threads. [18] is the phD dissertation of the same author as [17] but gave the method and steps more detailed. [19] analyzed the statistical characteristics of Slashdot and summarized the patterns of threads in it.

Our work is to query the tweets' content to find relevant, interesting, and fresh tweets via different threads. We use the whole TREC corpus to study global features of tweets' such as content, hashtags, urls, post time, etc. Because the tweets in twitter are very short (no longer than 140 characters), there is little useful information in a single tweet and the relevant between a query and a tweet is difficult to determine. So we also introduce other tweets in the same thread to gain more information to improve the search result. We employ SVM ranking model to rank our query results. The model is trained on pair-wise labeled data.

3   Ad-Hoc Search

The task of Microblog trec 2011 can be described as follow, given a query Q at time T, to find relevant tweets whose post times are no later than T to keep time reverse order(from newest to latest) to achieve high precision@30. A query is a simple topic described by some words.

The corpus is some specified users' tweets on twitter during Jun. 23, 2011 and Feb. 8, 2011. The whole data set is crawled from Twitter with specified seeds. After extracting from the crawled html pages, we got the tweet ID, screen name, http status code, post time, tweet content as a one-line record for each tweet. We only focus on English tweets. After preprocessing the tweets we got 5,650,490 English from 16,141,812 in total.

SVM Rank is an efficient machine learning method to rank. It uses Support vector machine (known as SVM) to rank and the kernel function should be linear. We study the distribution of each feature and take the logarithm of the user activeness and retweetness of the tweets to make them follow the power-law. Because if the distribution follows power-law, it is linear in log-log scale and it fits the SVM perfectly. Table 1 shows the features we used to train our model.

Table 1. Features used to train our model

Feature name	Description
Enhanced BM25	The content relevant
User activenss	The author's authority
Twitter Length	The number of words
Time of retweetness	Popularity
URL	Whether has urls
Hashtag	Hashtag relevant
Freshness	Resolve realtime

3.1 Conversational Thread

Thread originates from Email network with obviously reply relationships. However there are also conversational threads in forums, BBS and blogs. We can also define thread in twitter similarly. A conversational thread in twitter is a rooted tree whose nodes are tweets. The children of each node(if any) is the reply tweet of it. We consider the tweets in a conversational thread to be relevant to each other and thus be more relevant to a topic than other tweets outside the tree. Thread is the implicit structure which is helpful to determine the relevance to a particular topic.

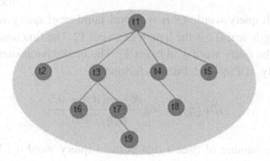

Fig. 1. Conversational thread

38,482 tweets in the corpus are labeled to be relevant or irrelevant to a particular topic. We crawled twitter data and get the conversational threads. We get 48,543 tweets in total to train and test our model. Figure 1 shows a typical conversation thread which has 9 tweets.

The height of a thread can be defined as the number of tweets in the longest reply-to links. Table.2 shows the height distribution of threads. It is not surprising that most of the thread is only one node. There are 38469 threads in total and about 90.34% of them have only 1 tweet. However, the average height of threads higher than 1 is about 3.80. That is to say we can have (if any) nearly 4 tweets to complement a single tweet.

Table 2. The height of threads

Height	Number of threads
1	34754
2	2172
3	553
4	298
5	185
6	124
7	84
8	62

3.2   Enhanced BM25

As for measuring the content relevance, we proposed a scoring method for tweets based on BM25.

Let Q be a query that consists of query words q1, q2, q3, etc. In the BM25 model, the score B(Ti, Q) for tweet Ti is calculated in the following way.

$$B(T_i, Q) = \sum_{j=1}^{|Q|} IDF(q_j) \cdot \frac{f(q_j, T_i) \cdot (k_1 + 1)}{f(q_j, T_i) + k_1 \cdot (1 - b + b \cdot \frac{|T_i|}{avgtl})}$$

where q_j is the jth query word, $|Q|$ is the total number of query words, avgtl is the average tweet length, and $|T_i|$ is the length of tweet T_i. The function $f(q_j, T_i)$ indicates the frequency of the query word q_j in tweet T_i. The model parameters k_1 and b are 2.0 and 0.75 separately. $IDF(q_j)$ is defined as follows.

$$IDF(q_j) = \log \frac{N - n(q_j) + 0.5}{n(q_j) + 0.5},$$

where $n(q_j)$ is the number of tweets that contain query word q_j. It is seen that the score of a tweet matching two different words is the same to that of another matching the same word twice, if the two query words have the same IDF value. Since tweets are extremely short, the word frequency actually does not count much, and matching different query words makes much more sense. Therefore, we need a way to boost such a case. The enhanced BM25 finally defines as follows with boosting where $h_{j,i}$ is a 0-1 binary that indicates whether query word q_j hits tweet i.

$$B(T_i, Q) = \left(\sum_{j=1}^{|Q|} IDF(q_j) \cdot \frac{f(q_j, T_i) \cdot (k_1 + 1)}{f(q_j, T_i) + k_1 \cdot (1 - b + b \cdot \frac{|T_i|}{avgtl})} \right) \cdot \sum_{j=1}^{|Q|} h_{i,j}$$

3.3   User Activeness

The activeness of Twitter users is measured by the number of tweets they posted during a period. An active user is more likely to share valuable information, since he got more used to collecting interesting information, and sharing it. Figure 2 illustrates distribution of the number of users over the number of tweets they posted. It shows that the majority of users posted only one tweet, while the most active one posted 570 tweets in the corpus. With the number of posted tweets less than 12, the distribution follows the power law.

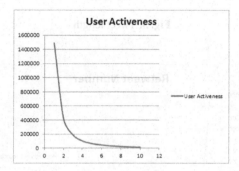

Fig. 2. User activeness

Let u_i represent the number of posted tweets of user i. Thus the feature score U_i for user i is calculated as follows.

$$U_i = \begin{cases} \ln u_i , & u_i \leq 12 \\ \ln 12, & u_i > 12 \end{cases}$$

3.4   Other Features

Tweet Length We use rule-based method known as WordNet is employed to deal with the stemming of nouns, verbs and adjectives. There are also lots of misspelled words in the tweets we also correct the words using method described in Wiktionary [20]. The length of a tweet is generally defined as the number of words after word stemming, skipping retweets, hashtags and urls. Figure 3 shows the tweets distribution over the length, which does not follow the power law. So we used the length directly as a feature.

The times of retweetness In Twitter, the times of retweetness reflect the popularity of a tweet. The relationship between the times of retweetness and the number of tweets is shown in Figure 4. We can see that a large number of tweets that are retweeted once, while only a few have been retweeted more than 20 times. The distribution for those retweeted follows the power law. Because of the power-law distribution, we use $\ln(r_i+1)$ to denotes where r_i is the times of retweetness of tweet i. r_i equals zero if it is not retweeted.

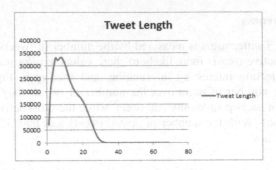

Fig. 3. Tweet length

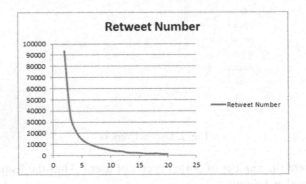

Fig. 4. Retweet Number distribution

URLs and Hashtags A tweet with urls usually contains more information than others. Thus the number of urls indicates the value of a tweet in our ranking model. Hashtags are designed to reflect the topic of a tweet, so we also consider them when determine the relevance. Since a hashtag maybe a combination of several words, we simply use substring match instead of whole word match as a query hits a hashtag. Similarly as the way we boost content match, the score is also boosted by multiplying the number of different query words got matched. The feature score H_i of hashtags for tweet i is calculated as formula Where q_j is the jth query word. $|Q|$ is the total number of query words. And $h_{j,i}$ is a 0-1 binary that indicates whether query word q_j hits the hashtags in tweet i.

$$H_i = (\sum_{j=1}^{|Q|} IDF(q_j) \cdot h_{j,i}) \cdot \sum_{j=1}^{|Q|} h_{j,i}$$

Freshness Because of we must keep the reverse time order, the freshness of a tweet is very important and may be more important than relevant. We use it as a feature. If the query time is T and the ith tweet's post time is $t_i \le T$, both are denoted by seconds from a given time point. We use $\ln(T-t_i+1)$ to measure the freshness of the tweet. The smaller the value is, the fresher the tweet.

4   Experiments and Evaluations

From the TREC Microblogging 2011's official result, we have got about labeled 38,482 tweets. We crawl twitter to get tweets they replied to and merge them up to get 48,543 tweets in total. Because we have the reply relationship among the tweets we can easily restore the conversational threads. We join up all the tweets in the same thread as a longer tweet to denote each single original tweet. So after that, we have 38,482 longer tweets. We compare the result of our method with simple method that determines the tweets relevance simply by whether they contain particular keywords or not. There are 50 topics in TREC's task and we use labeled tweets on 30 topics to train our model (topic-id from 1 to 30) and the other 20 topics (topic-id from 31 to 50) to test. For each test topic we give top30 results and compare. Table 3 shows the result. Using conversational threads can improve the precision@30 by about 29% in average. If we use tree thread height as a feature, the result will be even better. The second column show the result.

Table 3. Precision@30 compared with simple method and our method

Topic-id	Our method Presion@30	Using tree height a feature	Simple hit method
31	2	4	1
32	1	2	1
33	8	8	0
34	1	2	0
35	1	1	3
36	13	13	9
37	0	0	1
38	2	2	1
39	8	8	7
40	1	1	2
41	1	1	1
42	1	1	2
43	4	5	2
44	3	6	0
45	2	2	2
46	1	1	9
47	0	0	0
48	2	2	0
49	2	3	0
50	0	0	0
Total	53	62	41

5   Conclusion and Future Work

Microblog tweets are always very short, which hinders big difficulty to searching for relevant tweets. To handle the problem, this paper proposed to employ conversational thread structure to expand short tweets and use an enhanced BM25 formula to determine the relevant and thus can find relevant more easily.

In this preliminary work, we simply join up all the tweets in conversational thread without consider to merge up hashtags and urls. The further study on the whole frame of combine tweet words, hashtags and urls together seems to be a good try.

Acknowledgements. This work was supported by the National Natural Science Foundation of China with grant number 60933005 and the National Basic Research Program of China with grant number 2012CB316303. This work was also partly supported by the National Information Security Foundation of China with grant number 242-2011F45.

References

1. http://en.wikipedia.org/wiki/Microblogging
2. http://www.howardyermish.com/2009/08/10/drinking-from-the-fire-hydrant/
3. Broder, A.: A taxonomy of web search. IBM Research
4. Java, A., Song, X., Finin, T., Tseng, B.: Why We Twitter: Understanding Microblogging Usage and Communities. In: Joint 9th WEBKDD and 1st SNA-KDD Workshop (2007)
5. Ahn, Y.-Y., Han, S., Kwak, H., Moon, S., Jeong, H.: Analysis of Topological Characteristics of Huge Online Social Networking Services. In: WWW 2007 (2007)
6. Kwak, H., Lee, C., Park, H., Moon, S.: What is Twitter, a Social Network or a News Media? In: WWW 2010 (2010)
7. Weng, J., Lim, E.-P., Jiang, J., He, Q.: TwitterRank: Finding Topic-sensitive Influential Twitters. In: WSDM 2010 (2010)
8. Meeder, B., Karrer, B., Sayedi, A., Ravi, R., Borgs, C., Chayes, J.: We Know Who You Followed Last Summer: Inferring Social Link Creation Times In Twitter. In: WWW 2011 (2011)
9. Romero, D.M., Meeder, B., Kleinberg, J.: Differences in the Mechanics of Information Diffusion Across Topics: Idioms, Political Hashtags, and Complex Contagion on Twitter. In: WWW 2011 (2011)
10. Wu, S., Hofman, J.M., Maso, W.A., Watts, D.J.: Who Says What to Whom on Twitter. In: WWW 2011 (2011)
11. Castillo, C., Mendoza, M., Poblete, B.: Information Credibility on Twitter. In: WWW 2011 (2011)
12. Sakaki, T., Okazaki, M., Matsuo, Y.: Earthquake Shakes Twitter Users: Real-time Event Detection by Social Sensors. In: WWW 2010 (2010)
13. Lerman, K., Hogg, T.: Using Twitter to Recommend Real-Time Topical News. In: RecSys 2009 (2009)
14. Asur, S., Huberman, B.A.: Predicting the Future with Social Media. Arxiv preprint arXiv:1003.5699 (2010)

15. Lee, C., Kwak, H., Park, H., Moon, S.: Finding Influetntials Based on the Temporal Order of Information Adoption in Twitter. In: WWW 2010 (2010)
16. Suh, B., Hong, L., Pirolli, P., Chi, H.: Want to be Retweeted? Large Scale Analytics on Factors Impacting Retweet in Twitter Network. In: ICSC 2010 (2010)
17. Seo, J., et al.: Online Community Search Using Thread Structure. In: CIKM 2009 (2009)
18. Seo, J., et al.: Search Using Social Media Structures. Phd Dissertation
19. Gómez, V., et al.: Statistical Analysis of the Social Network and Discussion Threads in Slashdot. In: WWW 2008 (2008)
20. http://en.wiktionary.org/wiki/Wiktionary:Frequency_lists

Messages Ranking in Social Network

Bo Li, Fengxian Shi, and Enhong Chen

University of Science and Technology of China, China
{libo7,sfx,cheneh}@ustc.edu.cn

Abstract. Nowadays, people engage more and more in social networks, such as Twitter, FaceBook, and Orkut, etc. In these social network sites, users create relationship with familiar or unfamiliar persons and share short messages between each other. One arising problem for these social networks is information is real-time and updates so quickly that users often feel lost in such huge information flow and struggle to find what really interest them. In this paper, we study the problem of personalized ranking in social network and use the SSRankBoost algorithm, a kind of pairwise learning to rank method to solve this problem. We evaluate our approach using a real microblog dataset for experiment and analyze the result empirically. The result shows clear improvement compared to those without ranking criteria. *abstract* environment.

Keywords: ranking, social network, microblog.

1 Introduction

Nowadays, people engage more and more in social network services like Twitter, Facebook, Google+, Orkut, etc. Most of these social networks provide the short messages sharing service. People use this service heavily for staying in touch with friends and colleagues, receiving and publishing all kinds of real time information from all kinds of users with overlapping and disparate interests. On one hand, due to fast transmission rate and low cost of dissemination, the information flow updates quickly. According to recent reports, Twitter currently has over 73 million users, with more than 50,000 tweets being generated per minute. For China microblog Weibo, up to July 2010, the number of messages sent is more than 3 million per day and the average number of messages generated is close to 40 per second. On the other hand, people tend to idolize the information and have psychological dependence on it which also stimulates this information overload. For example, quite a lot of users follow hundreds or even thousands of people on microblogs.

The above two aspects bring in a problem that how users find what really interest them in this huge information flow. Imaging a Twitter lover who follows many people and these people post all kinds of messages from political to sports, from art to entertainment. To some topics the user likes, but to some other topics she/he dislikes. She/he is quite reluctant and feels frustrated if she/he has to read lots of uninteresting messages until what really catches her/his eyes and interests

Z. Shi, D. Leake, and S. Vadera (Eds.): IIP 2012, IFIP AICT 385, pp. 268–275, 2012.

her/him. Currently, most sites exhibit all the messages she/he receives in reverse chronological order when a user login in. This emphasis on time provides no guarantee that the most interesting messages appear on top, especially given that thousands of new messages are generated every minute. We think a more rational rank model is needed and thus we can better present messages to users.

In this paper, we study the message ranking problem in social network. There are some challenges for this problem. First, the text of these messages are short, usually less than 140 characters which is different from the traditional content based relevant ranking scenario. Second, due to fewer links among texts (we will use term text interchangeable to message), traditional link based ranking algorithms such as PageRank, HITS are not suitable to use. Third, the user behavior information we get is usually quite limit. For instance, in microblog, even we know user may have read a lot of messages, but we don't have explicit feedback information, like the specific extent of how user likes each message. What we have is a small portion of user behavior, such as whether she/he has replied the messages or retweeted the messages. At last, for the real-time applications, we require the time complexity of model is small. Considering all these factors, we use SSRankBoost (Semi-Supervised RankBoost) algorithm[1], a kind of learning to rank method, to solve this problem. Since user relationship is a main feature of social network, it can be mined for our ranking task. We utilize this underlying information in two perspectives: first, we analyze user authority on social network graph and exploit this as a feature for training. Second, according to the philosophy that similar users have similar interests, we measure user similarity and leverage user data to train the model. Using SSRankBoost, we modify the loss function of the model on the basis of user and his friends training sets. To evaluate the proposed approach, we use data from Sina Weibo, the biggest microblog site in China and empirically analyze the result which shows much improvement compared to those without ranking criteria.

In the remainder of the paper, first we briefly introduce some related work in section 2. Then in section 3 we apply SSRankBoost to message ranking and take microblogging as an example. At last, we present the experiment results in section 4 and conclude in section 5.

2   Related Work

Social network related problem has attracted many researchers' interests recently. Some researchers investigated the topological properties of the network [2] and user behaviors[3]. Many efforts have also been made for the influence analysis and information diffusion in social network. For example, [4] studied influence maximization on networks and proposed an algorithm to solve this problem.

Our work is different from the previous work in [5] and [6] which also tackle microblog ranking. [5] described several strategies for ranking microblogs in a real-time search engine, they proposed methods to re-rank the top-k tweets returned by an existing microblog search engine which was a non-personalized ranking while our work focus on personalized ranking. [6] studied the problem of designing a mechanism to rank items in forums by making use of the user reviews such as "thumb" and "star" ratings. Their ranking algorithm specifically required manual user input and only utilized this feedback information, while our work and [5] take into account social network properties and properties of the message itself.

3   Personalized Messages Ranking in Social Network

Before discusssing the detail of our approach, we first describe our setting and some notations here. Generally, for each target user, the system is given a set of some labeled message set M where each sample m_i in it is associated with a label $y_i \in \{0, 1\}$, representing different level of relevance or preference. In our experiment, we set $y_1 = 1$ or $y_0 = 0$ which means user prefers a sample m_i with $y_i = 1$ more compared with a sample m_j with label $y_j = 0$. As we have only two labels and there is no order between samples with the same label, the feedback information we used is called bipartite and thus our ranking is bipartite ranking [1]. We denote M_+, M_- to represent these two disjoint sample sets, the samples with label 1 and samples with label 0. For $\forall m_+ \in M_+, m_- \in M_-$, the function $\Phi(m_+, m_-) = 1, \Phi(m_-, m_+) = -1$, while for m_1, m_2 which are in the same set, $\Phi(m_1, m_2) = 0$. Finally, the goal of learning can be defined as the search of a scoring function $H : m \to R$ which assigns higher scores to more relevant instances than to less relevant or irrelevant ones.

3.1   SSRankBoost

Assume the target user is u_0, and his friend set is $F'^{(0)} = \{u_{k_1}, u_{k_2}, \ldots, u_{k_n}\}$. The similarity between u_0 and $u_{k_i}(i = 1, \ldots, n)$ is calculated through some kind of criteria (to be discussed in following section), and we select the top K most similar users and form set $F^{(0)} = \{u_{k_1}, u_{k_2}, \ldots, u_{k_K}\}$ for u_0, their corresponding similarities are $\{s_{k_1}, s_{k_2}, \ldots, s_{k_K}\}$. For convenience, we define $s_0 = 1$ and get the user set for u_0 as $U^{(0)} = \{u_0\} \bigcup F^{(0)}$. The training samples that we clean from the source data for each user in $U^{(0)}$ is $M_i'^{(0)}$, e.g. the training samples for u_0 in $U^{(0)}$ is $M_0'^{(0)}$. To utilize friends' information for the target user u_0, beside $M_0'^{(0)}$, we also use some of the data in each $M_i'^{(0)}(i = 1, \ldots, K)$, which is represented as $M_i^{(0)}, M_i^{(0)} \subset M_i'^{(0)}$ to train the model for u_0. We denote $M_0^{(0)} = M_0'^{(0)}$, then the data used for training for u_0 is $M^{(0)} = \bigcup_{i=0}^K M_i^{(0)}$. Furthermore, M_i^+, M_i^- as the corresponding sample set with label 1 and label 0 for each M_i which means messages in M_i^+ are all labeled with 1 and messages in M_i^- are all labeled with 0.

In this paper, we use SSRankBoost algorithm to model user interests. The final ranking function $H(m)$ that we need in SSRankBoost is in the form of $H = \sum_{t=1}^T \alpha_t h_t$. Let r_i be the respectively ranking loss function for set M_i which defines in equation (2), r_i minimizes the average numbers of irrelevant examples scored better than relevant ones in each M_i separately. Then by summing all the

r_i, we have the ranking loss $r(H)$ of H defined in equation (1). In the equation, s_i is the disfactor weight for each M_i, $D_i(m_-, m_+) = 1/(|\,M_i^+\,|\,|\,M_i^-\,|)$ (we will denote $P_i = |\,M_i^+\,|\,|\,M_i^-\,|)$ is the weight for pair (m_-, m_+) (at the t-th iteration, we use $D_{t,i}$ to denote weight D_i), $[\pi]$ is equal to 1 if the predicate π holds and 0 otherwise in equation(2). As r_i is the respectively ranking loss function for set M_i, using the upper bound $[x \geq 0] \leq e^x$, we have its bound shown in equation (3), so the total cost $r(H)$ also have a bound.

$$r(H) = r_0 + \sum_{i=1}^{k} s_i r_i = \sum_{i=0}^{k} s_i r_i \tag{1}$$

$$r_i = \sum_{m_-, m_+ \in M_i^- \times M_i^+} D_i(m_-, m_+)[H(m_-) - H(m_+)] \tag{2}$$

$$r_i \leq \sum_{m_-, m_+ \in M_i^- \times M_i^+} D_i(m_-, m_+)exp(H(m_-) - H(m_+)) \tag{3}$$

$$r(H) = \sum_{i=0}^{k} s_i r_i \leq \varepsilon = \sum_{i=0}^{k} \sum_{m_-, m_+ \in M_i^- \times M_i^+} s_i D_i(m_-, m_+)exp(H(m_-) - H(m_+)) \tag{4}$$

In the implement of SSRankBoost, the algorithm get a weak learner h_t according to pair weights at each bound. At the beginning of the algorithm, all sample pairs are considered to be uniformly distributed, that is, the pair weight $D_{0,i}(m_-, m_+) = 1/P_i, \forall (m_-, m_+) \in M_i$. In each iteration of SSRankBoost, the weight for each pair is therefore increased or decreased depending on whether h_t orders that pair correctly, leading to the following update rules on the t-th iteration:

$$D_{t+1,i}(m_-, m_+) = \frac{D_{t,i}(m_-, m_+)exp(\alpha_t(h_t(m_-) - h_t(m_+)))}{Z_{t,i}} \tag{5}$$

Where $Z_{t,i}$ is normalization factor such that $D_{t,i}$ remains probability distributions. For a fast implement, SSRankBoost replaces the $D_{t,i}(m_-, m_+)$ with $v_{t,i}(m_-)$ and $v_{t,i}(m_+)$ and the relationship between them is following:

$$D_{t,i}(m_-, m_+) = v_{t,i}(m_-)v_{t,i}(m_+) \tag{6}$$

The corresponding framework of SSRankBoost applied to messages ranking is given in algorithm 1.

In the above framework, the algorithm needs to train a weak learner at each iteration. We also adopt {0,1}-valued weak learner and as for how to find the weak learner $h_t(m)$ and choose the weight α_t, reader can refer to SSRankBoost algorithm[1].

Algorithm 1: SSRankBoost algorithm

Input: $M_i, i = 0, 1, ..., K$, M_i is the data set from u_i
Output: ranking function $H = \sum_{i=1}^{T} \alpha_t h_t$
for $i \leftarrow 0$ **to** K **do**

$\quad \forall m \in M_i^+, v_i(m) = \frac{1}{|M_i^+|}$;

$\quad \forall m \in M_i^-, v_i(m) = \frac{1}{|M_i^-|}$;

for $i \leftarrow 0$ **to** K **do**
$\quad Q_{0,i} = 1$;

for $t \leftarrow 1$ **to** T **do**
$\quad$ train the weak learner using $v_{t,i}(m), i = 0, 1, ..., K$;
$\quad$ choose the weight α_t ; **for** $i \leftarrow 0$ **to** K **do**
$\quad\quad \forall m \in M_i$, update;

$$v_{t+1,i}(m) = \begin{cases} \frac{v_{t,i}exp(-\alpha_t h_t(m))}{Z_{t,i}^+} & \text{if } m \in M_i^+ \\ \frac{v_{t,i}exp(\alpha_t h_t(m))}{Z_{t,i}^-} & \text{if } m \in M_i^- \end{cases} ;$$

$\quad\quad$ where $Z_{t,i}^+, Z_{t,i}^-$ normalized $v_{t,i}$ over M_i^+ and M_i^-;
$\quad\quad$ update $Q_{t,i}$: $Q_{t,i} = Q_{t-1,i}Z_{t,i}^+Z_{t,i}^-$;

3.2   Features

In this article, we use the Sina Weibo(http://weibo.com) data for experiment and evaluate the proposed model. In this section, we introduce all the features we used in this data set. Sina Weibo is the most popular microblog service in China which is much like the Twitter. We use the open API that it provides and crawl the data from December 2011 to February 2012. In this dataset, there are 224977 users and 196499 messages.

To learn a ranking function, it is necessary to design a suitable set of features which will impact performances significantly. In this dataset, we implement the following features for target user u_x and the specific message m:
1) Message aspect features: the tf-idf value, length of text, whether the text contains a URL, topic vector extracted by the LDA model, the retweet number of the message, the replied number of message.
2) Author aspect features: the author authority, the author's message number.
3) Combined features: the interactions ratio of u_x and u_y (u_x is the target user, u_y is the author of specific message m), the common friend ratio of u_x and u_y, the similarity of them. Here interaction means the behavior of user u_x retwweted/replied messages from u_y.

In the above features, we use the topic of the messages which is extracted by using unsupervised learning method LDA[7]. For features in terms of author, we use the author authority. Here, author authority means user's PageRank value in social network graph. The graph vertices represent users and edges representing the follower relationship between them, in another word, $w_{(i,j)}$ is

1 in the adjacency matrix of the graph if u_i follow u_j. Then we apply PageRank on this graph and get each user's authority.

3.3 User Similarity

To caculate the user similarity, We think there are some rules that can be obeyed: 1) the more common friends u_a and u_b have, the more similar they are; 2) the more interactions they have, the more similar they are; 3) the more similar their texts are, the more similar they are.

Let the common friends number of u_x and u_y be s, and their friends number be L_{u_x} and L_{u_y} respectively, the total interaction number that they have are I_{u_x} and I_{u_y} respectively, then we define $cf\_sim(u_x, u_y) = \frac{s}{L_{u_x}}, in\_sim(u_x, u_y) = \frac{I_{u_x u_y}}{I_{u_x}}$, where $I_{u_x u_y}$ denotes the number of interactions between u_x and u_y. At last, we also consider text similarity between users. We treat all the messages one user writes to be an article. Then we extract the topic by using LDA, whose each element represents the weight of the topic. We can also caculate the TF-IDF vector of each article. Assume the topic vector for u_x and u_y are $\vec{v_1}, \vec{v_2}$, while TF-IDF vectors are $\vec{t_1}, \vec{t_2}$, we define $text\_sim(u_x, u_y) = 0.5sim(\vec{v_1}, \vec{v_2}) + 0.5sim(\vec{t_1}, \vec{t_2})$ where $sim(\vec{x}, \vec{y})$ is two consine correlation between $\vec{x}, \vec{y}$. Combining the three similiarities, we finally get the similarity between u_x, u_y with following equation:

$$sim(u_x, u_y) = \alpha cf\_sim(u_x, u_y) + \beta in\_sim(u_x, u_y) + \gamma text\_sim(u_x, u_y) \quad (7)$$

Where $\alpha > 0, \beta > 0, \gamma > 0, \alpha + \beta + \gamma = 1$ and those three constants are the importance weights of the three kinds of similarities respectively.

3.4 Choose Microblog Samples

For a target user u, the behavior information we have is which messages she/he retweeted/replied, and which messages she/he didn't retweet/reply. It's obvious that if user retweets/replies a message, it indicates that user likes it. Assume that the messages user received from friends form a list $L = \{m_0, m_1, \ldots, m_t\}$ in which messages are arranged in reverse chronological order. If user retwcetes/replies m_k, then we set y_k to be 1 and for messages in set $N_5(m_k) = \{m_i, m_i \notin I(u) \land m_i \in L, i = k - 5, \ldots, k - 1, k + 1, \ldots, k + 5\}$, we set their lables to be 0. Here $I(u)$ is set of messages user have interacted with, $N_5(m_k)$ is the neighbor of m_k whose distance to m_k is no more than 5 in L. We didn't choose all the messages which u didn't have interactions because we don't know whether user likes it or not. But it's strongly believed that if two messages m_x and m_y appear at the successive positions and if u retweeted/replied m_x whilo didn't do it for m_y, then we can say that m_x is more preferred by u, and value $\Phi(m_x, m_y)$ can bet set to 1.

4 Experiment Result

We select 1023 users for experiments and use the metric precison and NDCG[8] to evaluate our results. We implement three schemes to analyze the impact of friends' data for users. The first scheme is to use all user's data for all users, which means we don't distinguish each user's data and it's not a personalized ranking. For instance, if there is 10 users, each has 100 samples, then we train the general model once with the 1000 samples and use this model for every user. This considers that every user has similar taste. The second scheme is to use user's own data for each user which suffers from the data spareness problem. The third scheme is to use k user's friend combined with user's own data to train the model, which is the method we discuss in this paper. Obviously, the second and third schemes are personalized ranking.

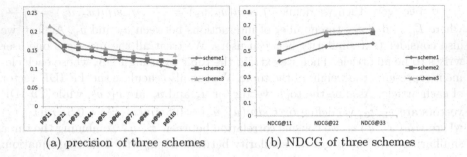

(a) precision of three schemes   (b) NDCG of three schemes

Fig. 1. Results of three schemes

Figure 1(a) and 1(b) show the result of precision, NDCG of the three schemes. Comparing the result of scheme 1,2 to 3, we can see the personalized ranking outperforms non-personalized ranking. It shows scheme 3 performs better than scheme 1,2 clearly. By comparing scheme 2 and scheme 3 in each figure, we can see the impact of friends data and the effects of our algorithm. From Figure 1(a) and 1(b), we see incorporating friend data using our algorithm really improves the result.

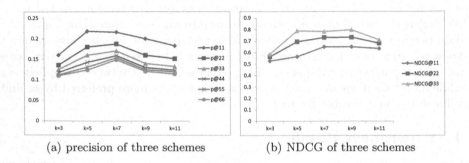

(a) precision of three schemes   (b) NDCG of three schemes

Fig. 2. Results of different K

We also study the impact of different K in the algorithm. For a target user and each of his K friends, we select C messages for training. Figure 2(a) and Figure 2(b) show the result of different K. From the figures, we can see when $K \geq 5$ and $K \leq 9$, the performance is best. When K is small, the performance improves as K grows. However, when $K \geq 9$, the benefit of friend data is decreasing, and the results become worse. To explain the results, we analyze the data and find that in the top $K(K \geq 9)$ friends lists, there are some users who have many followers and generate messages very frequently. These users mostly are celebrities or organizations and most of the common users have followed them for receiving news. In some sense, the data from these users could not characterize the target user's interest very well. For this reason, when too much data is used, the benefit of friend data is decreased.

5   Conclusion

Generally, this paper studies how to filter the real time web to find the most interesting messages for each user in social network. We apply SSRankBoost algorithm to rank the messages that user subscribed in social network. We utilize all kinds of features to evaluate the ranking method in the real data set. We think there is still a lot of room for improvement. One way is to incorporate more features into the ranking model, such as using the replied/retweeted author information of the message, the retweeted speed of the messages, the novelty of the message, etc.

References

1. Amini, M.R., Truong, T.V., Goutte, C.: A boosting algorithm for learning bipartite ranking functions with partially labeled data. In: SIGIR, pp. 99–106 (2008)
2. Mislove, A., Marcon, M., Gummadi, P.K., Druschel, P., Bhattacharjee, B.: Measurement and analysis of online social networks. In: Internet Measurement Comference, pp. 29–42 (2007)
3. Schneider, F., Feldmann, A., Krishnamurthy, B., Willinger, W.: Understanding online social network usage from a network perspective. In: Internet Measurement Conference, pp. 35–48 (2009)
4. Richardson, M., Domingos, P.: Mining knowledge-sharing sites for viral marketing. In: KDD 2002, pp. 61–70 (2002)
5. Nagmoti, R., Teredesai, A., Cock, M.D.: Ranking approaches for microblog search. In: IEEE/WIC/ACM International Conference on Web Intelligence and Intelligent Agent Technology, vol. 1 (2010)
6. Sarma, A.D., Sarma, A.D., Gollapudi, S., Panigrahy, R.: Ranking mechanisms in twitter-like forums. In: WSDM, pp. 21–30 (2010)
7. Blei, D.M., Ng, A.Y., Jordan, M.I.: Latent dirichlet allocation. Journal of Machine Learning Research 3, 993–1022 (2003)
8. Järvelin, K., Kekäläinen, J.: Cumulated gain-based evaluation of ir techniques. ACM Trans. Inf. Syst. 20(4), 422–446 (2002)

Diagnosis of Internetware Systems Using Dynamic Description Logic

Kun Yang, Weiqun Cui, Junheng Teng, and Chenzhe Hang

Chinese National Institute of Metrology, Beijing, China
yangkun@nim.ac.cn

Abstract. This paper proposes a kind of multi-agent based internetware system architecture, and introduces a diagnoser that can perform on-line diagnosis by observing the behaviors of it. This diagnoser brings semantic description to the procedure of states conversation of the system to be diagnosed by using dynamic description logic, which makes it possible to make use of more knowledge to analyze the failure further.

Keywords: diagnosis, internetware, dynamic description logic, discrete-event.

1   Introduction

The problem of failure diagnosis has received considerable attentions in the literature since detecting and isolating failures play an important role in building trustworthy software systems. There are two main methods in the research area of failure diagnosis: expert system-based method and model based method. The problem faced by the former is it is difficult to extract heuristic rule from expert knowledge and it is area-dependent that means the diagnostic rules have to change with the changing of the areas. As for the latter, it models the features of the object to be diagnosed and records the deep knowledge of them, and makes the diagnostic object model and the diagnostic process independent of each other.

Model based method doesn't depend on specific modeling language, and qualitative deviation, qualitative differential equations, process algebras, Bayesian networks, discrete event system method and Petri nets[1, 2] are all be used to model and analyze the diagnostic objects. Among them, the research about discrete event system modeling is the most active one. In [3], diagnostic object is modeled as a global finite state automata, and failure is modeled as series of state transferring that can't be observed, the diagnoser predicts current state of the system and judge whether a specific failure happens based on the event sequence observed. Based on the global model in [3], [4] proposes a distributed diagnostic strategy and protocol. [5] proposes a method to construct distributed diagnostic model for large scale discrete event system when global model can't be obtained and apply it to the diagnosis of telecommunications networks. Although the diagnostic objects of these works are physical system that is different from our internetware system, their modeling method for discrete event

Z. Shi, D. Leake, and S. Vadera (Eds.): IIP 2012, IFIP AICT 385, pp. 276–285, 2012.

system and the strategy of reducing the diagnostic searching space provide us a good reference and guidance.

As a novel software paradigm, Internetware's failure diagnosis problem is urgent with its environment becoming open, dynamic and difficult to control. As a dynamic system, it has a definite boundary and can be abstracted as a discrete event system to some extent, which makes it possible to observe its behaviors at some granularity and research its failure diagnosis under the discrete event system framework. Current discrete event system modeling methods focus more on the state transition, while more proper semantic to the process of   state transition will improve its diagnosis for following reasons:

- Software failure is semantically related with an application. People find that the software failures often happen for the components don't follow the design fully and execute the behaviors forbidden by designers or programmers, which means they violate the semantic of the system.
- Failure self-recovery needs semantic information. Because of the limitation of the observation method, the granularity of diagnosis result of traditional discrete event system is large, and the diagnostic searching space is huge. Lacking related semantic description makes it difficult to determine the exact failure state and recover that failure at the next step.

In order to bring proper semantic to the diagnostic process, this paper proposes a dynamic description logic based failure diagnosis method under the discrete event system framework, where event is modeled as an action, and state is modeled as a set of description logic simple formulas with their variables assigned specific values. Based on the analysis and reasoning to the formulas about a failure state, the failure location and type can be determined and its self-recovery becomes possible.

1.1   Multi-agent Based Internetware System

Ref [6] summarizes the software technology trend in the open, dynamic and variable internet environment and proposes the concept of internetware. Compared with classical software system, it has some distinctive features, such as the intelligentization of software entity, the separation of   software cooperation and the autonomization of system management and so on. Based on these features, we design an Internetware system whose architecture is illustrated as Fig.1 below.

There are two layers in the architecture above: the control layer and the target layer. The former mainly includes the sensor that perceives the changes of the environment and the effector that adjusts the actions of the system. The latter includes classic software system and its environment. Agent owns features needed by Internetware so it is the best basic component for Internetare system. While it is not a software development technique in mainstream currently, and most Web services doesn't developed by agent technique, so we package these web services by agent to map them to a multi-agent system, so as to combine them to finish a task by the cooperation of agents without modifying them. The diagnostic object here is a software system conforming to the architecture introduced.

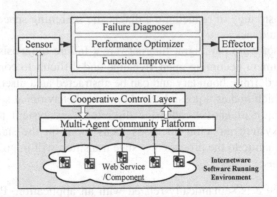

Fig. 1. The architecture of the Internetware System

2   D-ALCO Based Failure Diagnosis

2.1   D-ALCO Brief Introduction

Based on description logic ALCO, [7] proposes dynamic description logic D-ALCO by combining description logic, dynamic logic and theory of action. As a kind of formal tool for knowledge representing, D-ALCO has clear semantic feature and can provide not only decidable reasoning service, but also representing and reasoning for dynamic process and operating mechanism effectively.

The basic symbols in D-ALCO include ①Concept set N_C; ②Role set N_R; ③ Individual set N_I; ④Individual Variable set N_{IV}; ⑤Atom action set N_A; ⑥Concept constructors: { }, ¬, ⊔ and ∀; ⑦Formula constructors:¬, ∨ and < >; ⑧Action constructors: ∪, ;, * and ?; ⑨Some other symbols include≡, ( ) and ','. With these symbols, roles, concepts, formulas and actions can be constructed.

Def 1 for a given TBox $\mathcal{T}$, $\alpha(v_1, ..., v_n) \equiv (P_a, E_a) \equiv (P_a(v_1, ..., v_n), E_a(v_1, ..., v_n))$ is defined as an atom action. Where

1. $\alpha \in N_A$, is the name of the atom action defined;
2. $(v_1, ...,v_n)$ is the finite sequence of all individual variables appeared in P and E;
3. P_a is the finite set consisted by simple formulas, which denotes the precondition that must be satisfied before the action performs;
4. E_a is the finite set consisted by simple formulas, which denotes the result after the action's performing;
5. P_a and E_a satisfy the constraint that for any $\varphi \in E_a$, $\varphi^\neg \in P_a$ is satisfied.

Def 2 D-ALCO is a triple $M = (\Delta, W, I)$, where Δ is a non-empty set of individual which acts as the discourse domain of the model; W is the set of possible world; I gives an explanation $I(w)= (\Delta, \cdot^{I(w)})$ for each possible world w in W.

Def 3 for any $M = (\Delta, W, I)$, $\gamma : N_{IV} \to \mathcal{A}^I$ is called one designation based on M.

Any designation γ gives an assignment for each individual variable in N_{IV} and designates them as individuals in $\mathscr{d}$. If $v \in N_{IV}$ and $\gamma_1, \ldots, \gamma_n$ is all the designations based on M then the explanation domain for v can be denoted as $\mathscr{D}(v) = \cup_{1 \le i \le n} \{\gamma_i(v)\}$. For any $M = (\Delta, W, I)$, an explanation I and a designation γ can determine a mapping $\Gamma_{M,\gamma}$ from simple formula set $\varepsilon(v_1, \ldots, v_n) = \{\psi_1(v_1), \ldots, \psi_n(v_n)\}$ to Boolean value $\{True, False\}$, which can be denoted as: $\Gamma_{M,\gamma}(\varepsilon(v_1, \ldots, v_n)) = \varepsilon(v_1, \ldots, v_n)^{I,\gamma} = \{\psi_1(v_1), \ldots, \psi_n(v_n)\}^{I,\gamma} = \psi_1(v_1)^{I,\gamma} \wedge \ldots \wedge \psi_n(v_n)^{I,\gamma}$.

2.2　D-ALCO Based System Modeling

Under the framework of discrete event system, the multi-agent based internetware system can be modeled as a finite state automata $G=(X, \Sigma, \delta, X_0)$, where X denotes the finite set of state; Σ denotes the finite set of event; $\delta \subseteq X \times \Sigma \times X$ denotes the finite set of state transition; $X_0 \subseteq X$ denotes the set of initial states. The states and events in G have no semantic information, now we can model them by D-ALCO as follows.

Assume $(v_1, \ldots, v_n)$ is the finite sequence of individual variables involved in G, where $v_i \in N_{IV}$, $1 \le i \le n$. The state space X of G corresponds to the Cartesian produce of the explanation domain of all individual variables, that is $X := \prod_{i=1}^{n} D(v_i)$. Let Θ $(v_1, \ldots, v_n)$ is the set composed by simple formula set $\varepsilon(v_1, \ldots, v_n) = \{\psi_1(v_1), \ldots, \psi_n(v_n)\}$. For each $\varepsilon \in \Theta$ $(v_1, \ldots, v_n)$, there is only one designation γ for the finite sequence $(v_1, \ldots, v_n)$, which satisfies the function $\Gamma_{M,\gamma}(\varepsilon(v_1, \ldots, v_n)):X \to \{True, False\}$ is True. Let the explanation domain for the finite sequence $(v_1, \ldots, v_n)$ under any state $x_i \in X$ as a designation γ, then with this γ, the simple formula set ε that makes the $\Gamma M, \gamma(\varepsilon(v_1, \ldots, v_n)):X \to \{True, False\}$ is true corresponds to that state x_i, which means there is a one-to-one correspondence between simple formula set Θ and state set X. The simple formula set $\varepsilon \in \Theta(v_1, \ldots, v_n)$ with corresponding designation γ that corresponds to any state $x_i \in X$ can be recorded as $\varepsilon_\gamma(x_i)$, then the initial state x_0 in G can be recorded as $\varepsilon_\gamma(x_0)$.

An atom action α in D-ALCO is used to model an eventσ, and the finite set of atom action N_A is used to model the finite set of eventΣ. Based on the correspondence between simple formula set Θ and state set X, the transitions of the states can be extended to the changing of the formula set. A state transition in G can be described by using an atom action, and the form of atom action is as follows:
$$\alpha(v_1, \ldots, v_n) \equiv (P_\alpha, E_\alpha) \equiv (P_\alpha(v_1, \ldots, v_n), E_\alpha(v_1, \ldots, v_n))$$
where an action $\alpha \in N_A$ denotes an event; P_α and E_α are both finite sets of simple formula that denote the precondition before α performing and the effect after α performing respectively. The change caused by action α to the simple formula set can be described as $\rho_\alpha(\varepsilon)::= (\varepsilon \backslash E_\alpha \neg) \cup E_\alpha$, and that to the individual variables $(v_1, \ldots, v_n)$ can be described as $\Gamma_{M,\gamma}(\rho_\alpha(\varepsilon(v_1, \ldots, v_n)))::= \Gamma_{M,\gamma}((\varepsilon(v_1, \ldots, v_n) \backslash E_\alpha \neg) \cup E_\alpha)$. For an event σ denoted by an action $\alpha = (P_\alpha, E_\alpha)$, if $\delta(x_j, \sigma) = x_k$ is satisfied in G, then $\varepsilon_\gamma(x_k) \supseteq (\varepsilon_\gamma(x_j) \backslash E_\alpha \neg) \cup E_\alpha$ and $\Gamma_{M,\gamma}(\varepsilon_\gamma(x_j))::= \Gamma_{M,\gamma}(\varepsilon_\gamma(x_k)) = True$.

The process of modeling above can be illustrated intuitively as Fig.2 below.

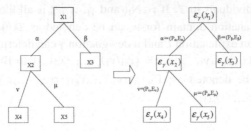

Fig. 2. Illustration of the D-ALCO modeling

2.3   D-ALCO Based Diagnoser Construction

Failure diagnoser is a finite state automata constructed based on $G_e=(X\times\Delta, \Sigma, \delta_e, X_{e0})$ that is obtained after extending the *D-ALCO* model of G introdued above. It is used to observe the system online and diagnose the failure based on those observations. Its model is as follows:

$$G_d = (Q_d, \Sigma_o, \delta_d, q_0)$$

Where $Q_d, \Sigma_o, \delta_d, q_0$ denote the state space, event set, transfer function and initial state of the finite state automata of failure diagnoser. G_d records all the states that the system can reach starting from the initial state and experiencing the same event observation sequence. For the existence of unobservable event, there may be several state transitions under the same event observation sequence, so $Q_d \subset 2^{\Theta\times\Delta}$, here Θ has the same meaning with the definition before, meaning all the sets of simple formula $\varepsilon_\gamma(x_i)$ that corresponding to the state set X in G; $\Delta=\{N\} \cup 2^{\Delta_f}$ which means the set of state label, where N means normal state and Δ_f means the failure category label set. Σ_o denotes the observable event set in the system G. $q_0=(\varepsilon_\gamma(x_0), \{N\})$, which is the same as the initial state X_{e0} of G_e.

For any state $q_d \in Q_d$, it can be denoted as $q_d=\{(\varepsilon_\gamma(x_1), l_1), \cdots, (\varepsilon_\gamma(x_n), l_n)\}$, where $x_i \in X, l_i \in \Delta$. δ_d can be defined as $\delta_d \subseteq (Q_d \times \Sigma_o \times Q_d)$. For any $\alpha \in \Sigma_o$, $(q_d, \alpha, \widetilde{q_d}) \in \delta_d$ is satisfied, if for any $(\varepsilon_\gamma(x_i), l_i) \in q_d$, there is a route s in G_e satisfying $\delta((\varepsilon_\gamma(x_i), l_i), s)=(\varepsilon_\gamma(x_j), l_j)$ and $M(s)=\alpha$, then $(\varepsilon_\gamma(x_j), l_j) \in \widetilde{q_d}$.

2.4   Distributed Diagnoser and Diagnostic Strategy

Web service system is distributed. Even for a centralized system, it is divided into several subsystems to solve problem cooperatively for the consideration of the communication between agents, so the distributed diagnostic strategy is important. The distributed diagnoser and its diagnostic strategy can be applied for several distributed sites, but here we only involve two sites for the illustrative purpose.

The event set $\Sigma=\Sigma_o \cup \Sigma_{uo}$ and the failure event $\Sigma_f \subseteq \Sigma_{uo}$. For the two distributed diagnoser, assume their observable event sets are Σ_{o1} and Σ_{o2} respectively and

$\Sigma_0 = \Sigma_{o1} \cup \Sigma_{o2}$, then their local unobservable sets are Σ/Σ_{o2} and Σ/Σ_{o1}. With the enlarging of the unobservable set, the possible state set that each diagnoser can reach under the same observable event sequence enlarges too, which makes the state space for diagnostic searching greater and makes it more difficult to locate failure and need higher computing cost. In order to limit the diagnostic searching space by making full use of known information, we use following strategies during the process of distributed diagnosis.

- *Recording the path of state transferring:* for the limitation of observation methods, different event tracks appear the same observation feature. Diagnoser records all the states a system can reach under the same observation, while if it can record the path of state transferring, then it can find impossible paths and delete the corresponding states on those paths, so as to reduce the diagnostic searching space.
- *The cooperation among distributed diagnosers:* although the information is local and limited for each distributed diagnoser, they can complement information and constrain mutually to limit the global diagnostic result to a smaller state space.

Now assume the model of a distributed diagnose is as follows:
$$G_{di}^e = (Q_d^e, \Sigma_{oi}, \delta_{di}^e, q_0^e)(\text{here } i \in \{1,2\})$$
Based on the first strategy, we record the path of state transferring in G_{di}^e, but only the direct precursor state of any state for the consideration of storage space and computational complexity. Q_d^e is obtained by extending Q_d introduced in previous section: let the state member of each state includes not only the state itself, but also its direct precursor state, that is $Q_d^e \subset 2^{(\Theta \times \Delta) \times (\Theta \times \Delta)}$. Let the direct precursor state of the initial state is itself, then $q_0^e = \{((\varepsilon_\gamma(x_0), \{N\}), (\varepsilon_\gamma(x_0), \{N\}))\}$ and for any $q_d^e \in Q_d^e$, $q_d^e = \{((\varepsilon_\gamma(x_1'), l_1'), (\varepsilon_\gamma(x_1), l_1)), \cdots, ((\varepsilon_\gamma(x_n'), l_n'), (\varepsilon_\gamma(x_n), l_n))\}$ $(l_i \in \Delta)$, where $(\varepsilon_\gamma(x_1'), l_1')$ and $(\varepsilon_\gamma(x_n'), l_n')$ correspond to the direct precursor states of the states corresponding to $(\varepsilon_\gamma(x_1), l_1)$ and $(\varepsilon_\gamma(x_n), l_n)$ in G_e respectively.

For the distributed diagnoser $G_{di}^e (i \in \{1,2\})$, its unobservable event set is Σ/Σ_{oi}, and for any event $\beta \in \Sigma/\Sigma_{oi}$, $M(\beta) = \emptyset$ is satisfied. δ_{di}^e is defined as follows:
$$\delta_{di}^e \subseteq (Q_d^e \times \Sigma_{oi} \times Q_d^e)$$
For any $\alpha \in \Sigma_{oi}$, $(q_d^e, \alpha, \widetilde{q_d^e}) \in \delta_d$ is satisfied. For any $((\varepsilon_\gamma(x_i'), l_i'), (\varepsilon_\gamma(x_i), l_i)) \in q_d^e$, if there is a event path s started from the state corresponding to $(\varepsilon_\gamma(x_i), l_i)$ that satisfies $\delta_e((\varepsilon_\gamma(x_i), l_i), s), (\varepsilon_\gamma(\widetilde{x_i}), \widetilde{l_i})$ in G_e, and $s \in (\Sigma/\Sigma_{oi})^* \alpha$, then $((\varepsilon_\gamma(\widetilde{x_i'}), \widetilde{l_i'}), (\varepsilon_\gamma(\widetilde{x_i}), \widetilde{l_i})) \in \widetilde{q_d^e}$. Where $(\varepsilon_\gamma(\widetilde{x_i'}), \widetilde{l_i'})$ corresponds to the precursor of the state corresponding to $(\varepsilon_\gamma(\widetilde{x_i}), \widetilde{l_i})$ in G_e.

The diagnoser introduced above records all the states that the system can reach driven by specific observable event. For the limitation of local information and observation methods, the state transferring of the system driven by unobservable event is not so definite, and according to the second strategy, it is needed to make that definite by cooperating with another diagnoser, so we consider the state transferring in a diagnoser driven by the event unobservable to itself while observable to another.

Def 4 for any state $q_d^e = \{((\varepsilon_\gamma(x_1'), l_1'), (\varepsilon_\gamma(x_1), l_1)), \cdots, ((\varepsilon(x_n'), l_n'), (\varepsilon_\gamma(x_n), l_n))\}$ $(l_i \in \Delta)$ in a distributed diagnoser $G_{di}^e (i \in \{1,2\})$, $S_i(q_d^e) = \Big\{ s \in (\Sigma/\Sigma_{oi})^*: s \in L_\sigma\Big(G_e, (\varepsilon_\gamma(x_k), l_k)\Big)\Big\}$ is an path set, where $\sigma \in \Sigma_{oj}$, $j \in \{1, 2\}/\{i\}$, $k \in \{1, \cdots, n\}$, and $s \in L_\sigma\Big(G_e, (\varepsilon_\gamma(x_k), l_k)\Big)$ denotes those event paths that start from the state corresponding to $(\varepsilon_\gamma(x_k), l_k)$ in G_e and end with specific event σ, then the state set that q_d^e can reach driven by the unobservable event in Σ/Σ_{oi} can be denoted as follows:

$$UR_i(q_d^e) = \{q_d^e\} \cup \bigcup_{s \in S_i(q_d^e)} \Big\{((\varepsilon_\gamma(\widetilde{x_k}'), \widetilde{l_k}'), (\varepsilon_\gamma(\widetilde{x_k}), \widetilde{l_k}))\Big\}$$

Where $k \in \{1, \cdots, n\}$, $(\varepsilon_\gamma(\widetilde{x_k}), \widetilde{l_k})$ is the state that $(\varepsilon_\gamma(x_k), l_k)$ reaches after the path $s \in S_i(q_d^e)$, and $(\varepsilon_\gamma(\widetilde{x_k}'), \widetilde{l_k}')$ is the precursor of $(\varepsilon_\gamma(\widetilde{x_k}'), \widetilde{l_k}')$ on that path.

Each distributed diangnoser determines the current state of the system based on the knowledge it owns and its observation method, and a coordination agent summaries their diagnostic results to obtain the global diagnostic result. Before introduce the strategy of summary, we introduce two operations definitions first.

Def 5 let $q_{d1}^e = \{((\varepsilon_\gamma(x_1'), l_{x_1}'), (\varepsilon_\gamma(x_1), l_{x_1})), \cdots, ((\varepsilon_\gamma(x_n'), l_{x_n}'), (\varepsilon_\gamma(x_n), l_{x_n}))\}$, $q_{d2}^e = \{((\varepsilon_\gamma(y_1'), l_{y_1}'), (\varepsilon_\gamma(y_1), l_{y_1})), \cdots, ((\varepsilon_\gamma(x_n'), l_{y_m}'), (\varepsilon_\gamma(x_n), l_{y_m}))\}$, the operation $\cap_e^i$ $(i \in \{1,2\})$ can be defined as follows:

$q_{d1}^e \cap_e^i q_{d2}^e \triangleq \{((\varepsilon_\gamma(z'), l_z'), (\varepsilon_\gamma(z), l_z)) \in 2^{(\Theta \times \Delta) \times (\Theta \times \Delta)}, (\varepsilon_\gamma(z), l_z) = (\varepsilon_\gamma(x_k), l_{x_k})$ $= (\varepsilon_\gamma(y_t), l_{y_t}), k \in \{1, \cdots, n\}, t \in \{1, \cdots, m\},$ if $i=1, (\varepsilon_\gamma(z'), l_z') = (\varepsilon_\gamma(x_k'), l_{x_k}'),$ if $i=2,$ $(\varepsilon_\gamma(z'), l_z') = (\varepsilon_\gamma(y_t'), l_{y_t}')\}$.

The operation $\cap_e^i$ is used to summary the diagnostic results submitted by every distributed diagnoser, and it denotes that for the different event paths to a specific state submitted by different distributed diagnosers, the summary result accepts the diagnostic result submitted by the distributed diagnoser that is more convinced about its path. This operation make it possible for all the distributed diagnosers to make full use of limited information and observation methods to constrain each other, so as to reduce the diagnostic space.

Def 6 let $q_1 = \{((\varepsilon_\gamma(x_1'), l_{x_1}'), (\varepsilon_\gamma(x_1), l_{x_1})), \cdots, ((\varepsilon_\gamma(x_n'), l_{x_n}'), (\varepsilon_\gamma(x_n), l_{x_n}))\}$, $q_0 = \{((\varepsilon_\gamma(y_1'), l_{y_1}'), (\varepsilon_\gamma(y_1), l_{y_1})), \cdots, ((\varepsilon_\gamma(x_n'), l_{y_m}'), (\varepsilon_\gamma(x_n), l_{y_m}))\}$, the operation $\cap_c$ can be defined as follows:

$q_1 \cap_c q_0 = \{((\varepsilon_\gamma(z'), l_z'), (\varepsilon_\gamma(z), l_z)) \in 2^{(\Theta \times \Delta) \times (\Theta \times \Delta)}, (\varepsilon_\gamma(z'), l_z') = (\varepsilon_\gamma(y_t'), l_{y_t})$ $= (\varepsilon_\gamma(x_k'), l_{x_k}'), k \in \{1, \cdots, n\}, t \in \{1, \cdots, m\}, (\varepsilon_\gamma(z), l_z) = (\varepsilon_\gamma(x_k), l_{x_k})\}$.

The operation $\cap_c$ is used to update the global diagnostic result in real time based on the summary result.

The global diagnostic result can be obtained based on the information submitted by all distributed diagnosers. Let q_1, q_2 and q denote the states of distributed diagnosers G_{d1}^e, G_{d2}^e and global diagnoser G_d^e respectively, the event path s = $s_1$aub, where $a, b \in \Sigma_o (\Sigma_o = \Sigma_{o1} \cup \Sigma_{o2}), u \in \Sigma_{uo}{}^*$, and q_{old} denotes the state of G_d^e after executing the event path of s_1a, then the state q after G_d^e executing the event path s can be obtained as follows:

1. If $b \in (\Sigma_{o1} \cap \Sigma_{o2})$ then

    (1). $q = (q_1 \cap_e^1 q_2) \cap_c q_{old}$ (when $a \in \Sigma_{o1}$)

    (2). $q = (q_1 \cap_e^2 q_2) \cap_c q_{old}$ (when $a \in \Sigma_{o2}$)

2. If $b \in (\Sigma_{o1}/\Sigma_{o2})$ then

    (1). $q = \left(q_1 \cap_e^1 UR_2(q_2)\right) \cap_c q_{old}$(when $a \in \Sigma_{o1}$)

    (2). $q = \left(q_1 \cap_e^2 UR_2(q_2)\right) \cap_c q_{old}$(when $a \in \Sigma_{o2}$)

3. If $b \in (\Sigma_{o2}/\Sigma_{o1})$ then

    (1). $q = (UR_1(q_1) \cap_e^1 q_2) \cap_c q_{old}$(when $a \in \Sigma_{o1}$)

    (2). $q = (UR_1(q_1)q_2) \cap_c q_{old}$(when $a \in \Sigma_{o2}$)

The distributed diagnosters can limit the failure states of the system by cooperation and bring a *D-ALCO* based description for every failure state, which lays a foundation for the failure recovery in the next step.

3   Knowledge Based Failure Recovery

3.1   Process of Failure Recovery

Based on the semantic description of failure state, we can recovery it by knowledge-based method. Its process is described as illustrated in Fig.3 below.

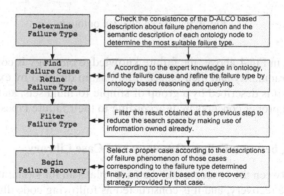

Fig. 3. Semi-automatic failure recovery flowchart

The green blocks diagram on the left denote the recovery steps, and the white block diagram on the right denotes the specific operations corresponding to each step. For the limitation of observation method, the description of failure state is not so exhaustive. People hope to resolve the failure problem in a smaller granularity and not limited to retrying or substitution, so step 2 and step 3 above repeat until there is no more expert knowledge can be used to filter the failure type.

3.2   Construction of Failure Knowledge Ontology

Failure knowledge ontology places an important role in the process of failure recovery. It lays a foundation for reusing expert knowledge automatically, analyzing the failure type accurately and reducing the failure searching space. It also acts as a backbone of knowledge to organize and manage the failure case library, so as to improve the efficiency of case searching and reduce the case searching space.

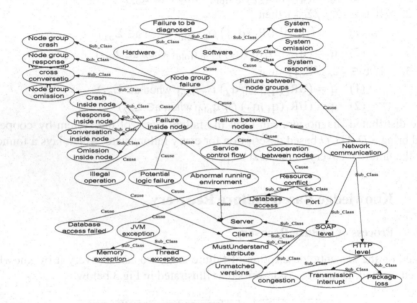

Fig. 4. Failure knowledge ontology

As illustrated in Fig.4 above, the failure knowledge ontology is constructed according the failure granularity and their causal relationship directed by experts. Each failure type node in it is described by a group of simple formula sets, and all failure cases are mapped with proper concepts in it according to their failure types.

3.3   Mapping between Failure Ontology and Case Library

The mapping between the concepts and the cases plays an important role during the process of failure recovery, and it is constructed as following code illustrated:

```
<O2DMappings srcOntology="DiagnosisOntology">
 <O2DMapping>
  <concept name="NodeFailure"/>
   <mappingExpression>
   <queryStatement name="query1"destSource="DiagnosisLibDB">
     <expression>
        Model.list ResourcesWith Property(symrec.FaultString,
           "NodeFailure")
     </expression>
```

```
      <objectKey >
        <keyAttribute name="recordid" type="string"/>
      </objectKey>
    </queryStatement>
  </mappingExpression>
</O2DMapping>
            . . .
</O2DMappings>
```

Where "NodeFailure" denotes a concept node in it, and "DiagnosisLibDB" denotes the failure case library. The code denotes constructing all the cases that match the query "query1" with the ontology concept "NodeFailure".

4   Conclusion

This paper proposes a D-ALCO based failure diagnostic method for a multi-agent based software architecture. It has some advantages as follows:

(1) It can be real-time and be transparent to the system to be diagnosed;

(2) It can find the running failure of the system and determine it type and location without manual intervention, so as to support the recovery next.

(3) It is general and independent. It describes the complex interactions between the components of a system by a concise model independent of it, so as to shorten the development cycle of the diagnostic system and improve the diagnostic efficiency by separating the reasoning kernel of the diagnostic system and the model of the system to be diagnosed.

References

1. Fabre, E., et al.: Fault detection and diagnosis in distributed systems: an approach by partially stochastic Petri nets. In: Discrete Event Dynamic Systems: Theory and Applications, pp. 203–231 (August 1998)
2. Genc, S., Lafortune, S.: Distributed Diagnosis of Discrete-Event Systems Using Petri Nets. In: van der Aalst, W.M.P., Best, E. (eds.) ICATPN 2003. LNCS, vol. 2679, pp. 316–336. Springer, Heidelberg (2003)
3. Sampath, M., et al.: Failure Diagnosis Using Discrete-Event Models. IEEE Transaction on Control Systems Technology 4(2), 105–124 (1996)
4. Debouk, R., Lafortune, S., Teneketzis, D.: Coodinated Decentralized Protocols for Failure Diagnosis of Discrete Event Systems. In: Discrete Event Dynamic Systems: Theory and Applications, pp. 33–86 (October 2000)
5. Pencole, Y., Cordier, M.-O.: A formal framework for the decentralised diagnosis of large scale discrete event systems and its application to telecommunication networks. Artificial Intelligence 164, 121–170 (2005)
6. Yang, F.-Q., et al.: Some Discussion on the Development of Software Technology. Acta Electronica Sinica 30, 1901–1906 (2002)
7. Chang, L., et al.: A tableau decision algorithm for dynamic description logic. Chinese Journal of Computers 31(6), 896–909 (2008)

Reasoning about Semantic Web Services
with an Approach Based on Temporal Description Logic

Juan Wang, Liang Chang, Chuangying Zhu, and Rongsheng Dong

Guangxi Key Laboratory of Trusted Software, Guilin University of Electronic Technology,
Guilin 541004, China
wangjuan_980644@163.com, changl@guet.edu.cn,
ccrsdong@guet.edu.cn

Abstract. Temporal description logic ALC-LTL not only has considerable expressive power, but also extends the description capability of description logic from the static domain to the dynamic domain. In this paper, ALC-LTL is applied for the composition of semantic Web services. We take the view that atomic process and composite process in the OWL-S ontology can be considered as atomic service and composited service respectively. Inputs, outputs, local variables, preconditions and results of atomic processes can all be described with ALC-LTL. Based on the models of services, the executability problem and the projection problem of Web services can be reasoned about effectively.

Keywords: temporal description logic; semantic Web services; OWL-S; executability problem; projection problem.

1 Introduction

The target of semantic Web service is to support automation of Web Services discovery, composition and execution by means of adding sufficient semantic information in the description of Web services [1]. OWL-S is an ontology represented by the Web Ontology Language OWL for describing Web services on the semantic Web environment. From the point of view of knowledge representation, the description of semantic Web services should contain not only static information but also dynamic information.

Description logic is the logic foundation of OWL. Although description logic provides considerable expressive power for describing knowledge of static domain, it cannot be used effectively to describe and reason about dynamic knowledge. According to the characteristics and application requirements of semantic Web, researchers have proposed kinds of extensions of description logics. Temporal description logic ALC-LTL is proposed by Baader et al. [2] as a combination of the description logic ALC and the linear temporal logic LTL. It not only has considerable expressive power for describing knowledge of both the static domain and the dynamic domain, but also is decidable and EXPTIME-complete for the satisfiability problem of formulas [2]. In this paper, we will use ALC-LTL to model and reason about semantic Web services.

Z. Shi, D. Leake, and S. Vadera (Eds.): IIP 2012, IFIP AICT 385, pp. 286–294, 2012.
© IFIP International Federation for Information Processing 2012

Reasoning problems investigated in this paper are the executability problem and the projection problem of semantic Web services. The executability problem will check whether a semantic Web services is executable or not w.r.t. a given initial state, and the projection problem will check whether a certain formula holds or not after the successful execution of the service [3].

In recent years, many formalisms have been proposed for these reasoning problems. McCarthy et al. [4] investigate these reasoning problems with the formalism of situation calculus. Calvanese et al. [5] investigate these reasoning problems based on LTL. Baader et al. [6] propose an action formalism based on description logic; the executability problem and the projection problem are investigated in this formalism and both of them are reduced to the consistency problem of ABoxes. As an application of that formalism, Baader et al. [3] model semantic Web services as actions and then the executability and the projection problems of semantic Web services can be investigated. Chang et al. [7] proposed a family of dynamic description logics named $DDL(X^{@})$ for representation and reasoning about actions.

In this paper, starting with the Process Model of OWL-S, we take the view that atomic process in OWL-S can be considered as atomic service and composite process in OWL-S as composited service. Therefore, the reasoning about composite process can be treated as the reasoning about composited service. Inputs, outputs, local variables, preconditions and results of the atomic processes are described with ALC-LTL. Based on the modeling of composited service, executability and projection problems of semantic Web services are reasoned about.

2   Temporal Description Logic ALC-LTL

The temporal description logic ALC-LTL combines description logic ALC with linear temporal logic LTL. Syntactically, ALC-LTL uses general concept inclusion axioms, concept assertions and role assertions of the description ALC instead of atomic propositions of LTL.

Concepts of ALC-LTL are constructed inductively as follows [2]:

$$C, D ::= C_i \mid \neg C \mid C \sqcup D \mid \forall R.\, C$$

where $C_i \in N_C$, $R \in N_R$. N_C and N_R respectively be disjoint sets of concept names and role names. Moreover, $C \sqcap D$ and $\exists R.C$ are introduced as abbreviations of $\neg(\neg C \sqcup \neg D)$ and $\neg(\forall R.\neg C)$.

An expression of the form of $C \sqsubseteq D$ for two concepts C and D is called a general concept inclusion axiom(GCI).Let N_I be a set of individual names. Expressions of the forms $C(p)$ and $R(p,q)$ are called concept assertion and role assertion respectively, where $p,q \in N_I$ and $R \in N_R$.

Formulas of ALC-LTL are constructed inductively as follows [2]:

$$\varphi, \psi ::= C \sqsubseteq D \mid C(p) \mid R(p,q) \mid \neg\varphi \mid \varphi \wedge \psi \mid X\varphi \mid \varphi \mathcal{U} \psi$$

where p, $q \in N_I$, $R \in N_R$ and C,D are concepts. Using these constructors, several other constructors can be defined as abbreviations: $false := \varphi \wedge \neg \varphi$, $true := \neg false$, $\varphi \vee \psi :=$ $\neg(\neg \varphi \wedge \neg \psi)$, $\varphi \rightarrow \psi := \neg \varphi \vee \psi$, $F\varphi := true \, \mathcal{U} \, \varphi$, $G\varphi := \neg F \neg \varphi$.

GCIs, concept assertions and role assertions are called ALC-assertion. ALC-assertion and its negation are called ALC-literals.

Semantically, the interpretation structure of ALC-LTL is similar with LTL, and states are organized by the progress of time. But the mapping of every state in ALC-LTL interpretation structure is not a set of atomic propositions, but an interpretation of ALC.

An *ALC-LTL interpretation structure* [2] is a pair $M = (\mathbb{N}, I)$, where,

(1) $\mathbb{N}$ is the set of natural numbers;

(2)   For every natural number $n \in \mathbb{N}$, description logic ALC interpretation of function I is $I(n) = (\Delta, \cdot^{I(n)})$, where interpretation function $\cdot^{I(n)}$ satisfies following conditions:

①   each concept name $C_i \in N_C$ is interpreted as a subset $C_i^{I(n)} \subseteq \Delta$ of Δ;

②   each role name $R_i \in N_R$ is interpreted as a binary relation $R_i^{I(n)} \subseteq \Delta \times \Delta$ of Δ.

③   each individual name $p_i \in N_I$ is interpreted as a element $p_i^{I(n)} \in \Delta$ of Δ, and for all natural number $m \in \mathbb{N}$ such that $p_i^{I(n)} = p_i^{I(m)}$.

Given an ALC-LTL interpretation structure $M = (\mathbb{N}, I)$, the semantics of concepts and formulas are defined inductively as follows [2].

Firstly, for any natural number $n \in \mathbb{N}$, each concept C is interpreted as a set $C^{I(n)} \subseteq \Delta$. The semantics of concepts of ALC-LTL are defined inductively as follows:

(1) $(\neg C)^{I(n)} := \Delta \setminus C^{I(n)}$, where $\setminus$ is the operator of set difference;

(2) $(C \sqcup D)^{I(n)} := C^{I(n)} \cup D^{I(n)}$, where $\cup$ is the operator of set union;

(3) $(\forall R.C)^{I(n)} := \{x \mid \text{for all } y \in \Delta, (x, y) \in R^{I(n)} \text{ implies } y \in C^{I(n)}\}$.

Secondly, for any natural number $n \in \mathbb{N}$, $(M,n) \vDash \varphi$ represents formula φ is true at an instant n. The semantics of formulas of ALC-LTL are defined inductively as follows:

(4) $(M,n) \vDash C \sqsubseteq D$ iff $C^{I(n)} \subseteq D^{I(n)}$;

(5) $(M,n) \vDash C(p)$ iff $p^{I(n)} \in C^{I(n)}$;

(6) $(M,n) \vDash R(p,q)$ iff $(p^{I(n)}, q^{I(n)}) \in R^{I(n)}$;

(7) $(M,n) \vDash \neg \varphi$ iff $(M,n) \nvDash \varphi$;

(8) $(M,n) \vDash \varphi \wedge \psi$ iff $(M,n) \vDash \varphi$ and $(M,n) \vDash \psi$;

(9) $(M,n) \vDash X\varphi$ iff $(M,n+1) \vDash \varphi$;

(10) $(M,n) \vDash \varphi \mathcal{U} \psi$ iff   there is $k \geq 0$ such that $(M,n+k) \vDash \psi$ and $(M,n+i) \vDash \varphi$ for all i, with $0 \leq i < k$.

A ALC-LTL formula φ is *satisfiable* iff there is an ALC-LTL interpretation structure $M = (\mathbb{N}, I)$ such that $(M,0) \vDash \varphi$.

The most basic reasoning problem in ALC-LTL is satisfiability problem. Baader et al. [2] has proved that this problem is EXPTIME-complete.

3   Modeling of Semantic Web Services

As a new generation of Semantic Web Services framework, OWL-S [8] has a broad perspective in application. OWL-S describes a semantic Web service by providing Service Profile, Service Model and Service Grounding. Service Profile describes what the service does; Process Model specifies how the service works; Service Grounding deals with the realization of services. Our work focuses more on Process Model.

Process Model has three kinds of processes: atomic process, composite process and simple process. Simple processes are non-invokable processes. Simple processes are not taken account of when modeling of the Web services in this paper. Flight reservation service BravoAir based on OWL-S1.2 has been described by OWL alliance. It is composed of two atomic processes, GetDesiredFlightDetails and SelectAvailableFlight, and a composite process, BookFlight. The process BookFlight is the composed of two atomic processes, LogIn and CompleteReservation. LogIn and BravoAir are described as Fig.1 and Fig.2 respectively.

(1) Modeling of Atomic Process

An atomic process is a description of a service which can be executed in single step. It composes of four parts: inputs, outputs, local variables, preconditions and results.

```
define atomic process LogIn
    (inputs:(LogIn_AcctName-AcctName
        LogIn_Password-Password),
    outputs:(LogIn_Output-boolean),
    results: ((hasPassword(LogIn_AcctName,LogIn_Password)
        |->LogIn_Output(true)
        &LoggedIn(LogIn_AcctName))
        (Correct_Password-Password
        &LogIn_Password -Password
        &hasPassword(LogIn_AcctName,Correct_Password)
        |->LogIn_Output(false)
        &NotLoggedIn(LogIn_AcctName))
        )
    )
```

Fig. 1. Atomic process LogIn

Inputs and outputs reflect messages exchange between the service invoker and the services. Local variables are used for depicting preconditions and results. This example doesn't involve it. Every input, output or local variables is comprised of variable name and type declaration. For example, in Fig.1, LogIn_AccName is the variable name; its type declaration is AccName. Let variable names and type declaration are x and C respectively, it can be described by means of the concept assertion $C(x)$, that is, LogIn_AccName-AccName can be described with LogIn_AccName(*AccName*).

The preconditions represented by logical formulas describe the satisfied condition before invoking the services. Results represent the effects after executing the services. Each result is composed of a condition and a set of effects and output bindings. They can be represented by logical formulas. When the Web services are executed, corresponding effects and outputs will be generated according to the condition. The preconditions, effects and the output bindings are depicted by means of Semantic Web Rule Language (SWRL). SWRL is a kind of rule language based on OWL and has considerable expressive power. SWRL-Condition is the conjunction of atoms in SWRL. Operator "$\wedge$" is used to construct the formulas. For example, LogIn_Output(true)&LoggIn(LogIn_AccName) can be represented as LogIn_Output($true$)$\wedge$LoggIn($LogIn\_AccName$).

```
Define composite process BravoAir
    (inputs:(DepartureAirport-Airport
            ArrivalAirport-Airport
            OutboundDate-FlightDate
            ...                    ),
    outputs:(FlightsFound-FlightList
            PreferredFlightItinerary-FlightItinerary
            ReservationID-ReservationNumber),
    results:(AlwaysTrue
            |->FlightsFound<=GetDesiredFlightDetails.GetDesiredFlightDetails_FlightsFound
            &PreferredFlightItinerary<=BookFlight.BookFlight_PreferredFlightItinerary
            &ReservationID<=BookFlight.BookFlight_ReservationID
            &HasFlightItinerary(TheClient,PreferredFlightItinerary))
    )
{GetDesiredFlightDetails(GetDesiredFlightDetails_DepartureAirport<=TheParentPerform.DepartureAirport,
                GetDesiredFlightDetails_ArrivalAirport<=TheParentPerform.ArrivalAirport,
                ...                                    );
SelectAvailableFlight(SelectAvailableFlight_FlightsAvailable<=GetDesiredFlightDetails.GetDesiredFlightDetails_FlightsFound);
BookFlight(BookFlight_SelectedFlight<=SelectAvailableFlight.SelectAvailableFlight_SelectedFlight,
            BookFlight_AcctName<=TheParentPerform.BookFlight_AcctName,
            BookFlight_Password<=TheParentPerform.Password)
}
```

Fig. 2. Composite process BravoAir

(2) Modeling of Composite Process

A composite process is composed of sub composite or atomic processes by control constructs. It mainly contains inputs, outputs, results and the body of the composite process. The body of a composite process is composed of control constructs and binding relationships. For example, in Fig.2, input DepartureAirport of GetDesiredFlightDetails binds with input GetDesiredFlightDetails_DepartureAirport of Get DesiredFlightDetails when composite process BravoAir invokes atomic process GetDesiredFlightDetails. The representations of inputs and outputs are similar with atomic processes. Results are the binding of inputs and outputs of the subprocesses. For example, output GetDesiredFlightDetails_FlightsFound of subporcess GetDesiredFlightDetails binds with output FlightsFound of composite process BravoAir.

We take the view that atomic process in OWL-S can be considered as atomic service and composite process in OWL-S as composited service. In the composited service, if the inputs or outputs has binding relationship with the inputs of composite process, the inputs that have binding relationship are replaced by the inputs of the composite processes; if the inputs or outputs has binding relationship with the outputs of the composite process, the outputs that exist binding relationship, are replaced by the outputs of the composite processes.

For each service s_i of the composited service $s_0, s_1, ... s_k$ ($0 \le i < k$, $i, k \in \mathbb{N}$), let P_i be a set of precondition formulas, representing the preconditions that should be satisfied before executing the services s_i. Let E_i be a set of effect formulas, representing the effects that should be satisfied after executing the services. P_i and E_i can be described as follows:

①   For each input or local variable of the atomic process, adding the form of formulas $C(x)$ in P_i;

②   For each precondition of the atomic process, adding the form of formulas $C(x)$ in P_i;

③   For each output of the atomic process, adding the form of formulas $C(x)$ in E_i;

④   For each result r_{ij} ($j \in \mathbb{N}$) of the atomic process, it combines a precondition and a group of effects and output constraints, let the precondition be φ_j, and the effects and output constraints transform to the formula $\psi_{j1}, \psi_{j2}, ..., \psi_{jt}$ ($t \in \mathbb{N}$), then adding φ_j in P_i, adding $\psi_{j1}, \psi_{j2}, ..., \psi_{jt}$ in E_i.

4   Reasoning about Semantic Web Services

Before trying to apply the service, we want to know whether it is indeed executable, this is an executability problem. If the service is executable, we may want to know whether applying it achieves the desired effect, this is a projection problem. These problems are basic inference problems considered in the reasoning about semantic Web services [3].

In order to describe reasoning tasks, composited service system should be described first of all. ALC-LTL literals are divided into two mutually disjoint sets, L_F and L_S, where L_F represents the set of ALC-LTL literals which are true on the current situation; each ALC-LTL literal in L_S corresponds to a service name s_i, and represents that the service s_i has just been performed. For any set Q composed of formulas, we use $Conj(Q)$ to denote the conjunction of all the elements of Q. The specification of dynamic system is as follows:

(1) For each service s_i in L_S, the effects of the services are specified by means of formulas of the form (1)

$$G(Conj(P_i) \rightarrow X(s_i \rightarrow Conj(E_i))) \tag{1}$$

Formula (1) shows that service s_i is executed under the conditions denoted by $Conj(P_i)$ brings about the conditions denoted by $Conj(F_i)$.

(2) For all ALC-LTL literal F in L_F, the frame axioms as formula (2) shows:

$$G(XF \leftrightarrow \bigvee_a (P_a \wedge Xa) \vee (F \wedge \bigwedge_b (\neg P_b \vee X \neg b))) \tag{2}$$

where $F \in L_F$. The atomic services a are those services that under the circumstances described by P_a make F become true, and b are those services that under the circumstance described by P_b make F become false. All ALC-LTL literal F can be depicted by using of frame axioms.

(3) The initial situation can be described with the expression, $Conj(P_{init})$. P_{init} is a set of concept assertions such that $P_{init} \subseteq L_F$.

Taking the process BookFlight as an example, it can be described as follows:

If AcctName is *Tom*, Password is *123456*, Confirm information is *confirmation* and SelectedFlight is *FlightItineraryList*.

First of all, atomic process LogIn is modeled. If the AcctName and the Password are correct, user logs in the system successfully. This service can be described as formula (3).

$$G(Conj(P_{\text{LogIn}}) \rightarrow X(\text{LogIn} \rightarrow Conj(E_{\text{LogIn}}))) \tag{3}$$

where $Conj(P_{\text{LogIn}})$=AcctName(*Tom*)$\wedge$Password(*123456*)$\wedge$hasPassword(*Tom,123456*), $Conj(E_{\text{LogIn}})$=Output(*true*)$\wedge$LoggedIn(*Tom*).

When the output of atomic process LogIn is true, the user chooses the flight from *FlightItineraryList* and confirms it. The description of ConfirmReservation is shown as formula (4).

$$G(Conj(P_{\text{ConfirmReservation}}) \rightarrow X(\text{ConfirmReservation} \rightarrow Conj(E_{\text{ConfirmReservation}}))) \tag{4}$$

where $Conj(P_{\text{ConfirmReservation}})$=SelectedFlight(*FlightItineraryList*)$\wedge$LoggedIn(T*om*)$\wedge$ Output(*true*)$\wedge$ Confirm(*comfirmation*),$Conj(E_{\text{ConfirmReservation}})$=preferredFlightItinerary (*BeiJing-Paris*)$\wedge$ReservationID(*20123039*).

Composite process (LogIn,ConfirmReservation) can be described by formula (3) and formula (4).

Next, frame axioms are used to describe the ALC-LTL literal in L_F, for example, LoggedIn(*Tom*) can be described as formula (5).

$$X\text{LoggedIn}(Tom) \leftrightarrow ((Conj(P_{\text{LogIn}}) \wedge X\text{LogIn}) \vee (\text{LoggedIn}(Tom) \wedge$$
$$(\neg Conj(P_{\text{LogIn}}) \vee X\neg \text{LogIn}) \tag{5}$$

Finally, the initial situation can be described as formula (6).

$$Conj(P_{init})\text{=AcctName}(Tom)\wedge\text{Password}(1253456)\wedge\text{hasPassword}(Tom,123456) \tag{6}$$

A service is executable if performing it does not contradict such a truth assignment, i.e., in current situation, if F is true, then the service s_i can be executed, else s_i can't be executed.

Let Γ be the formula descriptions of the services system, the expression $Occurs(s_0,s_1,...,s_k,rs)$ be the formula (7).

$$(s_0 \wedge X(s_1 \wedge X(...X(s_k \wedge rs)...)))\wedge G(rs \rightarrow XG\neg rs) \tag{7}$$

Formula (7) expresses that the sequence of services $s_0,s_1,...,s_k$ occurs, resulting in a situation denoted by the new ALC-LTL literal rs, and rs is true only once. rs acts a marker for the situation resulting by excuting $s_0, s_1, \ldots, s_k$ [5]. Projection problem can be reduced to the validity problem of formula (8):

$$\Gamma \rightarrow (Occurs(s_0, s_1, \ldots, s_k, rs) \rightarrow G(rs \rightarrow \varphi)) \tag{8}$$

The reasoning tasks above can be transformed to the satisfiability problem. Tableau decision algorithm for ALC-LTL can be used to verify the satisfiability problem. Consequently, it makes the verification of executability and projection problems of the composited service become true.

There are two problems need to be verified. First, composited service (LogIn, ConfirmReservation) is indeed executable; second, the formula LoggedIn(*Tom*) is indeed true after executing the services (LogIn,ConfirmReservation).

Let us introduce the formula $Occurs$(LogIn,ConfirmReservation,rs), which is formula (9).

$$(LogIn \wedge X(ConfirmReservation \wedge rs)) \wedge G(rs \rightarrow XG \neg rs) \tag{9}$$

Because in current situation, AcctName(*Tom*), Password(*123456*) and hasPassword(*Tom,123456*) in P_{init} are true, LogIn can be executed. In the same way, all elements in $P_{ConfirmReservation}$ are true, ConfirmReservation can be executed. Composited service (LogIn, ConfirmReservation) is executable.

Let Γ be the description of BookFlight, the projection problem can be described as formula (10).

$$\Gamma \rightarrow (Occurs(LogIn, ConfirmReservation, rs) \rightarrow G(rs \rightarrow LoggedIn(Tom))) \tag{10}$$

Formula (10) is satisfiable, so after executing the services (LogIn, ConfirmReservation), the formula LoggedIn(*Tom*) is true.

5   Conclusion

In this paper, we take the temporal description logic ALC-LTL proposed by Baader et al. as a tool for reasoning about semantic Web services. By treating each atomic process represented in OWL-S as an atomic service, the inputs, outputs, local variables, preconditions and results of atomic processes are all described by ALC-LTL. Based on modeling of the composited service, the reasoning problems of semantic Web services, such as executability and projection problems can be carried out.

One of our future works is to study the verification of composed Web services based on the reasoning mechanisms investigated in this paper. Another work is to design decision algorithm for ALC-LTL and develop a reasoning tool for it.

Acknowledgements. This work is supported by the National Natural Science Foundation of China (Nos. 60903079, 60963010, 61163041), the Natural Science Foundation of Guangxi Province (No.2012GXNSFBA053169), and the Innovation Project of Guangxi Graduate Education (No. 2011105950812M21).

References

1. MeIlraith, S.A., Son, T.C., Zeng, H.: Semantic web services. IEEE Intelligent Systems 16(2), 46–53 (2001)
2. Baader, F., Ghilardi, S., Lutz, C.: LTL over description logic axioms. In: Brewka, G., Lang, J. (eds.) Proceeding of the 11th International Conference on Principles of Knowledge Representation and Reasoning (KR 2008), pp. 684–694. AAAI Press, Cambridge (2008)
3. Baader, F., Lutz, C., Milicic, M., Sattler, U., Wolter, F.: A description logic based approach to reasoning about Web services. In: Vasiliu, L. (ed.) Proceeding of the WWW 2005 Workshop on Web Service Semantics: Towards Dynamic Business Integration, pp. 636–647. ACM Press, Chiba (2005)
4. McCarthy, J., Hayes, P.: Some philosophical problems form the standpoint of artificial intelligence. Machine Intelligence 4, 463–502 (1969)
5. Calvanese, D., De Giacomo, G., Vardi, M.: Reasoning about actions and planning in LTL action theories. In: Fensel, D., Giunchiglia, F., McGuinness, D., Willians, M. (eds.) Proceeding of the 8th International Conference on Principles of Knowledge Representation and Reasoning (KR 2002), pp. 593–602. Morgan Kaufmann, San Francisco (2002)
6. Baader, F., Lutz, C., Milicic, M., Sattler, U., Wolter, F.: Integrating description logics and action formalisms: first results. In: Veloso, M., Kambhampati, S. (eds.) Proceeding of the 12th National Conference on Artificial Intelligence (AAAI 2005), pp. 572–577. AAAI Press, Menlo Park (2005)
7. Chang, L., Shi, Z., Gu, T., Zhao, L.: A family of dynamic description logics for representing and reasoning about actions. Journal of Automatic Reasoning 49(1), 1–52 (2012)
8. The OWL Services Coalition. OWL-S: Semantic Markup for Web Services. Technical report (2002), http://www.daml.org/services

Constraint Programming-Based Virtual Machines Placement Algorithm in Datacenter

Yonghong Yu[1,2] and Yang Gao[1]

[1] State Key Lab for Novel Software Technology, Nanjing University,
Nanjing 210093, China
[2] College of Tongda, Nanjing University of Posts and Telecommunications,
Nanjing 210003, China
yuyh.nju@gmail.com, gaoy@nju.edu.cn

Abstract. As underlying infrastructure of cloud computing platform, datacenter is seriously underutilized, however, its operating costs is high. In this paper, we implement virtual machines placement algorithm in CloudSim using constraint programming approach. We first formulate the problem of virtual machines placement in virtualized datacenters as a variant of multi-dimensions bin packing problem, and then exploit constraint solver to solve this problem with the objective of minimizing number of physical machines that host virtual machines. Finally, we compare different virtual placement algorithms for evaluating constraint programming-based virtual machine placement algorithm including the built-in virtual machine placement algorithm in CloudSim and FFD algorithm. The experimental results show that constraint programming-based virtual machines placement algorithm can efficiently reduce the number of physical machines to achieve the goal of reducing datacenter operating costs and improving resource utilization.

Keywords: Datacenter; Virtual Machine Placement; Constraint Programming; CloudSim.

1 Introduction

As an emerging computing paradigm, Cloud Computing can provide users with on demand IT services and elastic computing platforms. Many institutes and companies recently focus on cloud computing technology. Some notable companies have designed and set up their cloud computing platforms respectively, such as Amazon EC2, IBM Blue Cloud, Microsoft Windows AZure, and Salesforce Sales Force. These cloud computing platforms offer customers storage, computing power, deploying environments and IT services at different layers. The development of cloud computing abilities is closely related to supports of underlying datacenters, which provide a powerful parallel computing power and massive data storage for cloud computing.

Operating costs of datacenters is rapidly increasing with scaling of datacenters. However, datacenters resources are seriously underutilized. Energy consumptions of IT devices in servers clusters is the largest contributor to operating

Z. Shi, D. Leake, and S. Vadera (Eds.): IIP 2012, IFIP AICT 385, pp. 295–304, 2012.

costs. It is estimated that in 2006 the total electricity consumptions of datacenters in US cost of about $4.5 billion. Furthermore, US energy consumption by datacenters could nearly doubled again in 2011 [1]. However, statistics in sports, e-commerce, and financial domains show the average server utilization varies between 11% and 50%[2]. The leading reason of underutilization is that administrator allocates hardware resources for hosting applications according to peak requirements of applications, although peak workloads of applications may not appear frequently. Furthermore, workloads of applications are dynamic, as well as there are stringent requirements for applications performance, such as applications throughput, response time, and cloud computing platform must meet applications SLA(Service Level Agreement). It is a big challenge for cloud datacenters to reduce operating costs and improve resource utilization while meet applications SLAs.

Server virtualization technology is an effective approach to improve the efficiency of resources and save energy while provides performance guarantees for applications. Using server virtualization technology, application runtime environment is encapsulated in a VM(Virtual Machine), and one or more virtual machines host on a PM(Physical Machine). Virtualization technology can provide performance isolation between VMs resided on the same PM, and avoid performance degradation caused by greedy or malicious applications . Moreover, it is helpful to improve resource utilization and save costs by multiplexing of datacenters resources across applications. Furthermore, virtualization technology enables VM live migration to handle dynamic workloads and meet application performance requirements.

However, adjusting the configuration of virtual machine and performing live migration manually obviously does not meet the needs of datacenters management. We need automate and intelligent approaches to address virtual machine management issue. Management of virtual machine consist of two phases: initial plan and runtime management. In initial plan stage, given a set of virtual machines and resource requirements of each virtual machine, as well as a set of physical servers, we need map each virtual machine onto physical machines. The goal of initial plan stage is to minimize the number of physical servers that host all virtual machines. In runtime management stage, we need to adjust configuration of virtual machines or perform live migration of virtual machine according to dynamic workloads. The purpose of runtime management is to minimize the number of virtual machines migration considered migration costs, while meet application SLAs. In this paper, we focus on the problem of virtual machine placement appeared in initialization plan stage of datacenters management with the goal of minimizing the number of physical machines. We first formulate the problem of virtual machine placement as a variant of multi-dimensions bin packing problem, and then exploit constraint solver to solve this problem with the objective of minimizing number of physical machines. Finally, we implement virtual machines placement algorithm using constraint programming technique in CloudSim[3]. Experimental comparison with other heuristic methods verifies the effectiveness of constraint programming-based approach.

The remainder of this paper is organized as follows. Section 2 describes related work. Then, Section 3 formulates virtual machine placement problem and describes in detail constraint programming-based virtual machines placement algorithm. Evaluation results are presented in Section 4. Finally,Section 5 concludes this paper and presents future work.

2   Related Work

Recently, several prototypes and methods have been proposed to address problems of virtual machine management by making a trade-off between applications performance, energy consumption and migration costs.

Verma et al.[4] have proposed pMapper, which is a power-aware application placement controller in heterogeneous server clusters. pMapper exploited heuristic to address virtual machine placement problem. Moreover, working set sizes of applications are considered while placing applications on physical machines. Wood et al.[5] presented Sandpiper, a system that automates the tasks of monitoring datacenters and detecting overloaded virtual machines, determining a new mapping relations between virtual machines and physical machines. Sandpiper implements a black-box approach which is fully OS- and application-agnostic as well as a gray-box approach that can leverage information collected from OS and application. To coordinate the actions taken by platform management and virtualization management in datacenters as well as support existing multivendor solutions, Kumar et al.[6] explored a practical coordination solution that loosely-couple platform and virtualization management in datacenters. In particularly, their solution designed a stabilizer component that prevent unnecessary virtual machine migrations by making a migration decision based on probability with which such a decision remains valid over certain duration in the future. In [7], Dhiman et al. present vGreen, a multitiered software system for energy-efficient virtual machine management in virtualized clusters environment. The most important feature of vGreen is that it implements policies for scheduling and power management of virtual machine using novel hierarchical metrics which capture energy consumption and performance characteristics of both virtual machines and physical machines. To handle local optimization problem of virtual machine placement in datacenters, Hermenier et al. investigated global optimization technique and proposed Entropy resource manager for homogeneous clusters, which utilizes constraint programming[8] to perform dynamic virtual machine consolidation with the aim of minimizing migration overhead. Our work draws upon the methods used in Entropy. On the contrary, we focus on virtual machine placement problem occurred in initial plan stage, and implement virtual machine placement algorithm based on constraint programming in CloudSim.

The above studies focus on configuration of virtual machine during runtime management of datacenters. Besides the above literatures, there're still some other research work similar to our work that is only concerned about initial plan task of datacenters. For example, Campegiani et al.[9] presented a genetic algorithm to search the optimal allocation strategy of virtual machines ,and designed penalty function to deal with unfeasible solutions. Bellur et al.[10] proposed linear programming and quadratic programming techniques to solve the problem. Jing XU et al.[11] presented a two level control system to find the mapping of workloads to virtual machines and virtual machines to physical resources using an improved genetic algorithm with fuzzy multi-objective evaluation.

3   Constraint Programming-Based Virtual Machine Placement Algorithm

The problem of virtual machine placement can be formulated as a variant of multidimensional bin packing problem, which is combinatorial NP-hard problem. For this problem, several heuristics have been developed, such as FFD(First Fit Decreasing), BFF(Best First Fit), which can quickly provide sub-optimal solution. In this section, we first formulate the problem of virtual machine problem, and then describe in detail virtual machine algorithm based on constraint programming.

3.1   Formalization of Virtual Machine Problem

In virtual machine placement problem, virtual machines are viewed as boxes, where various resource requests of each virtual machine are considered as dimensions of box with non-negative values. Physical servers are considered as bins, where CPU, memory and bandwidth capacities are regarded as properties of box. The goal of virtual machine placement problem is to determine the minimum number of physical machines required by the set of virtual machines.

The problem of virtual machine placement in datacenters is defined as : given a set of virtual machines $VM = \{vm_1, vm_2, ..., vm_n\}$ and a set of physical machines $PM = \{pm_1, pm_2, ..., pm_m\}$, where each vm_i is a triplet $vm_i = (cpu_i, ram_i, bw_i), 1 \le i \le n$ denoted cpu,memory and bandwidth requirements of virtual machine respectively, each pm_j is also a triplet $pm_j = (cpu_j, ram_j, bw_j), 1 \le j \le m$ denoted resource capacity of physical machine. In addition, $x_{ij}, 1 \le i \le m, 1 \le j \le n$ and $y_i, 1 \le i \le m$ are decision variables, $x_{ij} = 1$ if and only if vm_j is mapped onto pm_i, $y_i = 1$ if pm_i is used to host virtual machine. The objective is to minimize $\sum_{i=1}^{m} y_i$ while finding all values of x_{ij} .

There are several implicit constraints in the above definition: (1) each virtual machine can be hosted on only one physical machine; (2) for each type of resource, the amounts of resource requests of virtual machines sharing the same physical machine are smaller or equal to capacity of physical machine hosting them; (3)the number of physical machines that host virtual machines are not more than m, $\sum_{j=i}^{m} y_i \le m$.

3.2   Constraint Satisfaction Model of Virtual Machines Placement Problem

In this section, we describe how to tackle virtual machine placement problem in datacenters using constraint programming approach.

Constraint programming has been widely used in a variety of domains such as production planning, scheduling, timetabling and product configuration. The basic idea of constraint programming is that user formulates a real-world problem as a CSP(constraint satisfaction problem) and then a general purpose constraint solver calculates solution for it. Formally, a CSP is defined by a triplet $(Variables, Domains, Constraints)$, where $Constraints$ are the set of constraints being responsible for pruning the search space.

Combining the definition of virtual machine placement problem and the basic idea of constraint programming, the problem of virtual machine placement can by modeled by CSP in the following way:

Variables

- $X = \{x_{ij} | i \in [1, m], j \in [1, n]\}$ denote placement solution of virtual machine in datacenters.
- $Y = \{y_i | i \in [1, m]\}$, if the physical machine pm_i is used, then $y_i = 1$.
- $Load\_CPU = \{load\_CPU_i | i \in [1, m]\}$, $load\_CPU_i$ denotes the total of cpu requests of virtual machines hosted on pm_i.
- $Load\_RAM = \{load\_RAM_i | i \in [1, m]\}$, $load\_RAM_i$ denotes the total memory requests of virtual machines hosted on pm_i.
- $Load\_BW = \{load\_BW_i | i \in [1, m]\}$, $load\_BW_i$ denotes the total bandwidth requests of virtual machines hosted on pm_i.
- $solutionNum$ denotes the number of physical machines used in the solution for virtual machine placement problem.

Domains

- $\forall x_{ij} \in X, x_{ij} \in [0, 1]$
- $\forall y_j \in Y, y_j \in [0, 1]$
- $\forall load\_CPU_i \in Load\_CPU, load\_CPU_i \in [0, pm_i.cpu_i]$
- $\forall load\_RAM_i \in Load\_RAM, load\_RAM_i \in [0, pm_i.ram_i]$
- $\forall load\_BW_i \in Load\_BW, load\_BW_i \in [0, pm_i.bw_i]$
- $solutionNum \in [0, m]$

Constraints

- C1: the total of resource requests of virtual machines sharing the same physical machine are smaller or equal to resource capacity of physical machine hosting them.
  1. $\forall i, \sum_{j=1}^{n} x_{ij} \times vm_j.cpu_j \leq pm_i.cpu_i, i \in [1, m]$
  2. $\forall i, \sum_{j=1}^{h} x_{ij} \times vm_j.ram_j \leq pm_i.ram_i, i \in [1, m]$
  3. $\forall i, \sum_{j=1}^{h} x_{ij} \times vm_j.bw_j \leq pm_i.bw_i, i \in [1, m]$
- C2: assign $y_j = 1$ if there are some virtual machines packed in physical machine pm_j. In this paper, we only consider cpu load of physical machine. $\forall y_j \in Y, y_j = 1 \Leftrightarrow load\_CPU_j > 0$
- C3: each virtual machine can be packed in only one physical machine. $\forall j, \sum_{i=1}^{m} x_{ij} = 1, j \in [1, n]$
- C4: the number of physical machines that host virtual machines are not more than m. $solutionNum = \sum_{i=1}^{m} y_i \leq m$

3.3 Implementation of Constraint Programming-Based Virtual Machine Algorithm

After formulating the virtual machine placement problem as CSP, we now choose a constraint solver to solve the problem.

Constraint programming is often realized in imperative programming by software library, such as Geocode, JaCoP and Choco. We choose Choco as our constraint solver that is compatible with CloudSim. The algorithm for constraint programming-based virtual machine placement as follows:

Algorithm 1. CP-Based Virtual Machine Placement Algorithm.

Input:
 $List < PM > pmList$: the set of physical machines.
 $List < VM > vmList$: the set of virtual machines .

Output:
 $X[m][n]$: the allocation strategy for virtual machines.
 1: Defining variables $X[m][n]$, $Y[m]$, $Load\_CPU[m]$, $Load\_RAM[m]$, $Load\_BW[m]$ and $XT[n][m]$, which is a transpose of X.
 2: Creating model for CSP : $CPModel\, model = newCPModel()$
 3: Initialing variables defined in (1).
 4: // Adding constraint C1 in CPModel
 5: **for** $i = 0; i < m$; $i + +$ **do**
 6:     $model.addConstraint(eq(load\_cpu[i], scalar(vmList.cpu, X[i])));$
 7:     $model.addConstraint(eq(load\_ram[i], scalar(vmList.ram, X[i])));$
 8:     $model.addConstraint(eq(load\_bw[i], scalar(vmList.bw, X[i])));$
 9: **end for**
10: // Adding constraint C2 in CPModel
11: **for** $i = 0; i < m$; $i + +$ **do**
12:     $model.addConstraint(ifOnlyIf(eq(Y[i], 1), gt(load\_cpu[i], 0)));$
13: **end for**
14: // Adding constraint C3 in CPModel
15: **for** $i = 0; i < n$; $i + +$ **do**
16:     $model.addConstraint(eq(1, sum(XT[i])));$
17: **end for**
18: // Adding constraint C4 in CPModel
19: $model.addConstraint(eq(usedBin, sum(Y)));$
20: $model.addConstraint(leq(usedBin, m));$
21: //Creating constraint solver and reading CPModel
22: $Solver\, s = newCPSolver();$
23: $s.read(model);$
24: $s.minimize()$ and return $X[m][n]$;

In addition, we can set timeout for this algorithm to reduce the search time of constraint solver, and constraint solver will return local or global optimal solution until timeout expired.

4   Evaluations

To evaluate constraint programming-based virtual machine placement algorithm, we implement this algorithm in CloudSim and conduct two experiments to compare with FFD algorithm.

CloudSim is developed for simulation of cloud computing platform and support system modeling of datacenters, resource management and task scheduling. In CloudSim, the task of virtual machine placement is mainly completed by VmAllocationPolicySimple class which selects a physical machine with the most number of available process units to create virtual machine iteratively. The virtual machine placement policy will use up all physical machine as long as the number of virtual machines is greater than the number of physical machines in datacenter, and this motivated us to design a virtual machine placement policy to reduce the number of physical machines that can host all virtual machines.

We define four classes of virtual machines representing different types of virtual machines respectively. The characteristics of each type of virtual machine are described in Table 1.

Table 1. Virtual machine types

VM category	CPU(MIPS)	RAM(MB)	BW(Mbit/s)
VM1	2500	870	100000
VM2	2000	1740	100000
VM3	1000	1740	100000
VM4	500	613	100000

Among four classes of virtual machine, VM1 represents Computing-intensive applications, VM2 and VM3 represents IO-intensive middle scale applications, VM4 represents small scale applications. There are two classes of physical server in datacenter. Their hardware parameters are described in Table 2.

Table 2. Hardware parameters of physical servers

PM category	CPU(MIPS)	RAM(MB)	BW(Mbit/s)
PM1	2*1860	4096	1000000
PM2	2*2660	4096	1000000

Throughout experiments we use a machine with a Intel(R) Core(TM)2 Duo CPU E7500@2.93 and 2GB memory installed CloudSim 3.0, Choco2.1.3 and JDK 1.7. The simulated datacenter consists of 50 physical machines, and each type of physical machine occupies half of the total. In order to verify constraint programming-based virtual machine placement algorithm that can efficiently reduce the number of physical machines required, we compare our approach with the built-in virtual machine placement algorithm of CloudSim and FFD at

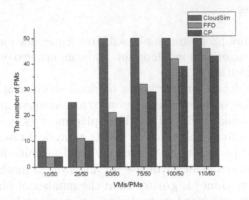

Fig. 1. The number of Physical machines used

different ratios of the number of virtual machines to physical machines, in which we set timeout = 1 second for constraint programming-based virtual machine placement algorithm. The experimental results are plot in Figure 1.

The results show that FFD algorithm and constraint programming algorithms is obviously better than CloudSim built-in algorithm in term of the number of physical machines used to host all request virtual machine in the same situations. When the ratio of number of virtual machines to number of physical machines equal to 1, CloudSim built-in algorithm occupies all physical machines, and other two algorithms only need about 20 physical machines. Comparing FFD algorithm with constraint programming-based virtual machine placement algorithm, the differences between FFD algorithm and constraint programming approach is small when the ratios are relatively lower, whereas with the increase of ratios of number of virtual machine to the number of physical machines, constraint programming approach can reduce more number of physical machines than FFD algorithm. In addition, when tmeout =1,2,3,4,5 minutes for constraint programming algorithm, we find the same result as Figure 1. It indicates that constraint programming algorithm can compute approximately global optimization solution within timeout.

To analyze the overhead of constraint programming-based virtual machine placement algorithm, we run each algorithm 10 times and average running time of each algorithms at different ratios, and setting timeout=1 minute for constraint programming approach. The experimental results are shown in Table 3.

From Table 3, we can find that the performance of FFD algorithm is optimal and valid that FFD algorithm can provide a fast and but often non-optimal solution. When the ratio of number virtual machines to number physical machines is 10/50, constraint programming-based virtual machine placement algorithm is able to return solution within timeout. However, when the number of virtual

Table 3. Average running time of three algorithms

VM/PM	CloudSim	FFD	CP
10/50	5.48ms	1.58ms	47ms
20/50	8.72ms	1.68ms	timeout
50/50	11.56ms	2.53ms	timeout
75/50	13.25ms	3.08ms	timeout
100/50	15.38ms	3.87ms	timeout

machine increase, constraint solver have not search whole space until timeout expired and only return approximately global optimal solution.

Our experiments conclude that constraint programming-based virtual machine placement algorithm can find better solution than FFD algorithm and CloudSim built-in placement algorithm with the goal of minimizing the number of physical machines hosting all requested virtual machine within timeout. Although constraint programming approach suffers from relatively long running times, we argue that time performance of virtual machine placement algorithm based on constraint programming is also acceptable because initial configuration of virtual machines is a task of planning and normally takes a long time. In practice, it is needed to trade-off between minimizing the number of physical machines and performance.

5   Conclusion and Future Work

In this paper, we focus on the problem of virtual machine placement appeared in initial plan stage of datacenters management and implement virtual machine placement algorithm exploiting constraint programming approach in CloudSim. The objective is to cut datacenters operating costs and improve resource utilization by reducing the number of physical machines and using virtualization technology. In order to achieve this objective, we formulate the virtual machine placement problem as CSP, and then choose Choco to solve this problem. We experimentally evaluate our approach and compare it with the built-in virtual machine placement algorithm in CloudSim and FFD algorithm. Experimental results show that constraint programming-based virtual machine placement algorithm can efficiently reduce the number of physical servers compared with FFD algorithm and CloudSim built-in placement algorithm, although suffers from relatively long search times.

A salient feature of cloud computing is to provide on-demand resource allocation strategy for applications in order to handle dynamic workloads. As part of future work, we are planning to study how to apply reinforcement learning and virtual machine live migration technologies to intelligently deal with dynamic peak workloads.

Acknowledgments. We would like to acknowledge the support for this work from the National Science Foundation of China (Grant Nos.61035003, 61175042,

61021062), the National 973 Program of China (Grant No. 2009CB320702), the 973 Program of Jiangsu, China (Grant No. BK2011005) and Program for New Century Excellent Talents in University (Grant No. NCET-10-0476).

References

1. Brown, R., et al.: Report to congress on server and data center energy efficiency: Public law, pp. 109–431 (2008)
2. Dasgupta, G., Sharma, A., Verma, A., Neogi, A., Kothari, R.: Workload management for power efficiency in virtualized data centers. Communications of the ACM 54(7), 131–141 (2011)
3. Calheiros, R., Ranjan, R., Beloglazov, A., De Rose, C., Buyya, R.: Cloudsim: a toolkit for modeling and simulation of cloud computing environments and evaluation of resource provisioning algorithms. Software: Practice and Experience 41(1), 23–50 (2011)
4. Verma, A., Ahuja, P., Neogi, A.: pmapper: power and migration cost aware application placement in virtualized systems. In: Proceedings of the 9th ACM/IFIP/USENIX International Conference on Middleware, pp. 243–264. Springer-Verlag New York, Inc. (2008)
5. Wood, T., Shenoy, P., Venkataramani, A., Yousif, M.: Sandpiper: Black-box and gray-box resource management for virtual machines. Comput. Netw. 53(17), 2923–2938 (2009)
6. Kumar, S., Talwar, V., Kumar, V., Ranganathan, P., Schwan, K.: vmanage: loosely coupled platform and virtualization management in data centers. In: Proceedings of the 6th International Conference on Autonomic Computing, pp. 127–136. ACM (2009)
7. Dhiman, G., Marchetti, G., Rosing, T.: vgreen: A system for energy-efficient management of virtual machines. ACM Transactions on Design Automation of Electronic Systems (TODAES) 16(1), 6 (2010)
8. Rossi, F., Van Beek, P., Walsh, T.: Handbook of constraint programming, vol. 35. Elsevier Science (2006)
9. Campegiani, P.: A genetic algorithm to solve the virtual machines resources allocation problem in multi-tier distributed systems. In: Second International Workshop on Virtualization Performance: Analysis, Characterization, and Tools (VPACT 2009), Boston, Massachusett (2009)
10. Bellur, U., Rao, C., SD, M.: Optimal placement algorithms for virtual machines. Arxiv preprint arXiv:1011.5064 (2010)
11. Xu, J., Fortes, J.: Multi-objective virtual machine placement in virtualized data center environments. In: 2010 IEEE/ACM Int'l Conference on & Int'l Conference on Cyber, Physical and Social Computing (CPSCom) Green Computing and Communications (GreenCom), pp. 179–188. IEEE (2010)

Recommendation-Based Trust Model in P2P Network Environment

Yueju Lei and Guangxi Chen

Guangxi Key Laboratory of Trusted Software Guilin University of Electronic Technology,
Guilin 541004, China
chgx@guet.edu.cn, leiyueju_422@sina.com

Abstract. Two kinds of peers' relationship are usually considered for reputation management in P2P network. One of them is direct trust relationship that the reputation is got with two peers interacting directly; the other is recommendation trust that the reputation is got by recommendation of the third party. In this paper, we presented an improved trust model based on recommended in P2P environment. In order to weight the transaction reputation and recommendation reputation, we introduced a risk value which resists the influence of false recommendation and collaborative cheating from malicious peers. The global reputation value of target peer can be calculated by using the different portions of transaction reputation, recommendation reputation and risk value. At last, we made the simulation experiments verifying the ability of resisting threats such as slandered by malicious peers, collaborative cheating, and so on.

Keywords: peer-to-peer network; transaction reputation; recommendation reputation; recommend.

1   Introduction

The trust relationship is a core part of interpersonal network relationships. The trust relationship often depends on the direct experience of entity itself to trustees or recommendation experience of other individuals. This relationship with mutual exchanges and mutual trust has formed a trusted network.

In the open P2P network, the relationship of all entities (peers) is coordinate. The open, dynamic and heterogeneous characteristics of P2P network make the entity can enter or leave freely. So, the relationship of peers in network is always dynamic changing. If there is no recommendation of trust third party or trust authority (such as the authentication center (CA)) to participate, it may be difficult to build trust relationships between entities. Therefore, it is very necessary to establish a trust mechanism to measure the credibility of peers in the P2P network.

2   Related Works

Researchers established numerous trust models, which cover the subjective trust and recommend trust [1-2]. W. T. Luke Teacy [3] et al established TRAVOS (Trust and

Z. Shi, D. Leake, and S. Vadera (Eds.): IIP 2012, IFIP AICT 385, pp. 305–310, 2012.

Reputation model for agent based on Virtual Organizations) trust model. In this model, trust value is calculated by probability theory, but lack personal experience between agents. Kamvar [4] et al presented a distrusted and secure method to compute global reputation values which is based on Power iteration. By having peers use these global reputation values to choose the peers from whom they download, the network effectively identifies malicious peers from the network. Guo Cheng [5] et al presented a trust evaluation model for quantifying the trustworthiness of peers based on recommendations. They introduced a probability method to relieve the influence of multipath propagation of trust and improved the stability of the result. Li Xiong [6] et al proposed the Peer Trust model(it is abbreviate to PTM)—a reputation-based trust supporting framework, which included a coherent adaptive trust model for quantifying and comparing the trust worthiness of peers based on a transaction-based feedback system over P2P network.

At present, trust model still has some problems and risks in P2P networks. In response to the above issues, a trust model based on recommended in the network environment is presented in this paper.

3   Trust Model

3.1   Notations

Here and after, we use the following notations in this paper:

- $i \rightarrow j$ The attitude of node i to node j; D_{ij} Direct trust value of $i \rightarrow j$
- $i \leftrightarrow j$ The transaction of node i to node j; $A(x)$ The Accuracy factor
- R_j Recommendation trust value of node j; d_x The offset factor
- $R_{x \rightarrow j}$ Feedback information of any node x to node j

3.2   Eigen Trust Model

The calculation method of trust values of node j   for local trust value in Eigen Trust model [7] is shown as follows:

1) Node i produce a local trust value for j   its direct deals with node j. Let
$$D_{ij} = S_{ij}/N_{ij} \tag{1}$$
Where S_{ij}, F_{ij} show the successful transaction times and failure times for node i; N_{ij} means the total transaction number of $i \leftrightarrow j$ in fixed time $[t_{start}, t_{end}]$; Sometimes, node i may have different opinion for a transaction. When i think the transaction is successful, but j may think it's not. So here, what we take is the attitude value of trustee to transaction. Obviously, if $N_{ij} = 0$, then $D_{ij} = 0$.

2) Set the $R_{x \rightarrow j}$ as the R_j feedback to other peers. Let
$$= (S_{xj} - F_{xj})/\sum_x S_{xj} \tag{2}$$
If $\sum_x S_{xj} = 0$ or $S_{xj} - F_{xj} < 0$, then $R_{x \rightarrow j} = 0$.

3) After iterating, we use their global reputation value as the peer's recommended trust in EigenRep model [7].

$$T_{i \to j} = \sum_x R_{x \to j} T_x \qquad (3)$$

In this model, there are two main problems. (1) The model ignores the influence of risk factor; thus $T_{i \to j}$ that we got may deviate from actual situation. (2) The model uses $T_{i \to j}$ iterative calculation as R_j; it doesn't consider the harm of malicious peers.

3.3   The Recommendation Trust Model

Direct trust Value

Definition 1. The local trust value is calculated by transaction experience directly. Take $i \to j$ for an example. The direct trust value is:

$$= \frac{\alpha S_{ij}}{N_{ij}} + \beta P(x) \qquad (4)$$

Where α, β are all the weighted factor; $0 \le \alpha, \beta \le 1$, their value depend on the importance of this transaction. $P(x)$ is punishment factor.

$$P(x) = \begin{cases} 0 & \text{normal trade} \\ -1 & \text{malicious behaviour} \end{cases} \qquad (5)$$

If D_{ij} and N_{ij} all above the system set the boundary value, then the node j can be identified as a credible node. It can make transaction without (R_j).

Recommendation Trust Value

We need to consider R_j when the peers don't trade directly or N_{ij} is too low or D_{ij} don't reach the threshold value of system. In order to response the recommendation trust value accurately, we introduce the accuracy factor $(A(x))$.

Definition 2. The calculation formula of R_j is:

$$R_j = \sum_x R_{x \to j} \times A(x) \qquad (6)$$

Where $\sum_x R_{x \to j}$ means a congregation that all the peers of recommendation information with node j transaction records.

Definition 3. Calculation formula of offset factor d_x is:

$$d_x = \left[\sum_{u \in IS(i) \cap IS(x)} |S_{iu}/N_{iu} - S_{xu}/N_{xu}| \right] / [IS_{iu} \cap IS_{xu}] \qquad (7)$$

Where IS_{iu} represents the congregation of $i \leftrightarrow u$, $IS_{iu} \cap IS_{xu}$ represents all the common peers of $i \leftrightarrow x$. Obviously, $0 \le d_x \le 1$. If $IS_{iu} \cap IS_{xu} = 0$, that means there is no common node of $i \to x$, let the initial value $d_x = 0.5$.

Definition 4. The calculation formula for accuracy factor of feedback node is:

$$A(x) = 1 - d_x^{1/s} \qquad (8)$$

Where $A(x) \le 1; s = 1,2,3 \cdots$

$A(x)$ is low enough, then the credibility of recommendation peer should be suspected. So, we set the boundary value, if $A(x) \le 0.1$, R_j can be neglected.

The risk Value

It exist a certain risk when any entity (node) interact with others in a virtual environment. So the calculation formula of risk value is:

$$R(x) = \xi D_{ij} + (1 - \xi) \sum_x R_{x \to j} \times A(x) \tag{9}$$

Where ξ is a weight, it should be set a relatively small value. Because of dynamic and heterogeneous of peers in P2P network, it is difficult to find the credible third party. The risk value of d_x is larger than D_{ij}.

The global reputation value

Definition 5. The calculation formula of global reputation value of node j is:

$$T_{i \to j} = \mu D_{ij} + (1 - \mu) R_j - R(x) \tag{10}$$

Where μ represents a weighted factor according to its trading experience.

We discuss the following cases:

1) In case, $\mu = 0$. The node itself has no definite view. Its trust value mainly depends on other peers' recommendation feedback.

2) In case, $\mu = 1$. The node only believes the history experience of itself. It is skeptical the feedback information of other recommendation peers.

3) In case, $0 < \mu < 1$.at beginning $T_{i \to j}$ is limited, so μ is a little proportion, with the accumulation of trading, the peers are more likely to trust their own experience. Therefore, the weighted factor is dynamic change. Let

$$\mu = 1 - \sigma^{k/n} \tag{11}$$

Where the value ranges of σ is $0.5 < \sigma < 1$; k shows the k time transactions; n shows total number of transactions and value range is $n \in \{1, 2, 3 \cdots\}$. Thus in this case, μ we get is fit for the situation and mode of actual transaction. Therefore, formula of $T_{i \to j}$ is:

$$T_{i \to j} = \mu D_{ij} + (1 - \mu) \sum_x R_{x \to j} \times A(x) - R(x) \tag{12}$$

Finally, we compare $T_{i \to j}$ with threshold value of system ($T_{i \to j0}$). If $T_{i \to j} > T_{i \to j0}$, then the node is credible and can be traded; otherwise, the node isn't credible.

4   The Experiment and Analysis

In this section, we will assess the performance of our recommendation trust model as compared to the Eigen Trust model and Peer Trust model in P2P network. Take the downloading files for instance.

4.1   The Simulation Experiment

Experimental environment: the number of peers is 100; total number of files is 500. We distribute 500 documents to 100 peers randomly, and make sure that each node has a document at least. The node has this file if it downloads successfully; otherwise, the peer downloads failing. We set that the number of honest peers is more than the number of malicious peers, and the first transaction request is launched by honest nodes.

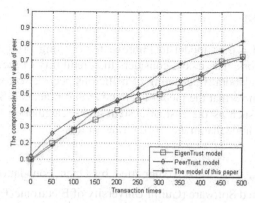

Fig. 1. The influence of transaction times to trust value

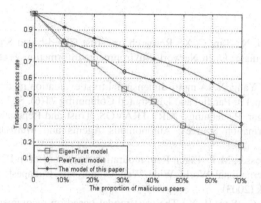

Fig. 2. Transaction success rate under different proportion of malicious peers

4.2   The Analysis

Figure 1: Eigen Trust model not consider the accuracy of recommendation peers, with the accumulation of transactions, $T_{i \to j}$ is increasing. PTM evaluates the reliability of peers by depending on other peer's recommendation completely, but it ignores the personal interactive historical experience. The model of this article considers transaction record and accuracy of recommendation peers. Therefore, $T_{i \to j}$ of target peer is real. Experimental results show that the accuracy and validity of the model.

Figure 2: After 50 transactions on average, we set the honest peer to evaluate other peers. From the results, the transaction success rate of each model has different degree dropping with gradually increasing in the proportion of the malicious peers. When the number of malicious peers is more than 40%, the error rate is high in Eigen Trust model and transaction success rate falls rapidly. For Eigen Trust model, it is almost hard to run if there are many malicious peers in environment. But no matter the size of proportion of malicious peers, PTM and the model of this article still remain relatively high transaction success rate. The results confirmed that the model has a certain safety and can play a role in the real application.

5   Conclusion

We have presented a trust model to improve for existing problems of Eigen Trust model in P2P network. The model considers the credibility of recommendation peers and prevents the collaborative cheating of malicious peers effectively. In P2P simulation, the results show that the model is effective and practical. But we not involved in some problems such as the cost of system about calculation trust value of peer, timeliness, etc. We hope that in the work further study.

Acknowledgement. This work is supported by Open Foundation of Guangxi Key Laboratory of Trusted Software (Guilin University of Electronic Technology).

References

1. Yu, D.-G., Chen, N., Tan, C.-X.: Research on Trust Cloud-based Subjective Trust Management Model under Open Network Environment. Information Technology 10(4), 759–768 (2011)
2. Li, X., Zhou, F., Yang, X.: A multi-dimensional trust evaluation model for large-scale P2P computing. Journal of Parallel and Distributed Computing 71(6), 837–847 (2011)
3. Teacy, W.T., Jigar, P., Jennings, N.R.: TRAVOS: Trust and Reputation in the Context of Inaccurate Information Sources. Autonomous Agents and Multi-agent Systems 12(2), 183–198 (2006)
4. Kamvarsd, Schlosser, M.T.: Eigen Rep: reputation management in P2P networks. In: Proc of the 12th International World Wide Web Conference Budapest, pp. 123–134. ACM Press, New York (2003)
5. Guo, C., Li, M.-C., Yao, H.-Y.: The trust model based on recommended under P2P network. Computer Engineering 34(24), 157–159 (2008)
6. Xiong, L., Liu, L.: Peer Trust: supporting reputation-based trust for peer-to-peer electronic communities. IEEE Trans. on Knowledge and Data Engineering 16(7), 843–857 (2004)
7. Dou, W., Wang, H.-M., Jia, Y., Zou, P.: A recommendation-Based Peer-to-Peer Trust Model. Journal of Software 15(4), 571–583 (2004)

Frequency-Adaptive Cluster Head Election in Wireless Sensor Network

Tianlong Yun[1], Wenjia Niu[1], Xinghua Yang[1], Hui Tang[1], and Song Ci[1,2]

[1] High Performance Network Lab, Institute of Acoustics, Chinese Academy of Science, Beijing
[2] University of Nebraska-Lincoln, Omaha, NE 68182, USA
{niuwj,yangxh,tangh,sci}@hpnl.ac.cn,
tyun@ieee.org

Abstract. Efficient information routing mechanism is a critical research issue for wireless sensor networks (*WSN*) due to the limit energy and storage resource of sensor nodes. The clustering-based approaches (e.g. *LEACH*, *Gupta* and *CHEF*) for information routing have been developed. Although these approaches did actually improve the routing efficiency and prolong the network lifetime, the clustering frequency f is usually pre-designed and fixed. However, the network context (e.g. Energy and load) often dynamically changes, which can provide the dynamical adjustment determination for f. Hence, in this paper, we propose a frequency-adaptive cluster-head election approach, which applies the network context for making corresponding f adjustment. Furthermore, an f -based clustering algorithm is presented as well. The case study and experimental evaluations demonstrate the effectiveness of the proposed approach.

Keywords: Wireless Sensor Networks, network context, frequency adaptive, clustering.

1 Introduction

The rapid development of wireless sensor network (WSN) enables the information gathering from many physical environments [1, 2]. In WSN, sensor nodes usually have very limited energy and memory space, but they are densely deployed to guarantee the reliability of sensing. The typical structure of a sensor network includes base station and a lot of homogeneous node [3]. In this structure, base station is in charge of initiating data requests and all the sensor node should communicate with base station to make the sensing data delivered. In order to reduce the overhead, some nodes will act as storage nodes and receive the data request for other common nodes in their vicinity [4], then compress the data request result from many nodes and send it to the base station. This method has reduced some energy consumption, however, "hotspot" problem may be generated [5, 6]. More specifically, one storage node will not be replaced by other high-energy storage node or even common node. Hence, some storage node will run out its energy and dies at an early stage, which will further affect the data transmission to the base station and effective sensing in some area.

Z. Shi, D. Leake, and S. Vadera (Eds.): IIP 2012, IFIP AICT 385, pp. 311–319, 2012.

Recently, to solve the hotspot problem, clustering based technique has begun to be used. Typical work involves the LEATH [7], Gupta [8] and CHEF [9]. In these methods, storage node will be replaced by new node which has more energy. The other nodes will select a storage node as their agent so that they don't have to communicate with base station directly. In all of these methods, clustering is done periodically. A completed operation of clustering for one time is called "round". All of the nodes will have to act as storage node sooner or later at some rounds and all of these methods can effectively utilize the overall energy of the sensor network.

However, in such methods, we think that a new round of clustering didn't always happen at a proper time. All of them have a fixed frequency of clustering so that a large amount of energy has been wasted because of the unnecessary re-clustering. For instance, in a WSN-based intelligent building application, the sensing frequency of temperature at working hours should be higher than that at off-duty hours. We will analysis a specific scenario to demonstrate the disadvantage of traditional method. First, we assume that clustering is performing faster than it should be. As we mention above, at off-duty hours when there are barely any data request, the frequency of clustering should be slow down. Otherwise, the new storage node will be elected anyway even if the energy distribution is still balanced. Another situation is that the frequency of clustering is lower than it should be. In this case, the change of network context cannot be utilized immediately. A storage node could run out its energy before the next round of clustering. Hence, in this paper, we propose a re-clustering frequency-adaptive routing approach which utilizes the current network context to achieve effective load balance between all the nodes. Our efforts mainly focus on two aspects: firstly, we define a function to determine when to start a new round of clustering for cluster-head election; secondly, based on this function, we design a corresponding algorithm to improve node's energy consuming.

The rest of the paper is organized as follows. Section 2 discusses the related work, followed by the proposed approach in Section 3. Section 4 presents experimental results that illustrate the benefits of the proposed scheme. Section 5 concludes the paper.

2   Related Work

Our approach has inspired by a variety of related work, the most famous one is the LEATH. In this approach, the author not only put forward the method itself, but also a completed clustering protocol for WSN. To better distribute the energy consumption among the nodes, LEACH employs a stochastic model which will select several nodes from the network to act as the storage node. The elected storage node is also called as cluster head. The cluster head will then advertise message and all of the other nodes will keep their receiver on to receive the advertisement. The common nodes will choose the cluster head which is nearest to them. Then the common nodes will notice the cluster head it joins. After the cluster head confirms its entire cluster member, it will build a schedule to tell the common node the time to transfer the data. Finally when the cluster gathers all the data of its member, it will compress it and send data packets to the base station.

Table 1. Characteristic between the 3 methods

	LEACH	Gupta	CHEF
Clustering Method	Stochastic	Fuzzy Logic	Fuzzy Logic
Base Station process	No	Yes	No

This method is fully distributed and the base station didn't have to involve in the clustering process. The nodes will generate the random number by themselves. And CSMA will be used to prevent the congestion when nodes exchange the message [10].

This method improves the use of the overall energy but didn't take the network context into consideration. Based on this method, Gupta has put forward a method using fuzzy logic. Three descriptor, energy, concentration and centrality of fuzzy logic are used to better balance the energy consumption. The process of the clustering is conducted in the base station while the CHEF think it is inefficient. CHEF improves Gupta's method by combining the LEATH and Gupta together as well as making the method distributed. Each node will first generate a random number to determine whether or not to elect for a cluster head. The node whose random number is lower than the pre-designed threshold will then do the fuzzy logic calculation, and some of them will be selected as cluster head (See Table.1).

Although above methods have prolonged the lifetime of the network, the clustering frequency is often fixed. Hence, we try making the clustering frequency dynamically changed based on network context, and aim to achieve better energy consumption.

3   Cluster-Head Election Model

In this paper, we propose an approach which can be aware of the data request frequency and then adjust the corresponding clustering frequency f. We call this approach frequency-adaptive cluster-head election. The aim of our approach is to balance and reduce the energy consumption by dynamically change the re-clustering frequency. The proposed clustering method is executed by the base station because it is not sensitive to energy consumption. At the initial state, all the cluster head will be assigned randomly. And later, the base station will only communicate with the cluster head for requesting data while other nodes will update their sensing data to their cluster head periodically. When it comes to the re-clustering part, the base station will query the energy information of all the nodes, and let the old cluster head pass their token to the new cluster head [11]. The operation of our method breaks into several rounds. The first round has slightly different because there is no cluster head existing for base station to communicate with. Prior to detail approach description, we present the network organization.

3.1   Network Organization

The WSN network consists of plenty of sensor nodes and a base station which is located far away from the nodes (See Fig.1). In our approach, we assume that all the

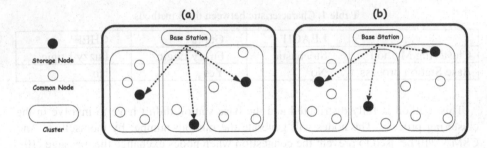

Fig. 1. The network organization of the method

sensor nodes are deployed in a square zone. The sensors themselves are homogenous and stationary. The base station will obtain the geographic locations of all the nodes using GPS technology [12]. Some nodes will be assigned as storage nodes which will in charge of communicating with the base station. The storage node will be initiated by clustering method based on the position and energy level of them. The new storage node will be assigned at another round of clustering while the happen time of the clustering is highly adaptive and dynamic.

3.2   System Architecture

Both the base station and the sensor node have their own system architecture. We will start with the description of the base station.

The base station is in charge of initiating a data request, communicating with sensor node and doing clustering. Fig.2 shows that there are five modules in the base station for clustering. Interface with application layer module will communicate with the application layer to get the need for sensing data and data request module will coordinate with other module to finish the data request task; interface with the node module will communicate with the nodes to get the sensing data and clustering information; the database module will maintain the data of the nodes and request result; the clustering module will conduct the calculation of the clustering.

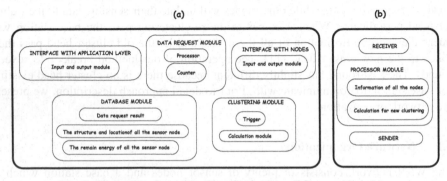

Fig. 2. The architecture of the Base Station(a) and Storage Node(b)

The structure of the storage node is less complicated (See Fig.2 (b)). Both sender and receiver are necessary parts for any node. For the storage node, it also maintains all the information of the member node and a calculation module which can be used to generate the index of changed nodes.

3.3   The Election Process

For the first round of clustering, some node have to be selected as the initial cluster head randomly based on the model in LEATH. First, all of the nodes will generate a random number between 0 and 1. If the number is less than a pre-designed chance parameter P, the percentage of storage nodes in the network, the node will elect itself as cluster head.

Then the elected cluster head should broadcast a notification message for other common nodes. All of the common node which its random number is higher than P should keep their receiver on. These common nodes should also count the number of the advertisement. Once the number exceeds h, the pre-designed cluster amount, it will choose the strongest one as its cluster head. Then the common node will send an answer message, in this time, the storage nodes must keep their receiver on. At the end, all of the storage nodes will have a sheet which includes the ID of all the members and the sheet will be sent back to the base station.

The description above is only applicable for the first time clustering. If there already some cluster heads exist, the cluster process will be different. The new round of clustering in base station is triggered by specific events that we will describe the algorithm to get the specific event in following part. Once the clustering event is triggered, the base station will send a notice to the entire old storage node. And the old storage node will notify its member node to ask for the energy level of them.

The common nodes will return their remaining energy level to the storage node and cluster head will return the information back to the base station. The base station will do the calculation which will be described specifically in the following fuzzy logic clustering controller part. After that, the base station will determine the new cluster head as well as its member nodes and will send messages to the old storage node with the index of new storage node and the member nodes they should include. And the old storage node will transfer its "head token" and the message from the base station to the new one.

The new storage node has to advertise its state as well. However, this advertisement will include the ID of its member node and the common node doesn't have to answer with another message because the storage already has the corresponding sheet. The index for the new storage node will only include the changed member node So that the overhead is reduced but the old storage should do some extra computation.

When the agent in application layer inquiry some data based on the need of service, the inquiry will be sent to the base station. After identifying the data request, the base station will look up the matching table to check which node should be checked and the corresponding storage node. Finally, the base station will directly get the data from the target storage node.

The calculation of the new cluster head uses the fuzzy logic controller [13, 14]. The controller has two descriptors: energy (the energy remain of the node) and distance (the sum of distances between the node and the nodes which is within r distance).A node has more chance to be selected as storage node if the value of the two descriptors is high. After the fuzzy logic operation, the base station will only notice the old storage node with index of nodes which have been changed.

A new round of cluster operation is triggered by specific events. The specific event will be defined as follows: A counter will count the incoming data request amount. Once the amount exceeds the threshold of the counter, a new round of clustering will start. The threshold of the counter is determined by the energy level of the overall network (The base station is aware of the energy level of each node, although it is delayed). Because the storage node is more likely to run out of its energy at a short time if the whole network is at a low energy level, so the next round of clustering should happen in time. Base on the overall remain energy and variance, counter threshold (T) for round t is showed below (k is a constant value depends on the network; N is the cluster number; X_{ij} is remain energy level of each node):

$$T_t = k \frac{\sum X_{it}}{\sqrt{\sum (X_{it} - \frac{1}{N} \sum X_{it})^2}} \tag{1}$$

Preliminaries :
N: the number of clusters;
C_i: cluster $i(i = 0, 1, \dots ,N\text{-}1)$;
S_i: the storage node of C_i;
X_{ji}: the sensor node in C_i;
$S(X_{ji})$: remain energy level of nodes
B: base station
T: threshold of clustering
M: counter number of base station
```
begin
while M exceed T do
for each storage node Si in the network do
Si=>B: state inform packet;
end for
while B receives all the state inform packet do
conduct clustering using fuzzy logic;
end while
for each storage node Si in the network do
B=>Si: clustering result inform packets;
for each node Xji in the cluster Ci do
if (Xji == new storage node) then
transfer index and token to the new storage node
end if
end for
end while
end
```

Our approach prevents the demerit we found in LEATH and other fuzzy logic-based approach at some specific situation. First, remain energy is used to balance the energy consumption of the nodes. Then, it is reasonable that the new round of clustering is based on the amount of incoming data request.

4   Experimental Evaluations

In previous sections, we have described the structure and method of our method. In this section, we will do some experiment and validate the effectiveness of the proposed approach. We implement the proposed method using MATLAB fuzzy control tool box and we use similar environment setup assumptions to test it to compare its performance with former methods.

The basic structure is built upon the LEATH protocol. In order to be compared to the energy consumption with former typical method, we use the energy consumption assumption in LEATH, CHEF and Gupta.   And we deployed 400 nodes into in 200*200 areas. The simulation time will be divided into several periods, in the peak periods, there are more data requests from the base station while in other periods, the frequency of data request is relatively low. And the data request obeys Poisson distribution.

Having collected 100 sets of data, we can see a slight improvement in the lifetime of the network if the incoming data request is largely depended on time (See Fig.4 (a)). However, if the data request is stable, the network lifetime of our approach is shorter than CHEF but still longer than LEACH.

Also we can compare another important parameter ΔT, the difference between the dead time of the first node and the last node. We can find that the time difference in our approach is shorter than the former method LEATH and CHEF. The first node fails at the 513rd round and the last node fail at the 678th round so that the ΔT for out method is 165 rounds. The ΔT for LEATH and CHEF is 203 and 187 respectively. This result shows that the energy consumption has a balanced distribution which is good because the intact network has a better performance. If some node fails at a early stage, the sensing cannot be performed in the corresponding zone.

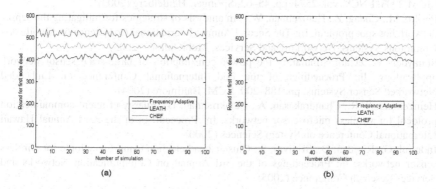

Fig. 3. Network Lifetime in Time-depend (a) and Stable (b) situation

5   Conclusion

This paper extends the existing clustering method to prolong the lifetime of WSN at some specific situation. The proposed approach can reduce energy consumption when major fluctuations exist in the data request frequency. This improvement is realized by making the clustering frequency f flexible. We put forward a simple but effective method with adaptive clustering frequency adjusting to better utilize and balance the overall energy of all the nodes. In addition, although most of the computation is carried on the base station, the message transmitted between node and the base station can be controlled at a reasonable level at the same time. The simulation result shows that our approach is a meaningful and a reasonable attempt for the improvement in fuzzy logic based approach. In the future, we will focus on extending our method so that it can be used for more complex situations.

Acknowledgements. This research is supported by the National Natural Science Foundation of China (No. 61103158), the National S\&T Major Project (No. 2010ZX03004-002-01), the Deakin CRGS 2011 and the Securing CyberSpaces Research Cluster of Deakin University, the Sino-Finnish International S\&T Cooperation and Exchange Program (NO. 2010DFB10570), the Strategic Pilot Project of Chinese Academy of Sciences (No.XDA06010302), and the Ultra-realistic Acoustic Interactive Communication on Next-Generation Internet (No. 11110036).

References

1. Akyildiz, I.F., et al.: Wireless sensor networks: a survey. Computer Networks 38(4), 393–422 (2002)
2. Yick, J., Mukherjee, B., Ghosal, D.: Wireless sensor network survey. Computer Networks 52(12), 2292–2330 (2008)
3. Sohrabi, K., et al.: Protocols for self-organization of a wireless sensor network. IEEE Personal Communications 7(5), 16–27 (2000)
4. Ghose, A., Grossklags, J., Chuang, J.: Resilient Data-Centric Storage in Wireless Ad-Hoc Sensor Networks. In: Chen, M.-S., Chrysanthis, P.K., Sloman, M., Zaslavsky, A. (eds.) MDM 2003. LNCS, vol. 2574, pp. 45–62. Springer, Heidelberg (2003)
5. Perillo, M., Cheng, Z., Heinzelman, W.: An analysis of strategies for mitigating the sensor network hot spot problem. In: The Second Annual International Conference on Mobile and Ubiquitous Systems: Networking and Services, MobiQuitous (2005)
6. Shnayder, V., et al.: Simulating the power consumption of large-scale sensor network applications. In: Proceedings of the 2nd International Conference on Embedded Networked Sensor Systems, pp. 188–200. ACM, Baltimore (2004)
7. Heinzelman, W.R., Chandrakasan, A., Balakrishnan, H.: Energy-efficient communication protocol for wireless microsensor networks. In: Proceedings of the 33rd Annual Hawaii International Conference on System Sciences (2000)
8. Indranil, G., Riordan, D., Srinivas, S.: Cluster-head election using fuzzy logic for wireless sensor networks. In: Proceedings of the 3rd Annual on Communication Networks and Services Research Conference (2005)

9. Jong-Myoung, K., et al.: CHEF: Cluster Head Election mechanism using Fuzzy logic in Wireless Sensor Networks. In: 10th International Conference on Advanced Communication Technology, ICACT 2008 (2008)
10. Dam, T.V., Langendoen, K.: An adaptive energy-efficient MAC protocol for wireless sensor networks. In: Proceedings of the 1st International Conference on Embedded Networked Sensor Systems, pp. 171–180. ACM, Los Angeles (2003)
11. Akkaya, K., Younis, M.: A survey on routing protocols for wireless sensor networks. Ad Hoc Networks 3(3), 325–349 (2005)
12. Mao, G., Fidan, B., Anderson, B.D.O.: Wireless sensor network localization techniques. Computer Networks 51(10), 2529–2553 (2007)
13. Ross, T.J.: Front Matter. In: Fuzzy Logic with Engineering Applications, p. i-xxi. John Wiley & Sons, Ltd. (2010)
14. Haining, S., Qilian, L.: Wireless Sensor Network Lifetime Analysis Using Interval Type-2 Fuzzy Logic Systems. In: The 14th IEEE International Conference on Fuzzy Systems, FUZZ 2005 (2005)

A Cluster-Based Multilevel Security Model for Wireless Sensor Networks

Chao Lee[1,2], Lihua Yin[2], and Yunchuan Guo[3]

[1] Institute of Computing Technology, Chinese Academy of Sciences
Beijing, China
lichao@software.ict.ac.cn
[2] Graduate University of Chinese Academy of Sciences
Beijing, China
[3] Institute of Information Engineering, Chinese Academy of Sciences
Beijing, China

Abstract. Wireless sensor network is one of the fundamental components of the Internet of Things. With the growing use of wireless sensor networks in commercial and military, data security is a critical problem in these applications. Considerable security works have been studied. However, the majority of these works based on the scenarios that the sensitivities of data in the networks are in the same. In this paper, we present a cluster-based multilevel security model that enforces information flow from low security level to high security level. The design of the model is motivated by the observation that sensor nodes in numerous applications have different security clearances. In these scenarios, it is not enough for just protecting the data at a single level. The multilevel security mechanism is needed to prevent the information flow from high level nodes to low level nodes. We give the formal description of the model and present a scheme to achieve it. In our model, sensor nodes are grouped into different clusters. In each cluster, the security clearance of sensor nodes must not be higher than the security clearance of the cluster head. We use cryptography techniques to enforce the information flow policy of this model. The higher level nodes can derive the keys of lower level nodes and use the derived key to get the information from lower-level nodes. *abstract* environment.

Keywords: wireless sensor network, multilevel security, information flow control.

1 Introduction

The Internet of Things has trended to growing use in commercial and military areas. It has been paid more and more attentions[1,2]. Wireless sensor network (WSN) is one of the fundamental components of the Internet of Things, which has attractive many researchers[3]. A typical wireless sensor network is composed of a large number of sensor nodes and one or several base stations, which are used to collect data from the sensor nodes. The base station broadcasts

Z. Shi, D. Leake, and S. Vadera (Eds.): IIP 2012, IFIP AICT 385, pp. 320–330, 2012.

control messages to engage sensor nodes to do some specific tasks. In response, the sensor nodes send back the collected data to the base station. They use radio frequency channels to broadcast messages for communication. Since the communication among sensors is via radio, which is exposed in the air, wireless sensor networks are highly vulnerable to security attacks. Security requirements for sensor networks have attractive many attentions[4,5,6]. However, the majority of these works are designed to provide uniform security across the network, which means that all the sensor nodes and information have the same security clearance and sensitivity. There are various scenarios that sensor nodes in WSNs play different security levels. For example, in a wireless sensor network operating in a battlefield, the data collected by platoon leader node can be read by battalion commander node but cannot be read by soldiers. The command broadcasted to all battalion commander nodes can be received by the nodes whose security clearances are higher than battalion commander, but can never be received by the nodes whose clearances are lower than it. Take metropolitan surveillance application as another example, the police can see all data, but citizens can only see a subset of the data. This type of applications with multiple priority groups demands different layers of sensed data and multilevel security model in sensor networks. This motivation is the main reason to develop multilevel security in WSNs.

In this paper, we propose a cluster-based multilevel security model to address the problem, in which all sensor nodes and cluster heads have different security clearances. The WSN is modeled as a tree, in which the base station is the root, and each cluster is the subtree. In each cluster, the security clearances of all nodes are lower than the clearance of the cluster head, and the clearances of nodes are decreased from the root to leaf. In our model, each information has a classification, and only the nodes whose clearance is higher than the classification can read and relay the information. We give the formal description of the model and achieve the prototype of it.

To achieve this model, we present a scheme to build the multilevel topology in each cluster, and a multilevel key computation scheme to enforce the information flow control. In this model, each node belongs to different security level. The cluster head election algorithm is used to elect cluster heads with different security levels. The key computation process is initialized by the base station. It computes the keys of all cluster heads according to the clearance of each cluster head. And the key of a sensor node in a cluster is based on the cluster head key and the sensor node's clearances. In other words, the keys are clearance-related. If an upper-level node S_H wants to read an information which flows from a lower-level node S_L and the information is encrypted by the key of S_L, then S_H must have the ability to derive the key of S_L from its own key to decrypt the information.

The rest of this paper is organized as follows: Section 2 presents the related work. In section 3, we describe the systems containing multiple sensitive data and users with different access privilege. Our proposed multilevel security model is described in Section 4. Section 5 provides the achievement scheme of the model.

Section 6 discusses our proposed scheme. Finally, the conclusion can be found in Section 7.

2   Related Work

In the past, the research of multilevel security (MLS) has mainly focused on operating systems, programs, and wired computer networks. For example, In [7], Bell and La Padula proposed the classical BLP model for operating system multilevel access control. In [8], Lu and Sundareshan proposed a model to describe the mechanism that enforces the security policy and requirements for computer network. In [9], Winjum and Berg described a MLS scheme for computer network routing information. Very recently, however, multilevel security communication in wireless sensor networks have begun to attract the attentions. In [10], Teng proposed a multi-layer encryption(MLE) scheme for multilevel access control in wireless sensor networks. In this work, users with different security clearance are assigned different group keys. Lower level users' key are computed by an one-way hash function from the keys of higher level users. However, the scheme doesn't enforce the information flow from low-security to high-security in networks. In [11], Panja proposed a scheme called role-based access in sensor networks which provides role-based multilevel security in sensor networks. Each group is organized in such a way that they can have different roles based on the context and thus can provide different levels of accesses. They organized the network using Hasse diagram then compute the key for each individual node and extend it further to construct the key for a group. In [12], Lee and Singhal introduced the concept of multiple security levels (MSL), which segregates different security levels by using a different computing infrastructure. They proposed an architecture to achieve the MLS property based on MSL concept. They divided the network to several domains and guards. A security domain is a discrete network consisting of a set of nodes having the same security level. The guards monitor and control the information flows among different security domains. However, their scheme assumes that all the same security level nodes can join the same group without the consideration of the communication range.

3   System Description

Before we describe the WSN with different access privilege, let us give the definitions of some terms.

- *Security Class*: Security class is a sensitivity level of an entity (e.g. UNCLASSIFIED, CONFIDENTIAL, SECRET, TOP SECRET). Let SC denote the set of security classes, which corresponds to a set of disjoint classes of sensitivity level. $SC = \{L_1, L_2, \ldots L_n\}$, where n is a finite integer.
- *Dominate Relation*: We denote $L_i \succeq L_j$ to say that the security class L_i dominates(or covers) L_j. $L_i \preceq L_j$ holds whenever $L_j \succeq L_i$. We then define the relation $\succ$ by $L_i \succ L_j \Leftrightarrow L_i \succeq L_j \wedge \nexists L_k \in SC.L_i \succeq L_k \succeq L_j$.

- *Clearance*: Clearance is the degree of trust associated with a subject. We define the clearance of a subject s is $CL(s) = [L_i, L_j] \in SC \times SC$, where $L_j \succeq L_i$, which means that s can read information at class L_j or lower, and write information at class L_i or higher. We denote $CL_\perp(s) = L_i$ and $CL_\top(s) = L_j$.
- *Classification*: Classification is a security class assigned to an object which specifies the sensitivity of the object. We denote $CF(o) = L_i \in SC$ as the classification of o.

In our model, we assume the WSNs consisting of sensor nodes, cluster heads and base station. Sensor nodes, which are grouped into clusters, probe the environment to track the target, and then send the collected data to the cluster heads. Each sensor has limited resources and short radio transmission range. If some of them have more than one hop from the cluster, they send their data to relaying nodes in the cluster and finally communicate with the cluster head. Cluster heads have more resources than sensors. They can execute relatively complicated numerical operations than sensors and have much larger radio transmission range than sensor nodes. Each cluster head is assumed to be reachable to all sensors in its cluster. Cluster heads can communicate each other directly and relay data between its cluster members and the base station.

In the system, we consider two modes. One is data collection, and the other is command distribution. A login user gets the collected data or distributes command via the base station. Each node and user have security clearances $[L_i, L_j]$. Any information transmitted over the WSN must be designated a security classification L_k where $L_i \preceq L_k \preceq L_j$ according to the sensor's task. For example, a sensor whose clearance is [*soldier, soldier*] can just generate information of *soldier* classification, while a sensor whose clearance is [*soldier, platoon_commander*] can generate information of *soldier* or *platoon_commander* according to the task or the collected data sensitivity. In the data collection mode, if the information collected by a sensor node is designated the classification *platoon_commander*, only the nodes whose clearances dominant the classification *platoon_commander* can get and relay the information, the login user on the base station whose clearance cover *platoon_commander* can read the data. In the distribution mode, a battalion commander whose clearance is [*soldier, battalion_commander*] wants to distribute a command to all platoon commanders, the command is designated the classification of *platoon_commander*, only the sensor whose clearance dominant *platoon_commander* can receive the command, but all the *soldier* nodes cannot get the information. In the system, the information transmitted over the network can only be allowed to send to the node whose security level is equal or higher than with information classification. We assume appropriate network communication protocols designed to ensure reliable information transmission across the network.

4   A Multilevel Security Model for WSN

We first give the definition of *information flow relation*.

Definition 1 (Information Flow Relation). *For subjects $S', S'' \in S$, and information $i \in O$. $S' \overset{i}{\rightsquigarrow} S''$ defines information i can flow from S' to S''. The information flow relation $\rightsquigarrow$ is defined as follows:*

$$S' \overset{i}{\rightsquigarrow} S'' \Leftrightarrow CL_{\bot}(S') \preceq CF(i) \preceq CL_{\top}(S'')$$

For example, in a cluster, there are two sensors S' and S''. The clearance of S' is $[soldier, soldier]$. Similarly, the clearance of S'' is $[soldier, commander]$. S' can send information with *soldier* classification to S''.

To build the multilevel security cluster, we present *completely dominate relation*.

Definition 2 (Completely Dominate Relation). *For two sensors $S', S'' \in S$, we say $CL(S'')$ completely dominate $CL(S')$ if and only if $CL_{\top}(S') \preceq CL_{\top}(S'') \wedge CL_{\bot}(S') \preceq CL_{\bot}(S'')$. We denote it $CL(S') \trianglelefteq CL(S'')$.*

Proposition 1. $\trianglelefteq$ *is a partial ordering relation.*

Proof. It easy to see that $\trianglelefteq$ is reflexive and antisymmetric. Below we just proof that $\trianglelefteq$ is transitive. Assume $[L_i, L_j] \trianglelefteq [L_m, L_n]$ and $[L_m, L_n] \trianglelefteq [L_r, L_s]$, according to Definition 2, we have $L_i \preceq L_m$, $L_j \preceq L_n$, $L_m \preceq L_r$ and $L_n \preceq L_s$. Since $\preceq$ is transitive, we can get $L_i \preceq L_r$ and $L_j \preceq L_s$. So $[L_i, L_j] \trianglelefteq [L_r, L_s]$, $\trianglelefteq$ is transitive. Therefore $\trianglelefteq$ is a partial ordering relation.

In our model, we represent each cluster by a tree $T(V, E)$, where V represents a set of sensors and E represents a set of communication links. Each node is denoted by $(ID, CL(ID))$.

Definition 3 (Multilevel Security Cluster). *We define a multilevel security cluster as tuples*

$$MLSC = (CH, S, CL, \trianglelefteq)$$

where CH is the cluster head. S is the set of sensor nodes (members) contained in the cluster. CL is clearance of subjects. $\trianglelefteq$ is the completely dominate relation. A cluster is a multilevel security cluster if and only if:

(1) For $\forall S', S'' \in S$, $S'.parentID = x$, $S'.parentID = y \Rightarrow x = y$.
(2) $S'.parentID = S''.ID \Rightarrow CL(S') \trianglelefteq CL(S'')$.

We denote the set of multilevel security cluster by MSLCs.

The Condition (1) points out that a sensor node only has a unique parent node, as only one next-hop node to be allowed from lower level to upper level. When a node S' joining a cluster, if there are more than one nodes S_i, S_j,...,S_m, and $S' \trianglelefteq S_i$, $S' \trianglelefteq S_j$, ..., $S' \trianglelefteq S_m$, then S' chooses the closest one to be its parent. The Condition (2) can ensure the secure information flow.

Lemma 1. *In a multilevel security cluster, $S', S'' \in S$ are two sensors, and S'' is the parent of S':*

(1) In the data collection mode, information i can flow from S' to S'' all the time.

(2) In the command distribution mode, information i can flow from S'' to S' only when $CF(i) \preceq CL_\top(S')$.

Proof. In a multilevel security cluster, there must be $CL(S') \trianglelefteq CL(S'')$. That is $CL_\perp(S') \preceq CL_\perp(S'') \wedge CL_\top(S') \preceq CL_\top(S'')$. In the data collecting mode, information i is generated by S', so $CL_\perp(S') \preceq CF(i) \preceq CL_\top(S')$. We can easily get $CL_\perp(S') \preceq CF(i) \preceq CL_\top(S'')$. According to Defintion 1, i can flow S' to S'' all the time. In the command distribution mode, we can get $CL_\perp(S'') \preceq CF(i) \preceq CL_\top(S'')$. If $CF(i) \preceq CL_\top(S')$, and $CL(S') \trianglelefteq CL(S'')$, then $CL_\perp(S'') \preceq CF(i) \preceq CL_\top(S'')$, i can flow S'' to S'. Otherwise, the flow is prevented.

Definition 4 (Cluster-based Multilevel Security Model for WSN). *The WSN cluster-based multilevel security model is defined by*

$$\text{WSN-CMLSM} = (MSLCs, BS, U, P, I, \rightsquigarrow)$$

where

- *MSLCs is the set of multilevel security clusters.*
- *BS is the base station. All the $C \in MSLCs$ connect to BS.*
- *U is the set of users. Each login user whose clearance is $[L_i, L_j]$ can read collected data whose classification is not higher than L_j, and write distribution command with classification not lower than L_i.*
- *P is the security policy $(SC, \preceq)$, which is defined by a lattice.*
- *I is the information transmitted over the WSN.*
- *$\rightsquigarrow$ is the information flow relation.*

In Figure 1, we illustrate an example of our scheme. The topology of the network is generated by the method in Section 5.

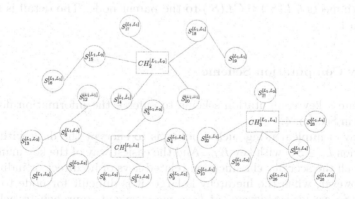

Fig. 1. Cluster-based multilevel security model exmamle

Let $S_i^{[L_i,L_j]}$ denote a sensor node i with the clearance $[L_i, L_j]$. For example, $S_1^{[L_1,L_2]}$ denotes sensor node 1 whose clearance is $[L_1, L_2]$, and $S_2^{[L_1,L_3]}$ presents a node whose identity is 2 and clearance is $[L_1, L_3]$.

If the node $S_8^{[L_1,L_2]}$ sends the data d_1 to the cluster head $CH_1^{[L_1,L_3]}$, it transmits the data through the path $S_8^{[L_1,L_2]} \rightarrow S_3^{[L_1,L_3]} \rightarrow CH_1^{[L_1,L_3]}$.(We will give the scheme of generating it in section 5).

Proposition 2. *The information flow in the WSN-CMLSM is secure.*

Proof. According to Lemma 1, we can easily prove it. Because of space constraint, we omit the details.

5 A Scheme to Achieve the Multilevel Model

We proposed a scheme to provide multilevel security for wireless sensor networks. The scheme consists of two parts. We first organized the sensors as a cluster-based multilevel security topology. And then the hierarchical keys on the topology are computed to enforce the information flow from low to high.

5.1 Multilevel Security Clusters Building

Each sensor node performs Algorithm 1 to build the multilevel security clusters. It takes as input $Range(S_i, CHs)$ and $Range(S_i, S)$, which means the cluster heads and sensor nodes in the communication range of S_i respectively. It outputs the routing information of S_i. In the algorithm, we assume that each sensor can detect the sensors and cluster heads in its communication range. It choose the nearest cluster head which satisfies the $CL(S_i) \trianglelefteq CL(CH)$ to the parent node. If it not exists the satisfied cluster head, the sensor node choose the nearest node which meets $CL(S_i) \trianglelefteq CL(S')$ to the parent node. The detail is shown in Algorithm 1.

5.2 Key Computation Scheme

We presente a key computation scheme to enforce the information flow policy discussed in Section 4.

One way of implementing such a policy is to encrypt the data with security classification $L_i \in SC$ with key K_{L_i}. And the easiest way of the key management is to hold all the security classification related keys of its direct or indirect child nodes. However, when the hierarchy is large, it is difficult for node to store all the keys. So we utilize *dependent keys management* approach to achieve the multilevel security policy.

Algorithm 1. Multilevel security clusters building algorithm

Data: $Range(S_i, CHs)$,$Range(S_i, S)$
Result: $S_i.parentID$

```
1  begin
2  |   if Range(Sᵢ, CHs) ≠ ∅ then
3  |   |   foreach CH ∈ Range(Sᵢ, CHs) do
4  |   |   |   if CL(Sᵢ) ⊴ CL(CH) then
5  |   |   |   |   Set_Candidate_CH ← CH;
6  |   |   |   end
7  |   |   end
8  |   |   if Set_Candidate_CH ≠ ∅ then
9  |   |   |   Chosen_CH = Nearest(Set_Candidate_CH);
10 |   |   |   Sᵢ.parentID = Chosen_CH.ID;
11 |   |   end
12 |   end
13 |   else
14 |   |   foreach S' ∈ Range(Sᵢ, S) do
15 |   |   |   if CL(Sᵢ) ⊴ CL(S') then
16 |   |   |   |   Set_Candidate_Node ← S';
17 |   |   |   |   Chosen_Node = Nearest(Set_Candidate_Node);
18 |   |   |   |   Sᵢ.parentID = Chosen_Node.ID;
19 |   |   |   end
20 |   |   end
21 |   end
22 end
```

Cluster Head Key Computation. The keys of cluster heads are computed in the base station. Assume the base station has the knowledge of the relation of security class before the WSN deployment is reasonable. $(SC, \preceq)$ are organized as a lattice. Each security class has many direct successors and predecessors. The base station uses one-way functions $H_1, H_2, \ldots H_m$ to compute the dependent keys, where m is the maximum number of children per node. If a security class L_j is directly covered by L_i whose key is K_i; and if L_j is the kth child of L_i, then $K_j = H_k(K_i)$. Moreover, if L_j has more than one direct parents $L_j^1, L_j^2, \ldots, L_j^m$, where L_j is the c_1th,$\ldots$, c_mth child of the parent $L_j^1, L_j^2, \ldots, L_j^m$, then $K_j = H_{c_1}(H_{c_1}(K_{L_j^1}), H_{c_2}(K_{L_j^2}), \ldots, H_{c_m}(K_{L_j^m}))$. According to the scheme, the key belongs to high security class can derived from the key of low security class. The key of L_i is denoted by K_{L_i}.Take Figure 2(a) as an example, let K_i is the key of L_i, where $1 \leq i \leq 7$, then we can compute $K_2 = H_1(K_1)$, $K_3 = H_2(K_1)$, $K_6 = H_3(H_3(K_2), H_1(K_3))$.

Sensor Node Key Computation. Since in each cluster, the sensor nodes are organized as a tree not a lattice, and the above cluster head key computation scheme is too energy consuming for sensor nodes, we compute the sensor node key as follows:

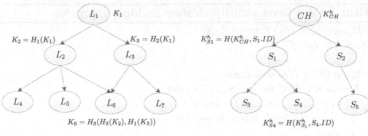

(a) security class key example   (b) sensor node key example

Fig. 2. Key computation scheme examples

1. Each node S_i computes the hierarchical key through a one-way hash function $K_{S_i}^h = H(K_{S_i.parent}^h, S_i.ID)$.

   Take Figure 2(b) as an example, $K_{S_4}^h$ can be computed from $K_{S_3}^h$ via $K_{S4}^h = H(K_{S_1}^h, S_4.ID)$.

2. In [13], Blundo et al, establish pair-wise keys using a bivariate t-degree polynomial $f(x, y) = \sum_{i,j=0}^{t} a_{ij}x^i y^j$, which has the property of $f(x, y) = f(y, x)$. For example, node S' and S'' compute the symmetric key to secure communication. let i and j are the ID of S' and S'' respectively. Node S' have $f(i, y)$, and it computes the key $f(i, j)$ to communicate with node S''. Node S'' can derive $f(j, i)$ through $f(j, y)$.

   The sensor node computes the communication key by $K_{S'}^c = K_{S'}^h \oplus f(S'.ID, y)$, where f is the polynomial in [13], y is the IDs of the nodes that connected to S'.

When sensor node S' sends the data to the cluster head. It encrypts the data by $K_{S'}^c$, and sends to its parent S''. S'' can compute the $K_{S'}^c$ through $f(S''.ID, S'.ID)$ and $K_{S'}^h = H(K_{S''}^h, S'.ID)$. S'' can get the information of S', and it forwards the message to its parent.

6   Theoretical Analysis

Theorem 3 *For security class L_i and L_j, if $L_i \preceq L_j$, then K_{L_i} can be derived from K_{L_j}.*

Proof. Assume $L_m, L_n, \cdots, L_r, L_s$ are between L_i and L_j, we can easily get $K_{L_m} = H(K_{L_i}), K_{L_n} = H(K_{L_m}), K_{L_{n+1}} = H(K_{L_n}), \cdots, K_{L_s} = H(K_{L_r})$, so $K_{L_j} = H(H(H(...H(K_i)...)))$. Therefore, if $L_i \preceq L_j$, K_{L_j} can derive K_{L_i}.

Proposition 4. *Information M is transited over the WSN, only the sensor node S' satisfied $CF(M) \preceq CL_T(S')$ can get the information.*

Proof. The information M $(CF(M) = L_m)$ is encrypted by K_{L_m}. When the information from the base station to sensors, only the cluster head satisfied

$L_k \preceq CL_\mathsf{T}(CH)$ can get the information, as K_{CH} can derive from K_{L_m} according to Theorem 3. The other cluster heads cannot get the information because of their inability to compute K_{L_m}. The cluster heads satisfied $L_m \preceq CL_\mathsf{T}(CH)$ send the information to their members. Since in MLSC, any sensor node S satisfies $CL(S) \trianglelefteq CL(S.parent)$ and the hierarchical key is computed by the topology, if $CF(M) \preceq CL_\mathsf{T}(S'),S'$ can derive K_{L_m} from $K_{S'}^h$. Therefore, S' can get the information of M. When the information is from the sensor to the base station, assume M is generated by a node S. In MLSC, the sensor's next hop must satisfy $CL(S') \trianglelefteq CL(S'.parent)$, so M can only send through the path $CF(M) \preceq CL_\mathsf{T}(S')$, according to Theorem 3, the sensors can derive K_{L_m}. the node $CF(M) \npreceq CL_\mathsf{T}(S')$ is not on the routing path of M, and the K_{L_m} cannot be derived, so only the sensor node S' satisfied $CF(M) \preceq CL_\mathsf{T}(S')$ can get the information.

Because of space constraint, the details of simulation will be presented in a separate paper.

7   Summary

In this paper, we proposed a cluster-based multilevel security model for wireless sensor networks. We divided the sensors in different clusters and modeled each cluster as a tree. In each cluster, the clearance of the cluster head completely dominated its member's clearance. Each sensor node has a unique parent, and the clearance of the node is completely dominated by its parent. The information flow policy is proposed to ensure the information flow low level to high level. We present a scheme to achieve it. In our scheme, cryptographic technologies are employed to enforce the information flow policy. The high level node can derive keys of the low level nodes, while the low-level node cannot derive keys of the high level nodes. Therefore, information can only flow from low to high that satisfies the requirement of the scenarios that the sensor nodes have different sensitivities.

Acknowledgment. This research is supported by the National High Technology Research and Development Program of China (863 Program) (2009AA01Z438), and the National Natural Science Foundation of China (61070186,61100186).

References

1. Atzori, L., Iera, A., Morabito, G.: The Internet of Things: A survey. Computer Networks 54, 2787–2805 (2010)
2. Weber, R.H.: Internet of Things C New security and privacy challenges. Computer Law & Security Review 26, 23–30 (2010)
3. Zhao, F., Guibas, L.J.: Wireless Sensor Networks: An Information Processing Approach. Morgan Kaufmann, San Francisco (2004)

4. Wang, Y., Attebury, G., Ramamurthy, B.: A survey of security issues in wireless sensor networks. IEEE Communications Surveys and Tutorials 8, 2–23 (2006)
5. Xiao, Y., Rayi, V.K., Sun, B., Du, X., Hu, F., Galloway, M.: A survey of key management schemes in wireless sensor networks. Computer Communications 30, 2314–2341 (2007)
6. Simplicio Jr., M.A., Barreto, P.S.L.M., Margi, C.B., Carvalho, T.C.M.B.: A Survey on Key Management Mechanisms for Distributed Wireless Sensor Networks. Computer Networks 54(15), 2591–2612 (2010)
7. Bell, D.E., LaPadula, L.J.: Secure computer systems: mathematical foundations and model.Technical Report M74-244, MTR (1973)
8. Lu, W.-P., Sundareshan, M.K.: A Model for Multilevel Security in Computer Networks. IEEE Trans. Softw. Eng. 16(6), 647–659 (1990)
9. Winjum, E., Berg, T.J.: Multilevel security for ip routing. In: Military Communications Conference 2008, pp. 1–8. IEEE Press, New York (2008)
10. Teng, P.-Y., Huang, S.-I., Perrig, A.: Multi-layer Encryption for Multi-level Access Control in Wireless Sensor Networks. In: Proceedings of The IFIP TC 11 23rd International Information Security Conference, pp. 705–709 (2008)
11. Panja, B., Madria, S.K., Bhargava, B.: A Role-based Access in a Hierarchical Sensor Network Architecture to Provide Multilevel Security. Comput. Commun. 31, 793–806 (2008)
12. Lee, J., Son, S.H., Singhal, M.: Design of an Architecture For Multiple Security Levels in Wireless Sensor Networks. In: 7th International Conference on Networked Sensing Systems (INSS), pp. 107–114. IEEE Press, New York (2010)
13. Blundo, C., De Santis, A., Herzberg, A., Kutten, S., Vaccaro, U., Yung, M.: Perfectly-Secure Key Distribution for Dynamic Conferences. In: Brickell, E.F. (ed.) CRYPTO 1992. LNCS, vol. 740, pp. 471–486. Springer, Heidelberg (1993)

A New Security Routing Algorithm Based on MST for Wireless Sensor Network

Meimei Zeng and Hua Jiang

Guilin University of Electronic Technology, Guilin, Guangxi, China
zmmmmz123@163.com

Abstract. In order to solve the general problems of information overlap, low energy utilization rate and network transmission security in routing protocols of wireless sensor networks, a scheme of energy-efficient and security-high routing for wireless sensor networks (EEASHR) based on improved Kruskal algorithms is proposed in this paper. The proposed scheme takes the energy size required for transmission and the value of reliability between nodes and nodes as the edges value of graph, then uses improved Kruskal algorithm to generate the minimum spanning tree (MST) in sink node, in other words, the optimal route path. The simulation in NS2 shows that the proposed algorithm improves network energy efficiency, reduces the node packet loss rate and prolongs the life cycle of wireless sensor networks.

Keywords: Kruskal, wireless sensor networks (WSNs), reliability, energy efficiency.

1   Introduction

With the development of Internet of Things, wireless sensor networks which are the ending net of Internet of Things meets more stringent requirements in security and energy consumption. It urges us to carry out more profound study on the development of wireless sensor networks.

Because of the resource constraints of sensor network nodes and the bad distributed environment, the tasks of improving the energy efficiency of sensor network nodes, and enhancing the network security are unusually formidable.

The following features should be included in a safe and efficient routing protocol. 1) Reduce the impact of the configuration error; 2) Reduce overall network energy consumption; 3) Ensure that only legitimate nodes participate in message transmission; 4) Prevent attackers from injecting spoofed routing information. All kinds of characteristics of wireless sensor networks make the routing protocols vulnerable for the attacks of spoofed routing information, such as selective forwarding, sinkhole, sybil attack, wormhole and so on[1-2]. Traditional security mechanisms based on the encryption system can't resolve the internal attacks from the nodes which are easily captured in an open environment[3-4]. A lot of research work has been done to prolong the network life cycle and improve the network security transmission index[5-6].

Z. Shi, D. Leake, and S. Vadera (Eds.): IIP 2012, IFIP AICT 385, pp. 331–336, 2012.
© IFIP International Federation for Information Processing 2012

According to the network topology, the routing protocols can be divided into flat and hierarchical routing protocols, and the relations between the nodes and the sink node are usually established in the form of multi-hop. The typical flat routing protocols include Flooding, Gossiping. Among them, there are the problems of information overlaps, abuses of resources, and poor expansibility in the Flooding[7] and Gossiping[8] protocol. Reference[9] shows an idea to send data by sending meta data to consult, but the reliability is poor. Besides the energy of nodes around the sink node is easy to run out. Nodes easily are captured. At present, the existing routing protocols are mainly focused on how to enhance the energy utilization of nodes or the credibility of network nodes[4][10].

2   System Model

2.1   Network Model

Assume that N sensor nodes are randomly distributed within a square area S, and will be no longer mobile after deployment. The unique sink node is deployed outside area S. It is the network that doesn't need human maintenance after the deployment. The network topology is relatively stable, and the mobility is small; All nodes are isomorphic with the same initial energy, data integration and unique identities (ID). These nodes do not require GPS equipment and do not need to know their specific location through methods of measurement; Wireless transmission power is controllable, namely nodes can adjust the size of the transmission power based on distance.

2.2   Wireless Energy Model [11]

Wireless energy model refers to the attenuation of the transmitted power decays exponentially with the transmission distance increases. For free space model, the transmit power was d^2 attenuation when the distance between the sending and receiving nodes is $d \prec d_0$, while multipath attenuation model the transmit power was d^4 attenuation under the same situation.

2.3   Wireless Transmission Energy Model [12]

Wireless transmission energy model as formula (1) and (2) shown, formula (1) indicates the energy loss of launching k bit data, and it is composed of the loss of launch circuit and the loss of power amplifier. E_{elec} means the needed energy which communication module receives or sends per bit data. ε_{fs}、 ε_{amp} respectively show the needed energy that launch amplifier transmits one bit data under two kind of channel models, d_0 is distance $d_0 = sqrt(\frac{\varepsilon_{fs}}{\varepsilon_{amp}})$. Formula (2) stands for the energy of receiving k bit data, which only depends on circuit loss. We assume the radio channel transmits information symmetrically, which means energy consumption from v to u equals that from u to v.

$$E_{T_x} = \begin{cases} k \times E_{elec} + k \times \varepsilon_{fs} d^2 & d \prec d_0 \\ k \times E_{elec} + k \times \varepsilon_{amp} d^4 & d \geq d_0 \end{cases} \tag{1}$$

$$E_{R_x} = k \times E_{elec} \tag{2}$$

2.4 The Method of Excluding Malicious Nodes

In order to void nodes with low confidence become members of transmission path nodes, excluding malicious nodes in networks. This paper improved Reliability evaluation mechanism proposed in paper [13]. As shown in formula (3), V_{I_b} stands for information offered by Intrusion detection system, 1 indicates good node, 0 indicates malicious node; V_{I_r} is an active observation value obtained by monitoring neighbor nodes actively; V_{S_r} is an indirect observation value obtained by exchanging information with neighbor nodes; f_1, f_2, f_3 can be set different values in accordance with the specific requirements of the network, we assume $f_1 + f_2 + f_3 = 1$ and normalize them.

$$E = [f_1(E_{T_x} + E_{R_x}) + f_2(1 - V_{I_r}) + f_3(1 - V_{S_r})] \times V_{I_b} \tag{3}$$

3 Algorithm Idea

Without considering the space difference, the proposed scheme abstracts nodes of the wireless sensor network into the vertices of the graph, and abstracts communication relation among network nodes into the edge of the graph between vertex and vertex. Thus, the wireless sensor network can be represented as graph G= (V, E), in addition, V represents the set of sensor network nodes and E represents communication relation among network nodes. The proposed scheme takes the energy required for transmission and the value of reliability between nodes and nodes as the edges of graph, using improved Kruskal algorithm to solve the problem existing in Flooding algorithm.

3.1 Kruskal Algorithm[14]

Kruskal algorithm is a classical algorithm to generate a minimum spanning tree. In this paper, the idea of minimum spanning tree is used to construct optimal routing paths in wireless sensor networks. The improved algorithm based on reference [13] as follows:

Assume that V is a set of vertices in the graph; E is the set of edges in the graph; RE is the set of edges of the optimal routing path. Optimal routing path can be achieved through the improved Kruskal algorithm.

1. Initialization: RV= {v_0, v_1,......, v_{n-1}},RV indicates a subgraph that includes n nodes but no edges.
2. If E={}, then output optimal path R, and the algorithm ended.
3. From the current position back to search (u,v) in sequential E (G) to get minimum weight edge (u,v). it should be noted that u and v are in different connected components. Namely (u,v) indicates the minimum weight edge. Move u and v into RE, merge them as one connected component and update E at the same time. If E{ (u, v) =0}, then move edge (u,v).
4. Turn to the step 2.

3.2   Optimal Route Path

The algorithm consists of two parts: building the optimal route and steady operation stage. In each round, the algorithm real-time updates the optimal route path through setting a time stamp on the sink node.

1. The node in network to broadcast request packets which carry the node information (ID) of senders and the E value of sender node orderly marked by node.
2. At last, request packet information can converge on sink node along with the transmission of the request packets in network. Sink node can obtain the topology of the entire network , ID, and the E value.
3. The sink node uses improved Kruskal algorithm to generate an optimal route path.
4. The sink node sends out query information based on the optimal path, then perception data converge to sink node along with the reverse path of query information. The parent node uses data integration to process data to reduce the data traffic, as Fig. 1 and Fig. 2 shown.

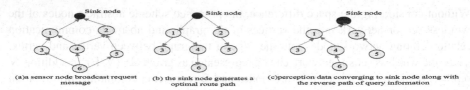

(a)a sensor node broadcast request message   (b) the sink node generates a optimal route path   (c)perception data converging to sink node along with the reverse path of query information

Fig. 1. The schematic plot for routing idea under the situation of no malicious nodes

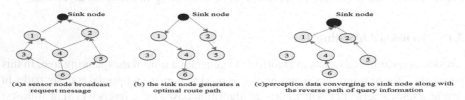

(a)a sensor node broadcast request message   (b) the sink node generates a optimal route path   (c)perception data converging to sink node along with the reverse path of query information

Fig. 2. The schematic plot for routing idea that No.3 is a malicious node

4   Simulation and Analysis

The proposed method is simulated via C++ program based on NS2 simulation platform. The main parameters of the sensor network are as follows:100 nodes are evenly distributed in 50*50 geographical area which is named S .The sink node is outside the S. The energy of every node is initially E_T =0.5J(E_T is the total energy.).The optimal routing path is generated by Kruskal algorithm in sink node. In order to validate the performance of proposed algorithm, the experiment made a comparison with Flooding algorithm .The network lifecycles of different protocols in the same initial energy are shown in Fig. 3.We can see that the lifecycle of proposed algorithm is nearly 80 times than Flooding algorithm (lifecycle ended until the death of the last node). The packet loss rates along with different number of malicious nodes are shown in Fig. 5. To simplify the experiment, malicious nodes are set manually. From Fig. 6 we can see that the packet loss rate of the EEASHR is generally reduced by about 25% compared with Flooding in the same circumstances.

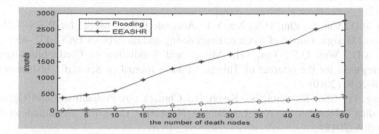

Fig. 3. network lifecycle of different protocol in the same initial energy

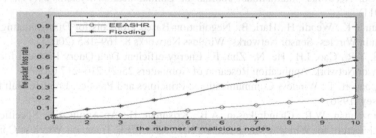

Fig. 4. network packet loss rate in different number of malicious nodes

5   Conclusions and Future Works

We have proposed EEASHR, an emerged efficient algorithm that uses the minimum spanning tree technique to provide data transport security in wireless sensor networks.

We have evaluated the cost, in terms of the network lifetime and the packet loss rate of the proposed Kruskal scheme and compared it with the Flooding protocol. The experiments showed that the proposed scheme can make full use of the node energy, and it also can significantly improve the WSN lifetime. At the same time, the proposed algorithm takes the reliability of nodes into consideration, thus it enhances network transition security by preventing malicious nodes from losing or interfering packets. In the future, we plan to study the hierarchical network topology which is suitable for large network nodes.

References

1. Kifayat, K., Merabti, M., Shi, Q.: Security in Wireless Sensor Networks. In: Staroulakis, P., Stamp, M. (eds.) Handbook of Information and Communication Security, pp. 513–552. Springer, Berlin (2010)
2. Jaydip, S.: A Survey on Wireless Sensor Network Security. International Journal of Communication Networks and Information Security 2, 55–78 (2009)
3. Boukerch, A., Xu, L., El-khatib, K.: Trust-based Security for Wireless Ad Hoc and Sensor Networks. Computer Communications 30, 2413–2427 (2007)
4. Wu, Y.F., Zhou, X., Feng, R.J., Wan, J.W., Xu, X.F.: Secure Routing based on Node Trust Value in Wireless Sensor Networks. Chinese Journal of Scientific Instrument 1, 221–227 (2012)
5. Kan, B.Q., Cai, L., Zhu, H.S., Xu, Y.J.: Accurate Energy Model for WSN Node and Its Optimal Design. Journal of Systems Engineering and Electronics 19(3), 427–433 (2008)
6. Hu, X.D., Wei, Q.F., Tang, H.: Model and Simulation of Creditability-based Data Aggregation for the Internet of Things. Chinese Journal of Scientific Instrument 31(11), 2636–2640 (2010)
7. Chang, D., Cho, K., Choi, N., Kwon, T., Choi, Y.: A Probabilistic and Opportunistic Flooding Algorithm in Wireless Sensor Networks. Computer Communications 35(4), 500–506 (2012)
8. Tang, J.H., Dai, S.S., Li, J.H.: Gossip-based Scalable Directed Diffusion for Wireless Sensor Networks. International Journal of Communication Systems 24(11), 1418–1430 (2011)
9. Joanna, K., Wendi, H., Hari, B.: Negotiation-Based Protocols for Disseminating Information in Wireless Sensor Networks. Wireless Networks 8, 169–185 (2002)
10. Cui, Y.R., Cao, J.H., He, N., Zhu, F.: Energy-efficient Data Query Protocol for Wireless Sensor Network. Application Research of Computers 25(2), 216–217 (2008)
11. Rappaport, T.: Wireless Communications: Principles and Practice. Prentice Hall Inc., New Jersey (1996)
12. Heinzelman, W.R., Chandrakasan, A.B., Krishnan, H.: An Application Specific Protocol Architecture for Wireless Micro Sensor Networks. IEEE Trans. on Wireless Communications, 660–670 (2002)
13. Zhang, J., Xu, L., Xu, D.W.: Secure Clustering Algorithm based on Trust Evaluation in Ad Hoc Network. Computer Application 10(27), 2426–2429 (2007)
14. Levin, M.S., Zamkovoy, A.: A Multicriteria Steiner Tree with the Cost of Steiner Vertices. Journal of Communications Technology and Electronics 5(12), 1527–1542 (2011)

A Novel Model for Semantic Learning and Retrieval of Images

Zhixin Li[1], ZhiPing Shi[2], ZhengJun Tang[1], and Weizhong Zhao[3]

[1] College of Computer Science and Information Technology,
Guangxi Normal University, Guilin 541004, China
[2] College of Information Engineering, Capital Normal University, Beijing 100048, China
[3] College of Information Engineering, Xiangtan University, Xiangtan 411105, China
lizx@gxnu.edu.cn, shizhiping@gmail.com, zjtang@gxnu.edu.cn,
zhaoweizhong@gmail.com

Abstract. In this paper, we firstly propose an extended probabilistic latent semantic analysis (PLSA) to model continuous quantity. In addition, corresponding EM algorithm is derived to determine the parameters. Then, we apply this model in automatic image annotation. In order to deal with the data of different modalities according to their characteristics, we present a semantic annotation model which employs continuous PLSA and traditional PLSA to model visual features and textual words respectively. These two models are linked with the same distribution over all aspects. Furthermore, an asymmetric learning approach is adopted to estimate the model parameters. This model can predict semantic annotation well for an unseen image because it associates visual and textual modalities more precisely and effectively. We evaluate our approach on the Corel5k and Corel30k dataset. The experiment results show that our approach outperforms several state-of-the-art approaches.

Keywords: semantic learning, automatic image annotation, continuous PLSA, aspect model, image retrieval.

1 Introduction

As an important research issue, Content-based image retrieval (CBIR) searches relative images of given example in visual level. Under this paradigm, various low-level visual features are extracted from each image in the database and image retrieval is formulated as searching for the best database match to the feature vector extracted from the query image. Although this process is accomplished quickly and automatically, the results are seldom semantically relative to the query example due to the notorious semantic gap [5]. As a result, automatic image annotation has emerged as a crucial problem for semantic image retrieval.

The state-of-the-art techniques of automatic image annotation can be categorized into two different schools of thought. The first one defines auto-annotation as a traditional supervised classification problem [4, 12], which treats each word (or semantic category) as an independent class and creates different class models for every word.

Z. Shi, D. Leake, and S. Vadera (Eds.): IIP 2012, IFIP AICT 385, pp. 337–346, 2012.
© IFIP International Federation for Information Processing 2012

This approach computes similarity at the visual level and annotates a new image by propagating the corresponding class words. The second perspective takes a different stand and treats image and text as equivalent data. It attempts to discover the correlation between visual features and textual words on an unsupervised basis [1, 2, 7, 8 10, 11, 13, 15], by estimating the joint distribution of features and words. Thus, it poses annotation as statistical inference in a graphical model.

As latent aspect models, both PLSA [9] and latent Dirichlet allocation (LDA) [3] have been successfully applied to annotate and retrieve images. PLSA-WORDS [15] is a representative approach, which acquires good annotation performance by constraining the latent space. However, this approach quantizes continuous feature vectors into discrete visual words for PLSA modeling. As a result, its annotation performance is sensitive to the clustering granularity. In our previous work [13], we propose PLSA-FUSION which employs two PLSA models to capture semantic information from visual and textual modalities respectively. Furthermore, an adaptive asymmetric learning approach is presented to fuse the aspects of these two models. Consequently, PLSA-FUSION can acquire higher accuracy than PLSA-WORDS. Nevertheless, PLSA-FUSION also needs to quantize visual features into discrete visual words for PLSA modeling. In the area of automatic image annotation, it is generally believed that using continuous feature vectors will give rise to better performance [2, 11]. However, since traditional PLSA is originally applied in text classification, it can only handle discrete quantity (such as textual words). In order to model image data precisely, it is required to deal with continuous quantity using PLSA.

This paper proposes continuous PLSA, which assumes that each feature vector in an image is governed by a Gaussian distribution under a given latent aspect other than a multinomial one. In addition, corresponding EM algorithm is derived to estimate the parameters. Then, as general treatment, each image can be viewed as a mixture of Gaussians under this model. Furthermore, based on the continuous PLSA and the traditional PLSA, we present a semantic annotation model to learn the correlation between the visual features and textual words. An asymmetric learning approach is adopted to estimate the model parameters. We evaluate our approach on standard Corel datasets and the experiment results show that our approach outperforms several state-of-the-art approaches.

The rest of the paper is organized as follows. Section 2 presents the continuous PLSA model and derives corresponding EM algorithm. Section 3 proposes a semantic annotation model and describes the asymmetric learning approach. Experiment results are reported and analyzed in section 4. Finally, the overall conclusions of this work are presented in section 5.

2   Continuous PLSA

Just like traditional PLSA, continuous PLSA is also a statistical latent class model which introduces a hidden variable (latent aspect) z_k ($k \in 1, \ldots, K$) in the generative process of each element x_j ($j \in 1, \ldots, M$) in a document d_i ($i \in 1, \ldots, N$). However, given this unobservable variable z_k, continuous PLSA assumes that elements x_j are sampled from a multivariate Gaussian distribution, instead of a multinomial one in traditional PLSA. Using these definitions, continuous PLSA [14] assumes the following generative process:

1. Select a document d_i with probability $P(d_i)$;
2. Sample a latent aspect z_k with probability $P(z_k|d_i)$ from a multinomial distribution conditioned on d_i;
3. Sample $x_j \sim P(x_j|z_k)$ from a multivariate Gaussian distribution $N(\mathbf{x}|\boldsymbol{\mu}_k, \Sigma_k)$ conditioned on z_k.

Continuous PLSA has two underlying assumptions. First, the observation pairs (d_i, x_j) are generated independently. Second, the pairs of random variables (d_i, x_j) are conditionally independent given the latent aspect z_k. Thus, the joint probability of the observed variables is obtained by marginalizing over the latent aspect z_k,

$$P(d_i, x_j) = P(d_i) \sum_{k=1}^{K} P(z_k \mid d_i) P(x_j \mid z_k). \tag{1}$$

A representation of the model in terms of a graphical model is depicted in Figure 1.

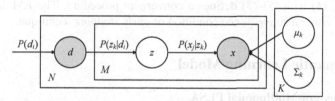

Fig. 1. Graphical model representation of continuous PLSA

The mixture of Gaussian is assumed for the conditional probability P(.|z). In other words, the elements are generated from K Gaussian distributions, each one corresponding a z_k. For a specific latent aspect z_k, the condition probability distribution function of elements x_j is

$$P(x_j \mid z_k) = \frac{1}{(2\pi)^{D/2} |\Sigma_k|^{1/2}} \exp(-\frac{1}{2}(x_j - \mu_k)^T \Sigma_k^{-1}(x_j - \mu_k)), \tag{2}$$

where D is the dimension, μ_k is a D-dimensional mean vector and Σ_k is a $D{\times}D$ covariance matrix.

Following the maximum likelihood principle, $P(z_k|d_i)$ and $P(x_j|z_k)$ can be determined by maximization of the log-likelihood function

$$\mathcal{L} = \sum_{i=1}^{N} \sum_{j=1}^{M} n(d_i, x_j) \log P(d_i, x_j),$$

$$= \sum_{i=1}^{N} n(d_i) \log P(d_i) + \sum_{i=1}^{N} \sum_{j=1}^{M} n(d_i, x_j) \log \sum_{k=1}^{K} P(z_k \mid d_i) P(x_j \mid z_k). \tag{3}$$

where $n(d_i, x_j)$ denotes the number of element x_j in d_i.

The standard procedure for maximum likelihood estimation in latent variable models is the EM algorithm [6]. In E-step, applying Bayes' theorem to (1), one can obtain

$$P(z_k \mid d_i, x_j) = \frac{P(z_k \mid d_i)P(x_j \mid z_k)}{\sum_{l=1}^{K} P(z_l \mid d_i)P(x_j \mid z_l)}. \tag{4}$$

In M-step, for any d_i, z_k and x_j, the parameters are determined as

$$\mu_k = \frac{\sum_{i=1}^{N} \sum_{j=1}^{M} n(d_i, x_j)P(z_k \mid d_i, x_j)x_j}{\sum_{i=1}^{N} \sum_{j=1}^{M} n(d_i, x_j)P(z_k \mid d_i, x_j)}, \tag{5}$$

$$\Sigma_k = \frac{\sum_{i=1}^{N} \sum_{j=1}^{M} n(d_i, x_j)P(z_k \mid d_i, x_j)(x_j - \mu_k)(x_j - \mu_k)^T}{\sum_{i=1}^{N} \sum_{j=1}^{M} n(d_i, x_j)P(z_k \mid d_i, x_j)}, \tag{6}$$

$$P(z_k \mid d_i) = \frac{\sum_{j=1}^{M} n(d_i, x_j)P(z_k \mid d_i, x_j)}{\sum_{j=1}^{M} n(d_i, x_j)}. \tag{7}$$

Alternating (4) with (5)–(7) defines a convergent procedure. The EM algorithm terminates by either a convergence condition or *early stopping* technique.

3   Semantic Learning Model

3.1   Gaussian-Multinomial PLSA

In order to deal with the data of different modalities in terms of their characteristics, we employ continuous PLSA and traditional PLSA to model visual features and textual words respectively. These two models are linked by sharing the same distribution over latent aspects $P(z|d)$. We refer to this semantic annotation model as *Gaussian-multinomial PLSA* (GM-PLSA), which is represented in Figure 2.

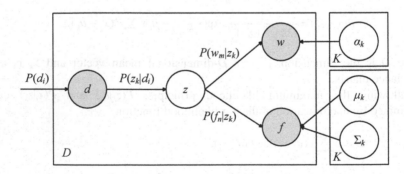

Fig. 2. Representation of the semantic annotation model GM-PLSA

GM-PLSA assumes the following generative process:

1. Select a document d_i with probability $P(d_i)$;
2. Sample a latent aspect z_k with probability $P(z_k|d_i)$ from a multinomial distribution conditioned on d_i;
3. For each of the words, Sample $w_m \sim P(w_m|z_k)$ from a multinomial distribution Mult($\mathbf{x}|\alpha_k$) conditioned on z_k.
4. For each of the feature vectors, Sample $f_n \sim P(f_n|z_k)$ from a multivariate Gaussian distribution $N(\mathbf{x}|\mu_k, \Sigma_k)$ conditioned on the latent aspect z_k.

Under this modeling approach, each image can be viewed as either a mixture of continuous Gaussian in visual modality or a mixture of discrete words in textual modality. Therefore, it can learn the correlation between features and words effectively and predict semantic annotation precisely for an unseen image.

3.2 Algorithm Description

We adopt asymmetric learning approach to estimate the model parameters because an asymmetric learning gives a better control of the respective influence of each modality in the latent space definition [15]. In this learning approach, textual modality is firstly chosen to estimate the mixture of aspects in a given document, which constrains the definition of latent space to ensure its consistency in textual words, while retaining the ability to jointly model visual features. The flow of learning and annotating is described in Figure 3.

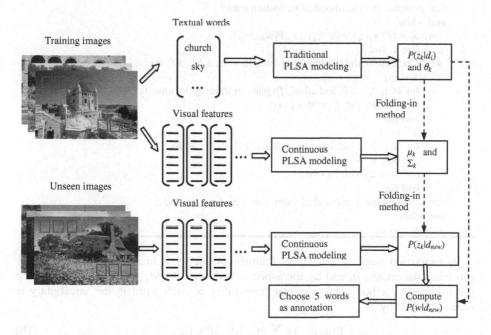

Fig. 3. The flow of learning and annotating algorithms of GM-PLSA

In training stage, each training image is processed and represented as a bag of visual features and textual words. The aspect distributions $P(z_k|d_i)$ are firstly learned for all training documents from textual modality only. At the same time, the parameter $P(w_m|z_k)$ (i.e. α_k) is determined too. Then we use folding-in method to infer the parameters μ_k and Σ_k for the visual modality with the aspect distributions $P(z_k|d_i)$ kept fixed. Consequently, we can get the model parameters α_k, μ_k and Σ_k, which remain valid in images out of the training set. The learning procedure is described in detail in Algorithm 1.

Algorithm 1. Estimation of the parameters: α_k, μ_k and Σ_k

Input: Visual features f_n and textual words w_m of training images.
Output: Model parameters α_k, μ_k and Σ_k.
Process:
1. random initialize the $P(z_k|d_i)$ and $P(w_m|z_k)$ probability tables;
2. **while** increase in the likelihood of validation data $\Delta L_t > T_t$ **do**
    (a) {E step}
        for $k \in 1, 2, ..., K$ and all (d_i, w_j) pairs in training documents **do**
        compute $P(z_k|d_i, w_j)$ with EM algorithm of traditional PLSA;
        end for
    (b) {M step}
        for $k \in 1, 2, ..., K$ and $j \in 1, 2, ..., M$ **do**
        compute $P(w_j|z_k)$ with EM algorithm of traditional PLSA;
        end for
        for $k \in 1, 2, ..., K$ and $i \in 1, 2, ..., N$ **do**
        compute $P(z_k|d_i)$ with EM algorithm of traditional PLSA;
        end for
    (c) compute the likelihood of validation data L_t;
    end while
3. save $\alpha_k = \{P(w_1|z_k), P(w_2|z_k), ..., P(w_{Mw}|z_k)\}$;
4. initialize μ_k and Σ_k;
5. **while** increase in the likelihood of validation data $\Delta L_c > T_c$ **do**
    (a) {E step}
        for $k \in 1, 2, ..., K$ and all (d_i, f_j) pairs in training documents **do**
        compute $P(z_k|d_i, f_j)$ with eq.(4);
        end for
    (b) {M step}
        for $k \in 1, 2, ..., K, i \in 1, 2, ..., N$ and $j \in 1, 2, ..., M$ **do**
        compute μ_k with eq.(7);
        compute Σ_k with eq.(8);
        end for
    (c) compute the likelihood of validation data L_c with eq.(3);
    end while
6. save μ_k and Σ_k.

In annotation stage, given visual features of each test image and the previously estimated parameters μ_k and Σ_k, the aspect distribution $P(z_k|d_{new})$ can be inferred using the folding-in method. The posterior probability of each word in the vocabulary is then computed by

$$P(w|d_{new}) = \sum_{k=1}^{K} P(z_k|d_{new})P(w|z_k). \tag{8}$$

As usual, we choose five words with the largest posterior probabilities as annotations of an unseen image.Having estimated the parameters of GM-PLSA, the stage of semantic retrieval can be put into practice directly. The retrieval algorithm takes as inputs a semantic word w_q and a database of test images. After annotating each image in the test database, the retrieval algorithm ranks the images labeled with the query word by decreasing posterior probability $P(w_q|d_{new})$.

4   Experimental Results

In order to test the effectiveness and accuracy of the proposed approach, we conduct our experiments on two standard datasets, i.e. Corel5k[7] and Corel30k[4].

The focus of this paper is not on image feature selection and our approach is independent of visual features. We simply decompose images into a set of 32×32 blocks, then compute a 36 dimensional feature vector for each block, consisting of 24 color features and 12 texture features.

4.1   Results on Corel5k Dataset

In this section, the performance is evaluated by comparing the captions automatically generated with the original manual annotations. We compute the recall and precision of every word in the test set and use the mean of these values to summarize the system performance.

We report the results on two sets of words: the subset of 49 best words and the complete set of all 260 words that occur in the training set. The systematic evaluation results are shown in Table 1. From the table, we can see that our model performs much better than the first three approaches. Besides, the model performs slightly better than MBRM. We believe that the application of the continuous PLSA is the reason for this result.

Table 1. Performance comparison on the task of automatic image annotation

Models	Translation	CMRM	CRM	MBRM	PLSA-WORDS	GM-PLSA
#words with recall > 0	49	66	107	122	105	125
Results on 49 best words, as in [7,8,10,11]						
Mean Recall	0.34	0.48	0.70	0.78	0.71	0.79
Mean Precision	0.20	0.40	0.59	0.74	0.56	0.76
Results on all 260 words						
Mean Recall	0.04	0.09	0.19	0.25	0.20	0.25
Mean Precision	0.06	0.10	0.16	0.24	0.14	0.26

Several examples of annotation obtained by our prototype system are shown in Table 2. Here top five words are taken as annotation of the image. We can see that even the system annotates an image with a word not contained in the ground truth, this annotation is frequently plausible.

344   Z. Li et al.

Table 2. Comparison of annotations made by PLSA-WORDS and GM-PLSA

Image				
Ground Truth	mule, deer, buck, doe	sphinx, stone, statue, sculpture	branches, grass, frost, ice	people, sand, water, sky
PLSA-WORDS Annotation	deer,mule,sculpture, stone, rabbit	sculpture, stone, sky, hut, statue	clouds, fox, frost, snow, ice	sky, beach, coast, sand, oahu
GM-PLSA Annotation	deer, mule, grass, sculpture, sand	stone, sculpture, statue, sky, sand	ice, frost, snow, clouds, branches	beach, water, sky, sand, coast

Mean average precision (mAP) is employed as a metric to evaluate the performance of single word retrieval. We only compare our model with CMRM, CRM, MBRM and PLSA-WORDS, because mAPs of other models cannot be accessed directly from the literatures.

The annotation results ignore rank order. However, rank order is very important for image retrieval. Our system will return all the images which are automatically annotated with the query word and rank the images according to the posterior probabilities of that word. Table 3 shows that GM-PLSA is slightly better than other models.

Table 3. Comparison of ranked retrieval results

Mean Average Precision on Corel5k Dataset		
Models	All 260 words	Words with recall ≥ 0
CMRM	0.17	0.20
CRM	0.24	0.27
MBRM	0.30	0.35
PLSA-WORDS	0.22	0.26
GM-PLSA	0.32	0.37

In summary, the experiment results show that GM-PLSA outperforms several previous models in many respects, which proves that the continuous PLSA is effective in modeling visual features.

4.2   Results on Corel30k Dataset

The Corel30k dataset provides a much large database size and vocabulary size compared with Corel5k. Since Corel30k is a new database, we only compare our model with PLSA-WORDS.

Figure 4 presents the precision-recall curves of PLSA-WORDS and GM-PLSA on the Corel30k dataset, with the number of annotations from 2 to 10. The precision and

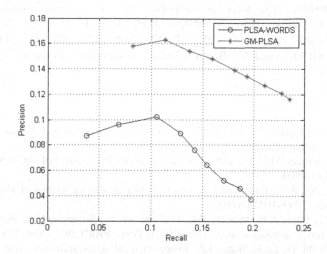

Fig. 4. Precision-recall curves of PLSA-WORDS and GM-PLSA

Table 4. Comparison of ranked retrieval results

Mean Average Precision on Corel30k Dataset		
Models	All 950 words	Words with recall ≥ 0
PLSA-WORDS	0.14	0.17
GM-PLSA	0.23	0.28

recall values are the mean values calculated over all words. From the figure we can see that GM-PLSA consistently outperforms PLSA-WORDS.

The superior performance of GM-PLSA on precision and recall directly results in its great semantic retrieval performance. From Table 4 we can also see the great improvements of GM-PLSA over PLSA-WORDS.

Overall, the experiments on Corel30k indicate that GM-PLSA is fairly stable with respect to its parameters setting. Moreover, since this annotation model integrates traditional PLSA and continuous PLSA, it has better robustness and scalability.

5   Conclusion

In this paper, we have proposed continuous PLSA to model continuous quantity and develop an EM-based iterative procedure to learn the conditional probabilities of the continuous quantity given a latent aspect. Furthermore, we present a semantic annotation model, which employ continuous PLSA and traditional PLSA to deal with the visual and textual data respectively. An adaptive asymmetric learning approach is adopted to estimate the model parameters. Experiments on the Corel dataset prove that our approach is promising for automatic image annotation. In comparison to previous proposed annotation methods, higher accuracy and effectiveness of our approach are reported.

Acknowledgments. This work is supported by the National Natural Science Foundation of China (Nos. 61165009, 60903141, 61105052), the Guangxi Natural Science Foundation (2012GXNSFAA053219, 2012GXNSFBA053166, 2011GXNSFD018026) and the "Bagui Scholar" Project Special Funds.

References

1. Barnard, K., Duygulu, P., Forsyth, D., et al.: Matching words and pictures. Journal of Machine Learning Research 3, 1107–1135 (2003)
2. Blei, D.M., Jordan, M.I.: Modeling annotated data. In: Proc. 26th Intl. ACM SIGIR Conf., pp. 127–134 (2003)
3. Blei, D.M., Ng, A.Y., Jordan, M.I.: Latent Dirichlet allocation. Journal of Machine Learning Research 3, 993–1022 (2003)
4. Carneiro, G., Chan, A.B., Moreno, P.J., Vasconcelos, N.: Supervised learning of semantic classes for image annotation and retrieval. IEEE Trans. PAMI 29(3), 394–410 (2007)
5. Datta, R., Joshi, D., Li, J., Wang, J.Z.: Image retrieval: ideas, influences, and trends of the new age. ACM Computing Surveys 40(2), article 5, 1–60 (2008)
6. Dempster, A.P., Laird, N.M., Rubin, D.B.: Maximum likelihood from incomplete data via the EM algorithm. Journal of the Royal Statistical Society 39(1), 1–38 (1977)
7. Duygulu, P., Barnard, K., de Freitas, J.F.G., Forsyth, D.: Object Recognition as Machine Translation: Learning a Lexicon for a Fixed Image Vocabulary. In: Heyden, A., Sparr, G., Nielsen, M., Johansen, P. (eds.) ECCV 2002. LNCS, vol. 2353, pp. 97–112. Springer, Heidelberg (2002)
8. Feng, S.L., Manmatha, R., Lavrenko, V.: Multiple Bernoulli relevance models for image and video annotation. In: Proc. CVPR, pp. 1002–1009 (2004)
9. Hofmann, T.: Unsupervised learning by probabilistic latent semantic analysis. Machine Learning 42(1-2), 177–196 (2001)
10. Jeon, J., Lavrenko, V., Manmatha, R.: Automatic image annotation and retrieval using cross-media relevance models. In: Proc. 26th Int'l ACM SIGIR Conf., pp. 119–126 (2003)
11. Lavrenko, V., Manmatha, R., Jeon, J.: A model for learning the semantics of pictures. In: Proc. NIPS, pp. 553–560 (2003)
12. Li, J., Wang, J.Z.: Automatic linguistic indexing of pictures by a statistical modeling approach. IEEE Trans. PAMI 25(9), 1075–1088 (2003)
13. Li, Z., Shi, Z., Liu, X., Li, Z., Shi, Z.: Fusing semantic aspects for image annotation and retrieval. Journal of Visual Communication and Image Representation 21(8), 798–805 (2010)
14. Li, Z., Shi, Z., Liu, X., Shi, Z.: Automatic image annotation with continuous PLSA. In: Proc. 35th ICASSP, pp. 806–809 (2010)
15. Monay, F., Gatica-Perez, D.: Modeling semantic aspects for cross-media image indexing. IEEE Trans. PAMI 29(10), 1802–1817 (2007)

Automatic Image Annotation and Retrieval Using Hybrid Approach

Zhixin Li[1], Weizhong Zhao[2], Zhiqing Li[2], and Zhiping Shi[3]

[1] College of Computer Science and Information Technology,
Guangxi Normal University, Guilin 541004, China
[2] College of Information Engineering, Xiangtan University, Xiangtan 411105, China
[3] College of Information Engineering, Capital Normal University, Beijing 100048, China
lizx@gxnu.edu.cn, zhaoweizhong@gmail.com,
lizhiqingchina@gmail.com, shizhiping@gmail.com

Abstract. We firstly propose continuous probabilistic latent semantic analysis (PLSA) to model continuous quantity. In addition, corresponding Expectation-Maximization (EM) algorithm is derived to determine the model parameters. Furthermore, we present a hybrid framework which employs continuous PLSA to model visual features of images in generative learning stage and uses ensembles of classifier chains to classify the multi-label data in discriminative learning stage. Since the framework combines the advantages of generative and discriminative learning, it can predict semantic annotation precisely for unseen images. Finally, we conduct a series of experiments on a standard Corel dataset. The experiment results show that our approach outperforms many state-of-the-art approaches.

Keywords: automatic image annotation, continuous PLSA, semantic learning, hybrid approach, image retrieval.

1 Introduction

Content-based image retrieval (CBIR) has been studied and explored for decades. Its performance, however, is not ideal enough due to the notorious *semantic gap* [18]. CBIR retrieves images in terms of their visual features, while users often prefer intuitive text-based image searching. Since manual image annotation is expensive and difficult to be extended to large image databases, automatic image annotation has emerged as a striking and crucial problem [5].

The state-of-the-art techniques of automatic image annotation can be categorized into two different schools of thought. The first one is based on discriminative model. It defines auto-annotation as a traditional supervised classification problem [3,4,12,17], which treats each semantic concept as an independent class and creates different classifiers for different concepts. This approach computes similarity at the visual level and annotates a new image by propagating the corresponding words. The second perspective takes a different stand. It is based on generative model and treats image and text as equivalent data. It attempts to discover the correlation between

Z. Shi, D. Leake, and S. Vadera (Eds.): IIP 2012, IFIP AICT 385, pp. 347–356, 2012.

visual features and textual words on an unsupervised basis by estimating the joint distribution of features and words. Thus, it poses annotation as statistical inference in a graphical model. Under this perspective, images are treated as bags of words and features, each of which are assumed generated by a hidden variable. Various approaches differ in the definition of the states of the hidden variable: some associate them with images in the database [8,10,11], while others associate them with image clusters [1,7] or latent aspects (topics) [2,14,15]. Both these two kind of approaches have their own advantages and disadvantages. This paper will show that it is feasible to combine the advantages of these two formulations.

As a latent aspect model, PLSA [9] has been successfully applied to annotate and retrieve images. PLSA-WORDS [15] is a representative approach, which achieves the annotation task by constraining the latent space to ensure its consistency in words. However, since traditional PLSA can only handle discrete quantity (such as textual words), this approach quantizes feature vectors into discrete visual words for PLSA modeling. Therefore, its annotation performance is sensitive to the clustering granularity. In the area of automatic image annotation, it is generally believed that using continuous feature vectors will give rise to better performance [2,3,11]. In order to model image data precisely, it is required to deal with continuous quantity using PLSA.

This paper presents continuous PLSA, which assumes that feature vectors of images are governed by a Gaussian distribution under a given latent aspect other than a multinomial one. In addition, corresponding EM algorithm is derived to estimate the parameters. Then, each image can be treated as a mixture of Gaussians under this model. Furthermore, we propose a hybrid framework to learn semantic classes of images. The framework employs continuous PLSA to model visual features of images in generative learning stage, and uses ensembles of classifier chains to classify the multi-label data in discriminative learning stage. We compare our approach with some state-of-the-art approaches on a standard Corel dataset and the experiment results show that our approach performs more effectively and precisely.

The rest of the paper is organized as follows. Section 2 presents the continuous PLSA model and derives corresponding EM algorithm. Section 3 proposes a hybrid framework and describes the training and annotation procedure. Experiment results are reported and analyzed in section 4. Finally, the overall conclusions of this work are presented in section 5.

2   Continuous PLSA

Just like traditional PLSA, continuous PLSA is also a statistical latent class model which introduces a hidden variable (latent aspect) z_k ($k \in 1, ..., K$) in the generative process of each element x_j ($j \in 1, ..., M$) in a document d_i ($i \in 1, ..., N$). However, given this unobservable variable z_k, continuous PLSA assumes that elements x_j are sampled from a multivariate Gaussian distribution, instead of a multinomial one in traditional PLSA. Using these definitions, continuous PLSA [13] assumes the following generative process:

1. Select a document d_i with probability $P(d_i)$;
2. Sample a latent aspect z_k with probability $P(z_k|d_i)$ from a multinomial distribution conditioned on d_i;
3. Sample $x_j \sim P(x_j|z_k)$ from a multivariate Gaussian distribution $N(\mathbf{x}|\mu_k,\Sigma_k)$ conditioned on z_k.

Continuous PLSA has two underlying assumptions. First, the observation pairs (d_i, x_j) are generated independently. Second, the pairs of random variables (d_i, x_j) are conditionally independent given the latent aspect z_k. Thus, the joint probability of the observed variables is obtained by marginalizing over the latent aspect z_k,

$$P(d_i,x_j) = P(d_i)\sum_{k=1}^{K} P(z_k \mid d_i)P(x_j \mid z_k). \tag{1}$$

A representation of the model in terms of a graphical model is depicted in Figure 1.

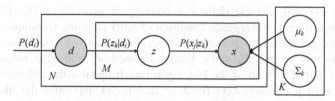

Fig. 1. Graphical model representation of continuous PLSA

The mixture of Gaussian is assumed for the conditional probability $P(.|z)$. In other words, the elements are generated from K Gaussian distributions, each one corresponding a z_k. For a specific latent aspect z_k, the condition probability distribution function of elements x_j is

$$P(x_j \mid z_k) = \frac{1}{(2\pi)^{D/2}|\Sigma_k|^{1/2}}\exp(-\frac{1}{2}(x_j - \mu_k)^T \Sigma_k^{-1}(x_j - \mu_k)), \tag{2}$$

where D is the dimension, μ_k is a D-dimensional mean vector and Σ_k is a $D \times D$ covariance matrix.

Following the maximum likelihood principle, $P(z_k|d_i)$ and $P(x_j|z_k)$ can be determined by maximization of the log-likelihood function

$$\mathcal{L} = \sum_{i=1}^{N}\sum_{j=1}^{M} n(d_i,x_j)\log P(d_i,x_j),$$
$$= \sum_{i=1}^{N} n(d_i)\log P(d_i) + \sum_{i=1}^{N}\sum_{j=1}^{M} n(d_i,x_j)\log \sum_{k=1}^{K} P(z_k \mid d_i)P(x_j \mid z_k). \tag{3}$$

where $n(d_i, x_j)$ denotes the number of element x_j in d_i.

The standard procedure for maximum likelihood estimation is the EM algorithm [6]. In E-step, applying Bayes' theorem to (1), one can obtain

$$P(z_k \mid d_i, x_j) = \frac{P(z_k \mid d_i)P(x_j \mid z_k)}{\sum_{l=1}^{K} P(z_l \mid d_i)P(x_j \mid z_l)}. \tag{4}$$

In M-step, for any d_i, z_k and x_j, the parameters are determined as

$$\mu_k = \frac{\sum_{i=1}^{N} \sum_{j=1}^{M} n(d_i, x_j)P(z_k \mid d_i, x_j)x_j}{\sum_{i=1}^{N} \sum_{j=1}^{M} n(d_i, x_j)P(z_k \mid d_i, x_j)}, \tag{5}$$

$$\Sigma_k = \frac{\sum_{i=1}^{N} \sum_{j=1}^{M} n(d_i, x_j)P(z_k \mid d_i, x_j)(x_j - \mu_k)(x_j - \mu_k)^T}{\sum_{i=1}^{N} \sum_{j=1}^{M} n(d_i, x_j)P(z_k \mid d_i, x_j)}, \tag{6}$$

$$P(z_k \mid d_i) = \frac{\sum_{j=1}^{M} n(d_i, x_j)P(z_k \mid d_i, x_j)}{\sum_{j=1}^{M} n(d_i, x_j)}. \tag{7}$$

Alternating (4) with (5)–(7) defines a convergent procedure. The EM algorithm terminates by either a convergence condition or *early stopping* technique.

As for the parameters, if parameter $P(x_j \mid z_k)$ is known, we could quickly infer the other parameters μ_k and Σ_k using folding-in method, and vice versa. Folding-in method is a partial version of the EM algorithm. It updates the unknown parameters with the known parameters kept fixed, so that it can maximize the likelihood with respect to the previously trained parameters.

3   Hybrid Generative/Discriminative Approach

3.1   Hybrid Framework

On the basis of continuous PLSA, we propose a hybrid framework which combines generative and discriminative learning. The framework firstly employs continuous PLSA to model visual features of images. As a result, each image can be represented as an aspect distribution. Then, this intermediate representation can be used to build ensembles of classifier chains, which can learn semantic classes of images and consider the correlation between the labels at the same time. The framework is shown in figure 2.

In training procedure, we firstly get the parameters μ_k and Σ_k given aspect z_k by modeling visual features of training images with continuous PLSA. At the same time, the aspect distribution $P(z_k \mid d_i)$ of each image is determined. This is the generative learning stage. The parameters μ_k and Σ_k are parameters of continuous PLSA. According to the independence assumption, these parameters remain valid for documents out of the training set. On the other hand, the aspect distribution $P(z_k \mid d_i)$ is only relative to the specific documents and cannot carry any prior information to an unseen image. This distribution, however, can represent each training image as a K-dimension vector. In addition, all the vectors can construct a simplex. Then, by making use of the aspect distribution and original annotation labels of each training image, we build a series of classifiers in which every word in the vocabulary is treated as an independent class. This is the discriminative learning stage. At this time, every image is

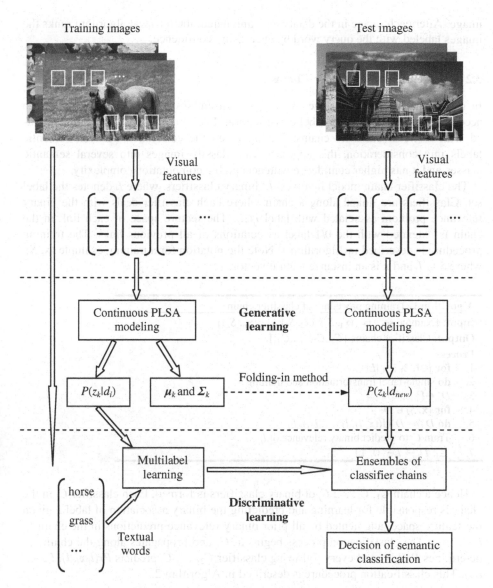

Fig. 2. Learning procedure of hybrid framework.

represented as an aspect distribution, but has several semantic labels. This circumstance is in conformity with multi-label learning, which can construct multiclass classifiers and integrate correlative information of textual words at the same time.

Correspondingly, there are two steps in annotation procedure. Firstly, since model parameters μ_k and Σ_k are determined in training procedure, we can compute the aspect distribution of each test image using folding-in method. Secondly, we classify the aspect distribution of each test image with the trained ensembles of classifier chains. Furthermore, we choose 5 words with highest confidence as annotations of the test image. After each image in the database is annotated, the retrieval algorithm ranks the images labeled with the query word by decreasing confidence.

3.2 Ensembles of Classifier Chains

In discriminative learning stage, we employ ensembles of classifier chains [16] to accomplish the task of multi-label classification. Each binary classifier is implemented with SVM in classifier chains. Having taken the correlation between semantic labels into consideration, this approach can classify images into several semantic classes and it has higher confidence with acceptable computation complexity.

The classifier chain model involves $|L|$ binary classifiers, where L denotes the label set. Classifiers are linked along a chain where each classifier deals with the binary relevance problem associated with label $l_j \in L$. The feature space of each link in the chain is extended with the 0/1 label associations of all previous links. The training procedure is described in Algorithm 1. Note the notation for a training example $(\mathbf{x}, S)$, where $S \subseteq L$ and $\mathbf{x}$ is an instance feature vector.

Algorithm 1. Training procedure of classifier chain

Input: Example set $D = \{(\mathbf{x}_1, S_1), (\mathbf{x}_2, S_2), ..., (\mathbf{x}_n, S_n)\}$.
Output: Classifier chains $\{C_1, C_2, ..., C_{|L|}\}$.
Process:
1. **for** $j \in 1, 2, ..., |L|$
2. **do** single-label transformation and training
3. $D' \leftarrow \{\}$
4. **for** $(\mathbf{x}, S) \in D$
5. **do** $D' \leftarrow D' \cup ((\boldsymbol{x}, l_1, l_2, ..., l_{j-1}), l_j)$
6. Train C_j to predict binary relevance of l_j
7. $C_j: D' \rightarrow l_j \in \{0, 1\}$

Hence a chain $C_1, C_2, ..., C_j$ of binary classifiers is formed. Each classifier C_j in the chain is responsible for learning and predicting the binary association of label l_j given the feature space, augmented by all prior binary relevance predictions in the chain l_1, $l_2, ..., l_{j-1}$. The classification process begins at C_1 and propagates along the chain: C_1 determines $Pr(l_1|\mathbf{x})$ and every following classifier $C_2, ..., C_j$ predicts $Pr(l_j|\mathbf{x}_i, l_1, l_2, ..., l_{j-1})$. This classification procedure is described in Algorithm 2.

Algorithm 2. Classifying procedure of classifier chain

Input: Test example $\mathbf{x}$.
Output: Results of all classifiers in the chain $Y = \{ l_1, l_2, ..., l_{|L|}\}$.
Process:
1. $Y \leftarrow \{\}$
2. **for** $j \in 1, 2, ..., |L|$
3. **do** $Y \leftarrow Y \cup (l_j \leftarrow C_j: (\boldsymbol{x}, l_1, l_2, ..., l_{j-1}))$
4. **return** $(\boldsymbol{x}, Y)$

This training method passes label information between classifiers, allowing classifier chain take into account label correlations and thus overcoming the label independence problem of binary relevance method. However, classifier chain still remains advantages of binary relevance method including low memory and runtime complexity.

The order of the chain itself clearly has an effect on accuracy. This problem can be solved by using an ensemble framework with a different random train ordering for each iteration. Ensemble of classifier chains trains m classifiers C_1, C_2, ..., C_m. Each C_k is trained with a random chain ordering of L and a random subset of D. Hence each C_k model is likely to be unique and able to give different multi-label predictions. These predictions are summed by label so that each label receives a number of votes. A threshold is used to select the most popular labels which form the final predicted multi-label set.

4   Experimental Results

In order to test the effectiveness and accuracy of the proposed approach, we conduct our experiments on an annotated image data set which was originally used in [7]. The dataset consists of 5000 images from 50 Corel Stock Photo cds. Each cd includes 100 images on the same topic. We divided this dataset into 3 parts: a training set of 4000 images, a validation set of 500 images and a test set of 500 images. The validation set is used to determine system parameters. After fixing the parameters, we merged the 4000 training set and 500 validation set to form a new training set. This corresponding to the training set of 4500 images and the test set of 500 images used by [7].

4.1   Parameters Setting

An important parameter of the experiment is the number of latent aspects for the PLSA-based models. Since the number of latent aspects defines the capacity of the model — the number of model parameters, it can determine the training time and system efficiency to a large extent. We choose three aspect numbers (90, 120 and 150) to do experiments. Through a series of experiments, we found that the system performs better when aspect number is 150. Therefore, we use 150 as aspect number, without ruling out the possibility that another aspect number would make the system performs much better. Furthermore, our approach constructs an ensemble including 90 classifier chains. Each classifier chain randomly chooses a subset of 500 images for training.

The focus of this paper is not on image feature selection and our approach is independent of visual features. So our prototype system uses similar features to [8] for easy comparison. We simply decompose images into a set of blocks (the size of each block is empirically determined as 16×16 through a series of experiments on the validation set), then compute a 36 dimensional feature vector for each block, consisting of 24 color features (auto correlogram computed over 8 quantized colors and 3 Manhattan Distances) and 12 texture features (Gabor energy computed over 3 scales and 4 orientations). As a result, each block is represented as a 36 dimension feature vector. Then each image is represented as a bag of features, that is, a set of 36 dimension vectors. All the feature vectors of training images compose the inputs of continuous PLSA. Therefore, this preprocessing procedure provides a uniform interface for continuous PLSA modeling.

4.2 Results of Automatic Image Annotation

In this section, the performance of our approach is compared with some state-of-the-art approaches — the Translation Model [7], CMRM [10], CRM [11], MBRM [8], PLSA-WORDS [15] and SML [3]. We evaluate the performance of image annotation by comparing the captions automatically generated with the original manual annotations. Similarly to [11], we compute the recall and precision of every word in the test set and use the mean of these values to summarize the system performance.

We report the results on two sets of words: the subset of 49 best words and the complete set of all 260 words that occur in the training set. The systematic evaluation results are shown in table 1. From the table, we can see that our approach performs significantly better than all other approaches. We believe that using hybrid framework to learn semantic classes is the reason for this result.

Table 1. Performance comparison on the task of automatic image annotation

Models	Translation	CMRM	CRM	MBRM	PLSA-WORDS	SML	Hybrid Approach
#words with recall > 0	49	66	107	122	105	137	137
Results on 49 best words, as in [7,8,10,11]							
Mean Recall	0.34	0.48	0.70	0.78	0.71	—	0.83
Mean Precision	0.20	0.40	0.59	0.74	0.56	—	0.78
Results on all 260 words							
Mean Recall	0.04	0.09	0.19	0.25	0.20	0.29	0.32
Mean Precision	0.06	0.10	0.16	0.24	0.14	0.23	0.28

Table 2. Comparison of annotations made by PLSA-WORDS and hybrid approach

Image				
Ground truth	grizzly, bear, meadow, grass	head, fox, snow, closeup	trees, sky, frost, ice	sand, water, people, sky
Annotations of PLSA-WORDS	bear, grizzly, horse, meadow, sand	clouds, sky, stone, sculpture, rabbit	grass, desert, ice, path, sculpture	beach, iceburg, snow, ice, water
Annotations of hybrid approach	bear, grizzly, grass, meadow, trees	fox, snow, sky, head, clouds	sky, trees, branch, ice, grass	water, sky, beach, people, snow

Several examples of annotation obtained by our prototype system are shown in Table 2. Here top five words are taken as annotation of the image. We can see that even the system annotates an image with a word not contained in the ground truth, this annotation is frequently plausible.

4.3   Results of Ranked Image Retrieval

In this section, mean average precision (mAP) is employed as a metric to evaluate the performance of single word retrieval. We only compare our model with CMRM, CRM, MBRM, PLSA-WORDS and SML, because mAP of the Translation model cannot be accessed directly from the literatures.

The annotation results ignore rank order. However, users always like to rank retrieval images and hope that the top ranked ones are relative images. In fact, most users do not want to see more than even 10 or 20 images in a query. Therefore, rank order is very important for image retrieval. Given a query word, our system will return all the images which are automatically annotated with the query word and rank the images according to the posterior probabilities of that word. Table 3 shows that our hybrid approach performs better than other models.

Table 3. Comparison of mAPs in ranked image retrieval

Models	CMRM	CRM	MBRM	PLSA-WORDS	SML	Hybrid Approach
All 260 words	0.17	0.24	0.30	0.22	0.31	0.35
Words with recall ≥ 0	0.20	0.27	0.35	0.26	—	0.41
Words with recall > 0	—	—	—	0.55	0.49	0.67

In summary, the experiment results show that our approach outperforms some state-of-the-art approaches in many respects, which proves that the continuous PLSA and our hybrid approach is effective in modeling visual features and learning semantic classes of images.

5   Conclusion

In this paper, we have proposed continuous PLSA to model continuous quantity and develop an EM-based iterative procedure to estimate the parameters. Furthermore, we present a hybrid generative/discriminative approach, which employs continuous PLSA to deal with the visual features and uses ensembles of classifier chains to learn semantic classes of images. Experiments on the Corel dataset prove that our approach is promising for automatic image annotation and retrieval. In comparison to some state-of-the-art approaches, higher accuracy and superior effectiveness of our approach are reported.

Acknowledgments. This work is supported by the National Natural Science Foundation of China (Nos. 61165009, 60903141, 61105052), the Guangxi Natural Science Foundation (2012GXNSFAA053219) and the "Bagui Scholar" Project Special Funds.

References

1. Barnard, K., Duygulu, P., Forsyth, D., et al.: Matching words and pictures. Journal of Machine Learning Research 3, 1107–1135 (2003)
2. Blei, D.M., Jordan, M.I.: Modeling annotated data. In: Proc. 26th Intl. ACM SIGIR Conf., pp. 127–134 (2003)
3. Carneiro, G., Chan, A.B., Moreno, P.J., Vasconcelos, N.: Supervised learning of semantic classes for image annotation and retrieval. IEEE Trans. PAMI 29(3), 394–410 (2007)
4. Chang, E., Goh, K., Sychay, G., Wu, G.: CBSA: content-based soft annotation for multimodal image retrieval using bayes point machines. IEEE Trans. CSVT 13(1), 26–38 (2003)
5. Datta, R., Joshi, D., Li, J., Wang, J.Z.: Image retrieval: ideas, influences, and trends of the new age. ACM Computing Surveys 40(2), article 5, 1–60 (2008)
6. Dempster, A.P., Laird, N.M., Rubin, D.B.: Maximum likelihood from incomplete data via the EM algorithm. Journal of the Royal Statistical Society 39(1), 1–38 (1977)
7. Duygulu, P., Barnard, K., de Freitas, J.F.G., Forsyth, D.: Object Recognition as Machine Translation: Learning a Lexicon for a Fixed Image Vocabulary. In: Heyden, A., Sparr, G., Nielsen, M., Johansen, P. (eds.) ECCV 2002. LNCS, vol. 2353, pp. 97–112. Springer, Heidelberg (2002)
8. Feng, S.L., Manmatha, R., Lavrenko, V.: Multiple Bernoulli relevance models for image and video annotation. In: Proc. CVPR, pp. 1002–1009 (2004)
9. Hofmann, T.: Unsupervised learning by probabilistic latent semantic analysis. Machine Learning 42(1-2), 177–196 (2001)
10. Jeon, J., Lavrenko, V., Manmatha, R.: Automatic image annotation and retrieval using cross-media relevance models. In: Proc. 26th Int'l ACM SIGIR Conf., pp. 119–126 (2003)
11. Lavrenko, V., Manmatha, R., Jeon, J.: A model for learning the semantics of pictures. In: Proc. NIPS, pp. 553–560 (2003)
12. Li, J., Wang, J.Z.: Automatic linguistic indexing of pictures by a statistical modeling approach. IEEE Trans. PAMI 25(9), 1075–1088 (2003)
13. Li, Z., Shi, Z., Liu, X., Shi, Z.: Automatic image annotation with continuous PLSA. In: Proc. 35th ICASSP, pp. 806–809 (2010)
14. Liu, J., Li, M., Liu, Q., et al.: Image annotation via graph learning. Pattern Recognition 42(2), 218–228 (2009)
15. Monay, F., Gatica-Perez, D.: Modeling semantic aspects for cross-media image indexing. IEEE Trans. PAMI 29(10), 1802–1817 (2007)
16. Read, J., Pfahringer, B., Holmes, G., Frank, E.: Classifier Chains for Multi-label Classification. In: Buntine, W., Grobelnik, M., Mladenić, D., Shawe-Taylor, J. (eds.) ECML PKDD 2009. LNCS, vol. 5782, pp. 254–269. Springer, Heidelberg (2009)
17. Wang, C., Yan, S., Zhang, L., et al.: Multi-label sparse coding for automatic image annotation. In: Proc. CVPR, pp. 1643–1650 (2009)
18. Smeulders, A.W.M., Worring, M., Santini, S., et al.: Content-based image retrieval at the end of the early years. IEEE Trans. PAMI 22(12), 1349–1380 (2000)

Double Least Squares Pursuit
for Sparse Decomposition

Wanyi Li, Peng Wang, and Hong Qiao

Institute of Automation, Chinese Academy of Sciences
No. 95 of Zhongguancun East Road, Haidian District, Beijing, China 100190
{wanyi.li,peng_wang,hong.qiao}@ia.ac.cn
http://www.ia.ac.cn

Abstract. Sparse decomposition has been widely used in numerous applications, such as image processing, pattern recognition, remote sensing and computational biology. Despite plenty of theoretical developments have been proposed, developing, implementing and analyzing novel fast sparse approximation algorithm is still an open problem. In this paper, a new pursuit algorithm Double Least Squares Pursuit (DLSP) is proposed for sparse decomposition. In this algorithm, the support of the solution is obtained by sorting the coefficients which are calculated by the first Least-Squares, and then the non-zero values over this support are detected by the second Least-Squares. The results of numerical experiment demonstrate the effectiveness of the proposed method, which is with less time complexity, more simple form, and gives close or even better performance compared to the classical Orthogonal Matching Pursuit (OMP) method.

Keywords: Sparse decomposition, Sparse representation, Sparse approximation algorithm, Double Least-Squares Pursuit.

1 Introduction

The sparse decomposition problem (also referred to as sparse approximation) is one of the main problems for sparse representation and compressed sensing. Given a full rank matrix $\mathbf{A} \in \mathbb{R}^{n \times m}$ with $n < m$ and a vector $\mathbf{b} \in \mathbb{R}^n$, the sparse decomposition problem can be stated as follows [1]:

$$(P_0): \quad \min_{\mathbf{x}} \|\mathbf{x}\|_0 \qquad subject\ to \qquad \mathbf{b} = \mathbf{Ax} \qquad (1)$$

i.e., find a sparsest representation for $\mathbf{b}$ over $\mathbf{A}$, or

$$(P_0^\epsilon): \quad \min_{\mathbf{x}} \|\mathbf{x}\|_0 \qquad subject\ to \qquad \|\mathbf{b} - \mathbf{Ax}\|_2 \leq \epsilon \qquad (2)$$

i.e., find a sparsest approximation for $\mathbf{b}$ with error ϵ.

(P_0) is the exact case and (P_0^ϵ) is the error-tolerant version of (P_0), with error tolerance $\epsilon > 0$, and $\|\mathbf{x}\|_0$ represents the number of nonzero entries in vector $\mathbf{x}$.

Z. Shi, D. Leake, and S. Vadera (Eds.): IIP 2012, IFIP AICT 385, pp. 357–363, 2012.

In both (P_0) and (P_0^{ϵ}) , matrix $\mathbf{A}$ is often referred to as dictionary while vector $\mathbf{b}$ is referred to as observation. In practice, (P_0^{ϵ}) is more suitable for real world problems than (P_0).

It has been proven that, given an arbitrary redundant dictionary $\mathbf{A}$ and an observation $\mathbf{b}$, to solve the sparse representation (P_0) and (P_0^{ϵ}) is a NP-hard problem [2]. As a result, researchers turn to find the approximate solutions for this problem.

The methods for solving (P_0) and (P_0^{ϵ}) mainly include two categories: greedy algorithms [3,4] and convex relaxation techniques [5-8]. Greedy algorithms for approximating the solution of $l_0 - norm$, such as orthogonal matching pursuit (OMP) [3] and matching pursuit(MP) [4], make a sequence of locally optimal choices in hope of determining a globally optimal solution. Although these methods are simple and efficient, the solutions are sub-optimal. Convex relaxation techniques replace the combinatorial sparse approximation problem with a related convex program in hope that the solutions coincide, i.e., replacing the highly discontinuous $l_0 - norm$ by a continuous or even smooth approximation, such as l_p norms for $p \in (0, 1]$ or even by smooth function. A lot of algorithms have been proposed for $l_p - norm$ sparse decomposition [5-8]. Although these methods have made some success in solving many practical problems, the computational cost of the methods still needs to be further treated, and to develop, implement and analyze novel fast sparse approximation algorithms is still an open problem [9].

In this paper, we propose a new pursuit algorithm — Double Least Squares Pursuit (DLSP) motivated by Least-Squares. In the proposed DLSP algorithm, the support of the solution is obtained by sorting the coefficients which are calculated by the first Least-Squares, and then the non-zero values over this support are detected by the second Least-Squares. The results of numerical experiment demonstrate that the proposed method is with less time complexity, more simple form and quite good performance.

The paper is organized as follows. Section 2 introduces the basic Orthogonal Matching Pursuit (OMP) method. Section 3 is devoted to the details of our proposed algorithm — Double Least-Squares Pursuit (DLSP). Section 4 presents the numerical results and comparison experiment results with other methods. Section 5 contains the conclusions.

2   Orthogonal Matching Pursuit (OMP)

The OMP [3] is the most classical greedy algorithm for approximating the solution of (P_0) or (P_0^{ϵ}). Its basic procedure is described in Fig. 1. OMP selects the column which is most correlated with the current residuals at each step, and the selected column is added into the support set. Then, the residuals are updated by projecting the observation onto the linear subspace which is spanned by the columns that have already been selected, i.e., the support set, and the algorithm then iterates. The algorithm does not stop until the $l_2 - norm$ of the residual reaches a pre-specified value ϵ. Compared with other alternative methods, OMP is with more simple form and faster implementation.

Thresholding-Algorithm is a simplification of the OMP, which selects the k largest inner product as the desired support, and only the first projection is used. Besides, many improved or extended versions of OMP have been developed [11-13].

Task: Approximating the solution of (P_0^ϵ) : $\min_\mathbf{x} \|\mathbf{x}\|_0 \ subject\ to\ \|\mathbf{b} - A x\|_2 \leq \epsilon$
Parameters: Given the matrix $\mathbf{A}$, the vector $\mathbf{b}$, and the error threshold ϵ.
Initialization: k=0, and

- the initial solution $\mathbf{x}^0 = 0$,
- the initial residual $\mathbf{r}^0 = \mathbf{b} - A\mathbf{x}^0 = \mathbf{b}$,
- the initial solution support $S^0 = support(\mathbf{x}^0) = \Phi$.

Main Iteration: k=k+1 and performing the following steps:

- **Greedy selection.** Find atom $\mathbf{a}_{j_0}, j_0 = arg\max_j \mathbf{a}_j^T \mathbf{r}^{k-1}$
- **update Support:**Update $S^k = S^{(k-1)} \cup j_0$.
- **Update Provisional Solution:** Compute $\mathbf{x}_k$, the minimizer of $\min \|\mathbf{b} - A\mathbf{x}\|_2^2 \ subject\ to\ Support\{\mathbf{x}\} = S^k$.
- **Update Residual:** Compute $\mathbf{r}_k = \mathbf{b} - A\mathbf{x}^k$.
- **Stopping Rule:**If $\|\mathbf{r}^k\|_2 \leq \epsilon$,stop,Otherwise apply another iteration.

Output: The proposed solution is $\mathbf{x}_k$ obtained after k iterations.

Fig. 1. Orthogonal-Matching-Pursuit, a greedy algorithm for approximating the solution of (P_0^ϵ)

3 Double Least Squares Pursuit

3.1 Motivation

Euclidean or $l_2 - norm$ problem is one of the most common norm approximation problems [10], which is also called the least-squares approximation problem, i.e.,

$$\min \|\mathbf{b} - A\mathbf{x}\|_2^2 \tag{3}$$

and the objective is the sum of squares of the residuals.

The problem has the unique solution,

$$\mathbf{x}_{opt} = \mathbf{A}^T(\mathbf{A}\mathbf{A}^T)^{-1}\mathbf{b} = \mathbf{A}^+\mathbf{b} \tag{4}$$

From above we have

$$\mathbf{b} \approx \sum_{i=1}^{n} \mathbf{a}_1 x_{opt}^1 + \mathbf{a}_2 x_{opt}^2 + \dots + \mathbf{a}_m x_{opt}^m \tag{5}$$

where $\mathbf{a}_i$ represents the i-th column of matrix $\mathbf{A}$, and x_{opt}^i is the i-th entry of x_{opt}.

The absolute value of x_{opt}^i represents the contribution of the i-th column of $\mathbf{A}$, i.e., $\mathbf{a}_i$, for the representation of $\mathbf{b}$. So we can select the k columns of $\mathbf{A}$ correspond with the k largest entries of x_{opt} as the desired support.

3.2   Double Least-Squares Pursuit

Fig. 2 presents a formal description of Double Least Squares Pursuit.

Note that the stage " set $\mathbf{x}_{opt} = \mathbf{A}^+\mathbf{b}$ " in "Initialization " is our first Least-Squares and the Updating Provisional Solution stage is another Least-Squares, so we called the proposed algorithm as Double Least Squares Pursuit.

After getting the solution of the least-squares approximation problem $\mathbf{x}_{opt}$, we select the k columns of $\mathbf{A}$ correspond with the k largest entries of $\mathbf{x}_{opt}$ as the desired support, and then the non-zero values over this support are detected by the second Least-Squares.

Task: Approximating the solution of (P_0^ϵ) : $\min_{\mathbf{x}} \|\mathbf{x}\|_0$ *subject to* $\|\mathbf{b} - \mathbf{A}x\|_2 \le \epsilon$
Parameters: Given the matrix $\mathbf{A}$, the vector $\mathbf{b}$, and the error threshold ϵ.
Initialization: k=0, and

- the initial solution $\mathbf{x}^0 = 0$,
- the initial residual $\mathbf{r}^0 = \mathbf{b} - \mathbf{A}\mathbf{x}^0 = \mathbf{b}$,
- the initial solution support $S^0 = support(\mathbf{x}^0) = \Phi$.
- the solution of the least-squares approximation problem, $\min \|\mathbf{b} - \mathbf{A}\mathbf{x}\|_2^2$
  $\mathbf{x}_{opt} = \mathbf{A}^+\mathbf{b}$

Main Iteration: Incrementing k by 1 and performing the following steps:

- **Updating Support.** find the maxima, j_0 *of* $x_{opt}^j : \forall j \notin S_{k-1}, x_{opt}^{j_0} \ge x_{opt}^j$, and update $S^k = S^{(}k-1) \cup j_0$
- **Update Provisional Solution:** Compute $\mathbf{x}_k$, the minimizer of $\min \|\mathbf{b} - \mathbf{A}\mathbf{x}\|_2^2$ *subject to* $Support\{\mathbf{x}\} = S^k$.
- **Update Residual:** Compute $\mathbf{r}_k = \mathbf{b} - \mathbf{A}\mathbf{x}^k$.
- **Stopping Rule:** If $\|\mathbf{r}^k\|_2 \le \epsilon$, stop, Otherwise apply another iteration.

Output: The proposed solution is $\mathbf{x}_k$ obtained after k iterations.

Fig. 2. Double Least Squares Pursuit for approximating the solution of (P_0^ϵ)

In Equation (5), when $\mathbf{A}$ is an orthogonal matrix, we have

$$\mathbf{x}_{opt} = \mathbf{A}^T(\mathbf{A}\mathbf{A}^T)^{-1}\mathbf{b} = \mathbf{A}^T\mathbf{I}\mathbf{b} = \mathbf{A}^T\mathbf{b} \tag{6}$$

$\mathbf{x}_{opt}^i$ becomes the inner product of $\mathbf{a}_i$ and $\mathbf{b}$, and Double Least-Squares Pursuit is equivalent to the Thresholding-Algorithm. Both of them find the support of the solution by choosing the k largest inner products.

In the above algorithm description, the number of the required non-zeros, k, is assumed to be already obtained. Alternatively, we can increase k until the error $\|\mathbf{b} - \mathbf{A}\mathbf{x}^k\|_2$ reaches a pre-specified value ϵ.

3.3 Time Complexity

For DLSP, due to the pseudo-inverse of matrix $\mathbf{A}$ can be computed and stored in advance, the searching for the k elements of the supports amounts to a simple sort of the entries of the vector $\mathbf{A}^+\mathbf{b}$. OMP needs to compute the inner products in each iteration. So if the proposed solution has k_0 non-zeros, the OMP method requires $2k_0mn$ flops, while DLSP and Thresholding-Algorithm methods require only k_0mn flops. Obviously, DLSP is faster and simpler than OMP.

4 Numerical Experiments

We compared the Double Least-Squares Pursuit (DLSP) with OMP and the Thresholding-Algorithm on a simple date set to demonstrate their comparative behavior. Experimental data and measurement method in [1] are adopted in this paper: Random matrix $\mathbf{A}$ with size 30×50 is created with entries drawn from the norm distribution, and the columns of this matrix are normalized to have a unit $l_2 - norm$. The sparse vector $\mathbf{x}$ is with independent and identically-distributed random supports of cardinalities in the range $[1,10]$, and its non-zero entries are drawn as random uniform variables in the range $[-2,1] \cup [1,2]$. Once $\mathbf{x}$ is generated, we compute $\mathbf{b} = \mathbf{A}\mathbf{x}$, and then apply the above mentioned algorithms to seek for $\mathbf{x}$. We perform 1,000 such tests per cardinality, and the average results are used. According to the Uniqueness-Spark Theorem[1], in all of our tests the original solution is also the sparsest, as the spark of $\mathbf{A}$ is 31.

In this experiment, the distance between the solution $\mathbf{x}_k$ and the ground truth $\mathbf{x}$ is measured by $l_2 - error$ and recovery of the support. The $l_2 - error$ is computed by $\|\mathbf{x} - \mathbf{x}^k\|^2/\|\mathbf{x}\|^2$. The recovery of the support is computed as the distance between the supports of the two solutions, denoting the two supports as S' and S, we define the distance by

$$dist(S', S) = \frac{\max\{|S'|, |S|\} - |S' \cap S|}{\max\{|S'|, |S|\}} \tag{7}$$

where, $|S|$ represents the size of S.

All these algorithms seek the proper solution until the residual is below a certain threshold (i.e., $\|\mathbf{r}^k\|_2 \le 1e-4$. The experimental results are summarized in Fig. 3 and Fig. 4.

Fig. 3 and Fig. 4 show that the performance of DLSP is close to OMP and better than Thresholding-Algorithm. DLSPs performance in terms of relative l_2 recovery error is better than OMPs, while in terms of the success rate in detecting the true support is close to OMPs.

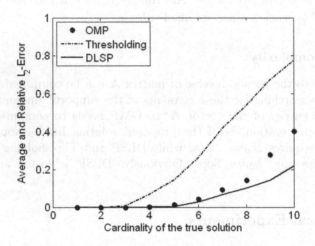

Fig. 3. Algorithms performance in terms of relative l_2 recovery error

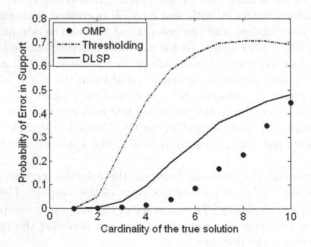

Fig. 4. Algorithms performance in terms of the success rate in detecting the true support

5 Conclusion

This paper presents a new pursuit algorithm Double Least-Squares Pursuit (DLSP) for sparse decomposition. This method finds the support of the solution and gets the non-zero values over this support by applying Least-Squares twice. Experimental results demonstrate that the proposed method is with more simple form, less time complexity and quite good performance.

Our future work will focus on deep theoretical analyze of DLSP which will help in understanding the results obtained in this paper and making more theoretical improvement.

Acknowledgments. This work was partly supported by the NNSF (National Natural Science Foundation) of China under the grant 61100098 and the Knowledge Innovation Program of the Chinese Academy of Sciences under the grant GY-110502.

References

1. Elad, M.: Sparse and Redundant Representations: From Theory to Applications in Signal and Image Processing. Springer (2010)
2. Natarajan, B.K.: Sparse approximate solutions to linear systems. SIAM Journal on Computing 24, 227–234 (1995)
3. Pati, Y.C., Rezaiifar, R., et al.: Orthogonal matching pursuit: recursive function approximation with applications to wavelet decomposition. In: 1993 Conference Record of The Twenty-Seventh Asilomar Conference on Signals, Systems and Computers (1993)
4. Mallat, S.G., Zhang, Z.: Matching pursuits with time-frequency dictionaries. IEEE Transactions on Signal Processing 41(12), 3397–3415 (1993)
5. Candes, E., Tao, T.: Decoding by linear programming. IEEE Trans. Inf. Theory 51, 4203–4215 (2005)
6. Candes, E., Romberg, J.: l1-magic:Recovery of sparse signals via convex programming. California Institute of Technology (2005), http://www.acm.caltech.edu/l1magic/
7. Cands, E., Romberg, J., Tao, T.: Stable signal recovery from incomplete and inaccurate measurements. Communications on Pure and Applied Mathematics 59, 1207–1223 (2006)
8. Candes, E., Tao, T.: Near-optimal signal recovery from random projections: Universal encoding strategies? IEEE Transactions on Information Theory 52(12), 5406–5425 (2006)
9. Sparse Representations for Signal Processing and Coding, http://www.see.ed.ac.uk/~tblumens/Sparse/Sparse.html (Visited March 22, 2012)
10. Boyd, S., Vandenberghe, L.: Convex Optimization. Cambridge University Press (2004)
11. Temlyakov, V.N.: Weak greedy algorithms. Advances in Computational Mathematics 5, 173–187 (2000)
12. Cai, T.T., Lie, W.: Orthogonal Matching Pursuit for Sparse Signal Recovery with Noise. IEEE Transactions on Information Theory 57(7), 4680–4688 (2011)
13. Tong, Z.: Sparse Recovery with Orthogonal Matching Pursuit under Rip. IEEE Transactions on Information Theory 57(9), 6215–6221 (2011)

Ensemble of k-Labelset Classifiers
for Multi-label Image Classification

Dapeng Zhang[1] and Xi Liu[2]

[1] Institute of Information Science and Engineering, Yanshan University,
Qinhuangdao, 066004, China
daniao@ysu.edu.cn
[2] Key Laboratory of Intelligent Information Processing, Institute of Computing Technology,
Chinese Academy of Sciences, Beijing, 100190, China
Liux@ics.ict.ac.cn

Abstract. In the real world, images always have several visual objects instead of only one, which makes it difficult for traditional object recognition methods to deal with them. In this paper, we propose an ensemble method for multi-label image classification. First, we construct an ensemble of k-labelset classifiers. A voting technique is then employed to make predictions for images based on the created ensemble of k-labelset classifiers. We evaluate our method on Corel dataset and demonstrate the precision, recall and F_1 measure superior to the state-of-the-art methods.

Keywords: Ensemble learning, k-labelset classifier, multi-label, image classification, voting.

1   Introduction

The images in real world are usually associated with multiple labels. With regard to their semantic understanding, each image will be assigned with one or more labels from a predefined label set. However, most of traditional image classification methods are concerned with learning from a set of images with only one label. It is hard for them to directly handle with the multi-label learning problem. Multi-label image classification is still a challenging research issue.

Many methods have been developed to solve multi-label image classification. In general, they transform the multi-label classification task to a set of independent two-class classification problems. Early work by Boutell et al. [1] built individual binary classifiers for each label to perform multi-label scene classification. The labels of an image are determined by the outputs of these individual classifiers. The solution is theoretically simple and intuitive but it ignores label correlations that exist in the images.

To exploit these correlations, researchers have made modifications to feature representations or existing discriminative methods. Godbole et al. [2] presented a new feature representation by extending the original features with relationships between classes. The new heterogeneous features were used to train a new SVM ensemble. Qi et al. [3]

Z. Shi, D. Leake, and S. Vadera (Eds.): IIP 2012, IFIP AICT 385, pp. 364–371, 2012.

simultaneously classified labels and modeled the correlations between them by using a unified Correlative Multi-Label Support Vector Machine. In [4], a Max-Margin factorization model was created to learn label classifiers. The regularization term in the model forced label classifiers to share a low dimensional representation, which enabled the algorithm to use correlation between labels for the label prediction task. In multi-label learning, one fundamental method of using label correlations is the label combination method, or label powerset method. The basis of this method is to consider each different subset of labels as a single label to form a single-label binary classifier. We call this classifier as k-labelset classifier if the number of the labels in the subset is k. Although the label powerset method suffers from high time complexity and few training examples, it is simple and directly takes into account label correlations.

This paper proposes a novel approach to multi-label image classification. It constructs an ensemble of k-labelset classifiers and gives the final labels by voting. To avoid label combination complexity and few examples for some k-labelsets, we abandon those k-labelset classifiers whose training samples are below a specified threshold. Besides, the images are always associated with three labels so we only consider $k = 1$, 2, 3 labelset. For the 1-labelset classifiers which focus on only one label, we use local features such as SIFT while for the other k-labelset classifiers, we use global features such as GIST. The final ensemble combination is accomplished by summing the votes of all the k-labelset classifiers for each label. Thresholding all the label votes gives the classified labels. Our approach is experimentally tested on the Corel datasets.

2   Ensemble of k-Labelset Classifiers

In this section, we first describe the construction of k-labelset binary classifiers and then detail how to make ensemble combination of the generated k-labelset classifiers for multi-label image classification. The whole procedure is illustrated in Fig. 1.

Given a collection of training images x_i, $i = 1, ..., n$, each example x_i is represented as a vector of d dimensions and is annotated by a subset of label set L. The image dataset is therefore represented as $(x_1, Y_1), ..., (x_n, Y_n)$. The aim is to train a classifier that can predict a subset of labels for each image.

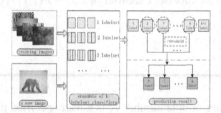

Fig. 1. The workflow of our approach

2.1 k-Labelset Classifiers

Let the image label set $L = \{l_i\}$, $i = 1, ..., |L|$. A set Y_i with $|Y_i| = k$ is called k-labelset. We use the term L^k to denote the set of all distinct k-labelsets on L and the size of L^k is given by the binomial coefficient: $\mid L^k \mid = \begin{pmatrix} \mid L \mid \\ k \end{pmatrix}$. For each distinct k-labelset, we regard it as a single label. The number of class values for the single label will be 2^k. To ease the computational complexity, in this paper we only consider the class in which all k labels appear in an image. Therefore the images which contain all the k labels are considered as positive images while the other images are considered negative for this single label. A binary classifier can then be created accordingly and we call it k-labelset classifier.

We iteratively construct an ensemble of k-labelset classifiers. Since most of the images are often possessed with no more than 3 labels, we only create 1-labelset, 2-labelset and 3-labelset classifiers. For k-labelset classifiers ($k = 1, 2, 3$), the algorithm selects a k-labelset Y_i from L^k at each iteration. If the number of the training images which have the selected k-labelset is above the specified threshold, a k-labelset classifier for the k-labelset will be created. Finally, we can obtain an ensemble of k-labelset classifiers and their corresponding k-labelsets. The pseudocode is given in Algorithm 1.

Algorithm 1. Gen Ensemble of K-labelset Classifier

1.   Input: training set D = {(x_i, Y_i), ... ,(x_n, Y_n)}, set of labels L, minimum training sample number T_n.

2.   Output: an ensemble of k-labelset classifiers $h_{k,i}$ and corresponding to k-labelsets Y_i

3.   For k =1, 2, 3 do

4. $R \leftarrow L^k$

5.   For $i = 1$ to $\mid L^k \mid$ do

6. $Y_i \leftarrow$ a k-labelset selected from R

7.   If the number of training images which contain Y_i is larger than T_n

8.   Train a k-labelset classifier $h_{k,i}$: $x \rightarrow P(Y_i) \ \square \{0,1\}$

9.   End

10. $R \leftarrow R \setminus \{ Y_i \}$

11.   End

The minimum training sample number T_n is a user specified parameter. It is set to be a fixed number such as 20 or a fixed ratio of the total training samples. This can avoid the cases in which there exist few samples for some k-labelset classifiers and also many k-labelset classifiers will thus not be created, which greatly reduce the computation. The few training number of some k-labelsets also means that the k labels have little correlation, and we can hypothesize that our approach will manage to model label correlations by using the minimum training sample threshold. It is also easily seen that we will suffer imbalanced data problem during the training of a k-labelset classifier because of the relatively small number of positive images for the k-labelset. To tackle with this, the negative images for the k-labelset will be sampled by a rate (i.e. 1:2) so as to train a good k-labelset classifier. Furthermore, for 1-labelset classifier, we will use the local image features and for 2-labelset and 3-labelset classifiers, we will use the global images features because 1-labelset classifiers focus on one local label while 2 or 3-labeset classifiers consider 2 or 3 labels as a whole.

2.2   Ensemble Voting

The labels of an image will be predicted based on the voting outputs of the obtained ensemble of the k-labelset classifiers. Given a new image, each k-labelset classifier $h_{k,i}$ provides binary decisions for each label in the corresponding k-labelset Y_i. Sum the binary decisions of all the k-labelset classifiers for each label and then calculate the average decision. A label will be assigned to the image if the label's average decision is larger than a user-specified threshold t. Also we can directly give top n (i.e. 3) labels with high average precision for the image. Algorithm 2 illustrates the detailed procedure of the ensemble voting prediction. The mk in the algorithm represents the number of k-labelset classifiers.

Algorithm 2. EnsembleVoting_Prediction

---

1.   Input:a new image x, an ensemble of k-labelset classifiers and the corresponding set of k-labelsets, set of labels L, the threshold for label prediction t.

2.   Output: multi-label image classification vector Result

3.   Initialize: Votesj=0, Sumj=0, j=1, ..., $|L|$

4.   For k= 1 to 3 do

5.   For i = 1 to mk do

6.   For each label l_j in the corresponding k-labelset of $h_{k,i}$

7.   Votesj = Votesj + $h_{k,i}$ (x, l_j);

8.   Sumj = Sumj + 1;

9.   End

10.   End

11.   End

12.   For j=1 to $|L|$ do

13   Avgj = Sumj / Votesj;

14   if Avgj > t then

15   Resultj = 1;

16   else Resultj = 0;

17.   End

---

3   Experiment

3.1   Experiment Setup

In this work, we use bag-of-local feature descriptors as image representations for 1-labelset classifiers and use a global image feature gist for 2, 3-labelset classifiers. In the BoW framework, a set of local feature descriptors are extracted from each image. All the local descriptors are then clustered and the prototype of each cluster is treated as a "visual word". Assigning each local feature to its nearest visual word and counting the occurrence number, we can represent an image by a histogram of visual words. The cluster number is set 500 in the experiment. For each image, 12*12 pixel local patches over a grid with spacing of 6 pixels are extracted and the local patches are described by a 200-dimensional texton histogram descriptor [5], which encodes both texture and color information. As for image gist [6], it is a holistic image representation that describes global structure of an entire image. We compute the gist by use of Oliva and Torralba's implemention [7]. Linear kernel based support vector machines (SVM) are employed to train the k-labelset classifiers and we will use SVMlight [8] to implement these SVMs.

We use the example-based multi-label classification evaluation measures in [9] as our experimental evaluation metric. Let D be a multi-label image test set, which consists of $|D|$ multi-label images (x_i, Y_i), $i = 1, ..., |D|$, $Y_i \subset L$. Y_i is the true set of image labels for image i in D and let Z_i be the predicted set of labels for image i. The precision, recall and F_1 measures are calculated respectively as follows:

$$Precision(D) = \frac{1}{|D|} \sum_{i=1}^{|D|} \frac{|Y_i \cap Z_i|}{|Z_i|} \tag{1}$$

$$Recall(D) = \frac{1}{|D|} \sum_{i=1}^{|D|} \frac{|Y_i \cap Z_i|}{|Y_i|} \tag{2}$$

$$F_1(D) = \frac{1}{|D|} \sum_{i=1}^{|D|} \frac{2|Y_i \cap Z_i|}{|Z_i| + |Y_i|} \tag{3}$$

3.2   Evaluation on Corel Dataset

The Corel dataset [10] has been extensively used as a standard benchmark dataset for annotation prediction tasks and we will evaluate our multi-label image classification approach on it. In the dataset, there are altogether 5000 images in 50 different sets (CDs). Its vocabulary size is 374 and all the images are associated with 1~5 labels. We will conduct the experiments on a randomly selected subset of the dataset, which contains 25 labels such as "bear", "forest", "mountain", and "sky" and about 4000 images. 80 percent of the new image set is kept as training set, and the left images are made as test set.

T_n in Alg.1 affects our method a lot. Too large will lead to a small number of k-labelset classifiers while too small will cause the few samples problem. We experiment with several t_n (15, 20, 25, 30, 35), the number of k-labelset classifiers for each t_n is shown in Table 1.

Table 1. The number of *k*-labelset classifiers for different T_n

t_n	15	20	25	30	35
$m1$	25	25	25	25	25
$m2$	89	83	71	60	53
$m3$	78	63	48	33	25
$avgL$	17.48	15.20	12.44	9.76	8.24

The 2-labelset classifiers provide classification for 2 labels and the 3-labelset classifiers provide classification for 3 labels. We define the metric $avgL = (m1+2*m2+3*m3)/|L|$, $|L|=25$, which measures the average number of classifiers for each label. From Table 1, we can see $t_n =25$ is a reasonable value under which $avgL$ is adequate for ensemble voting for each label, and the number of 2,3-labelset classifiers is large enough for capturing the label correlations and also the 25 least number of training sample make the *k*-labelset classifiers sufficient for training.

The different value of the threshold t in Alg.2 will lead to different precision, recall and F_1 measure. Note that when calculating these measures the ground-truth labels for test images are confined to the 25 labels. Table 2 gives the results of $t =0.35, 0.40$, and 0.45 respectively. From the perspective of F_1 measure, $t = 0.40$ presents the best results.

Table 2. Comparison of the performances of our method with other approaches

Method	Precision	Recall	F_1-measure
Trans[10]	0.06	0.04	0.05
CRM[11]	0.16	0.19	0.17
Independent SVMs[4]	0.22	0.25	0.23
Ours, t=0.35	0.49	0.57	0.527
Ours, t=0.40	0.52	0.54	0.530
Ours, t=0.45	0.56	0.43	0.486

To further demonstrate the effectiveness of our method, we compare it with several word annotation prediction models including Translation Model (Trans) [10], Continuous Relevance Model (CRM) [11], and independent SVMs [4] on the PicSOM features. Although these models use all 4500 training examples, we think that they are partly comparative. As shown in Table 2, our approach gives the best performance.

Fig.2. shows four images which are labeled by our method. The predicted labels are black bold under each image while the ground-true labels are shown in blue italic font. The results are quite satisfying on the whole.

Fig. 2. Some examples of the labeling results

4   Conclusions and Future Work

This paper presents a novel ensemble method for multi-label image classification. Through a simple ensemble voting of a set of k-labelset classifiers, our method can predict labels for any image. The measures evaluated on Corel dataset demonstrate the effectiveness of the method. The use of 1,2,3-labelset classifiers makes it natural to capture label correlations and that the k-labelset classifiers with few training images are discarded ensures the computational efficiency. For the future, we intend to consider using unlabeled or partly labeled images to improve the performance since fully labeled images are insufficient and hard to obtain. One possible means is to extend the k-labelset classifiers to semi-supervised learning classifiers.

References

1. Boutell, M.R., Luo, J., Shen, X., Brown, C.M.: Learning multi-label scene classification. Pattern Recognition 37, 1757–1771 (2004)
2. Godbole, S., Sarawagi, S.: Discriminative Methods for Multi-labeled Classification. In: Dai, H., Srikant, R., Zhang, C. (eds.) PAKDD 2004. LNCS (LNAI), vol. 3056, pp. 22–30. Springer, Heidelberg (2004)
3. Qi, G.J., Hua, X.S., Rui, Y., Tang, J., Mei, T., Zhang, H.J.: Correlative multi-label video annotation. In: ACM International Conference on Multimedia, pp. 17–26 (2007)
4. Loeff, N., Farhadi, A.: Scene Discovery by Matrix Factorization. In: Forsyth, D., Torr, P., Zisserman, A. (eds.) ECCV 2008, Part IV. LNCS, vol. 5305, pp. 451–464. Springer, Heidelberg (2008)
5. Li, T., Kweon, I.S.: A semantic region descriptor for local feature based image classification. In: ICASSP (2008)
6. Oliva, A., Torralba, A.: Modeling the shape of the scene: A holistic representation of the spatial envelope. IJCV 42(3), 145–175 (2001)
7. Oliva, A., Torralba, A.: Spatial envelop (2001),
   http://people.csail.mit.edu/torralba/code/spatialenvelop
8. Joachims, T., Mlight, S.V.: Learning to classify text using support vector machines - methods, theory, and algorithms. Kluwer, Dordrecht (2002)

9. Tsoumakas, G., Vlahavas, I.P.: Random *k*-Labelsets: An Ensemble Method for Multilabel Classification. In: Kok, J.N., Koronacki, J., Lopez de Mantaras, R., Matwin, S., Mladenič, D., Skowron, A. (eds.) ECML 2007. LNCS (LNAI), vol. 4701, pp. 406–417. Springer, Heidelberg (2007)

10. Duygulu, P., Barnard, K., de Freitas, J.F.G., Forsyth, D.: Object Recognition as Machine Translation: Learning a Lexicon for a Fixed Image Vocabulary. In: Heyden, A., Sparr, G., Nielsen, M., Johansen, P. (eds.) ECCV 2002. LNCS, vol. 2353, pp. 97–112. Springer, Heidelberg (2002)

11. Lavrenko, V., Manmatha, R., Jeon, J.: A model for learning the semantics of pictures. In: NIPS (2003)

Robust Palmprint Recognition Based on Directional Representations

Hengjian Li[1], LianHai Wang[1], and Zutao Zhang[2]

[1] Shandong Provincial Key Laboratory of Computer Network, Shandong Computer Science Center, Jinan, 250014, China
{lihengj,wanglh}@keylab.net
[2] School of Mechanical Engineering, Southwest Jiaotong University, Chengdu, 610031 China
zzt@swjtu.edu.cn

Abstract. In this paper, we consider the common problem of automatically recognizing palmprint with varying illumination and image noise. Gabor wavelets can be well represented for biometric image for their similar characteristics to human visual system. However, these Gabor-based algorithms are not robust for image recognition under non-uniform illumination and noise corruption. To improve the recognition performance under the low quality conditions, we propose novel palmprint recognition approach using directional representations. Firstly, the directional representation for palmprint appearance is obtained by the anisotropy filter, which is robust to drastic illumination changes and preserves important discriminative information. Then, the PCA is employed to reduce the dimension of image feature. At last, based on a sparse representation on palmprint feature, the compressed sensing is used to distinguish palms from different hands. Experimental results on the PolyU palprint database show the proposed algorithm have better performance. And the proposed scheme is robust to varying illumination and noise corruption.

Keywords: palmprint recognition, directional representation, compressed sensing, noise corruption, illumination changes.

1   Introduction

Biometrics is the automated method of recognizing a person based on a physiological or behavioral characteristic. Biometric technologies are becoming the foundation of an extensive array of highly secure identification and personal verification solutions. Among different biometric authentication, palmprint recognition is one of the most promising approaches since it has several special advantages such as stable line features, rich texture features, low-resolution imaging, low-cost capturing devices, easy self positioning, and user-friendly interface[1]. And various palmprint high performance recognition algorithms have been to proposed to improve the recognition rate[2]. For example, in early times, Lu et al. [3] and Wu et al. [4] proposed two methods based on Principal Components Analysis (PCA) and Linear Discriminant Analysis (LDA), respectively. Ekinci et al, proposed Kernel PCA to improve the palmprint

Z. Shi, D. Leake, and S. Vadera (Eds.): IIP 2012, IFIP AICT 385, pp. 372–381, 2012.

recognition performance and proposed a palmprint recognition approach integrating the Gabor wavelet representation and kernel PCA[5-6]. However, because of the illumination influence, the recognition performance of appearance based approaches mentioned above is not very satisfying in some aspects.

Orientation based approaches are deemed to have the best performance in palmprint recognition field, because orientation in palmprint contains more discriminative information than other features, and is insensitive to illumination changes. A.Kong and D.Zhang were the first authors who investigated the orientation information of the palm lines for palmprint verification and extracted the orientations of the palm lines by using the Winner-take-all Rule, which was defined as Competitive Code (CompC) [7]. W.Jia designed a matching algorithm based on pixel-to-area comparison, which improve the palmprint performance via reducing the Equal Error Rate(EER)[8]. However, by using only one dominant orientation to represent a local region, we may lose some valuable information because there are cross lines in the palmprint. By directly using multiple orientation features, H.Li used a the matching score level fusion strategy to fuse of different directional Palm-Codes to obtain the high performance[9]. Those above palmprint recognition algorithms imply that directional representations have certain robustness.

Traditional Gabor-based image representation it has got two important drawbacks. First, it is computationally very complex. Second, a lot of memory is needed to store Gabor features [10]. And what is important, Gabor-based image representation is not robust to varying illumination and noise corruption. To improve the implemented efficiently and robustness in the palmprint recognition, we propose novel palmprint recognition approach using directional representations and compressed sensing in this paper. The novel proposed approach can effectively overcome the Gabor-based shortcoming and can significantly improve the performance of appearance based approaches. Perhaps, the simplest classification scheme is a nearest neighbor (NN)classifier to distinguish different biometric traits using Subspace-based features [11]. Under this classifier, an image in the test set is recognized (classified) by assigning to it the label of the closest point in the learning set, where distances are measured in the image space. However, it does not work well under varying lighting conditions. Based on a sparse representation computed by l1-minimization, J. Wright et al propose a general classification algorithm for face object recognition. In this classification, the testing image is coded as a sparse linear combination of the training samples. Compared with the traditional recognition algorithm, a sparse representation provides deep insights into feature classification and has proven its superior performance [12].

The rest of this paper is organized in the following. In Section 2, the directional representations of palmprint images using multiple anisotropy filters will be introduced. Feature extraction and dimension reduction using PCA and classification using compressed sensing will be proposed in Section 3. Experimental results on PolyU Palmprint Database are given in Section 4. Finally, conclusions are made in Section 5.

2   The Directional Representations

From some literatures, it can be seen that different representations of palmprint can be adopted for Subspace-based approaches. These representation algorithms include

Gabor, wavelet transform[5], dual-tree complex wavelet transform[13] et,al. General-ly speaking, the commonly used strategy is applying the corresponding filter to con-volute with the original palmprint image. However, these mentioned above still have some drawbacks. Let us take the most popular Gabor filters as an example. The Gabor filters, which could effectively extract the image local directional features at multiple scales, have been successfully and prevalently used in palmprint recognition, leading to better results. Since the Gabor features are extracted in local regions, they are less sensitive to variations of illumination. Also, the computational burden of Gabor-based representations is very heavy.

An image can be modeled as a piecewise smooth 2D signal with singularities. These properties need anisotropic refinement representations. However, the tradition-al Gabor isotropic refinement can't efficiently represent the palmprint structure. The Anisotropic Filter (AF) is to obtain sparse representation by the idea of efficiently approximating contour-like singularities in 2-D images. The AF is a smooth low resolution function in the direction of the contour, and behaves like a wavelet in the orthogonal (singular) direction. That is, the AF is built on Gaussian functions along one direction, and on second derivative of Gaussian functions in the orthogonal direc-tion. The structure of AF is very special for capturing the orientation of palmprint image[14]. The AF has the following general form

$$G(u,v) = (4u^2 - 2)\exp(-(u^2 + v^2)) \tag{1}$$

where (u,v) is, in this case, the plane coordinate and can be obtained in the following way.

$$\begin{bmatrix} u \\ v \end{bmatrix} = \begin{bmatrix} 1/\alpha & 0 \\ 0 & 1/\beta \end{bmatrix} \begin{bmatrix} \cos\theta & \sin\theta \\ -\sin\theta & \cos\theta \end{bmatrix} \begin{bmatrix} x - x_0 \\ y - y_0 \end{bmatrix} \tag{2}$$

Where $[x_0, y_0]$ is the center of the filter, the rotation θ, to locally orient the filter along palm contours and α and β are to adapt to contour type.

The choice of the Gaussian envelope is motivated by the optimal joint spatial and frequency localization of this kernel and by the presence of second derivative-like

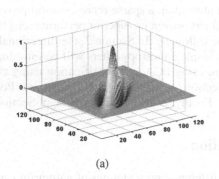

(a)   (b)

Fig. 1. (a)Appearance of anisotropic filter(2D); (b) Sample anisotropic filter with a rotation of 2*pi/3 radians and scales of 3 and 9

filtering in the early stages of the human visual system. It is also motivated by the presence of second derivative-like filtering in the early stages of the human visual system. Usually, $\beta > \alpha$ is set to better obtain the line orientation of palmprints. A 3D visualization of an AF can be seen in Fig.1. The competitive rule is a Winner-take-all rule defined as:

$$j = \arg \min_{p} \iint I * G\left(\theta_p\right) dxdy \qquad (3)$$

where j is called the winning index. The orientations of the six filters, θ_p are $p\pi / 12$, where p=0,1,2,...11.

Fig. 2 shows three palmprint images and their directional representations. Among them, Fig 2(a) and (c) come from the same palm, but were captured in different illumination conditions. Although the illumination conditions changed drastically, their directional representations are still very similar (see Fig 2(b) and (d)). From this example, it can be concluded that the directional representations is also robust for the change of illumination.

Fig. 3 shows three palmprint images and their directional representations. Among them, Fig 3(a) comes from Fig 2(a) which adds Gaussian white noise(noise level is 25) . Fig 3(c) comes from Fig 2(c) which adds Gaussian white noise(noise level is 25) .Although the noise made the image quality and understandability worse, their directional representations are still very similar. From this example, it can be concluded that the directional representations is also robust for the noise corruption.

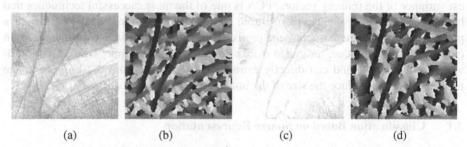

| (a) | (b) | (c) | (d) |

Fig. 2. Plmprint images and their directional representation and directional representation s are robust for illumination changes

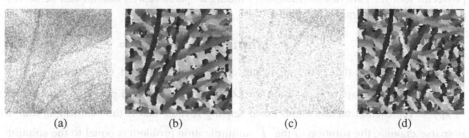

| (a) | (b) | (c) | (d) |

Fig. 3. Plmprint images and their directional representation and directional representation s are robust for noise corruption

3   The Proposed Robust Recognition Algorithm

This section describes the proposed palmprint verification algorithm using a compressed sensing as a classifier. We first extract from the original palmprint images a region of interest (ROI), as illustrated in Fig.4. The directional representations using multiple anisotropic filter will be obtained on the ROIs of palmprint images via winner-take-all rule. The PCA is used to extract the feature and reduce dimension of At last, the compressed sensing is used to classify the palms from different hands.

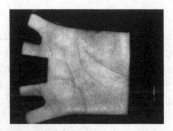 

Fig. 4. (a) The determination of ROI. (b) A cropped ROI image of the palmprint image in (a).

PCA has been widely used as linear feature extraction in computer vision. It computes the basis of a pace which is a space which is represented by its training vectors yields projection directions that maximize the total scatter across all classes. These basis vectors, actually eigenvectors, computed by PCA are in the direction of the largest variance of the training vectors. PCA is one of the most successful techniques that have been used in biometric recognition. The 2-DPCA is introduced to generate a projection space while extracting the projected feature of each image on the space. In classical PCA technique, an image matrix should be mapped into 1-D vector in advance. 2-DPCA method can directly extract feature matrix from the original image matrix, which can reduce the size of the image covariance matrix[15].

3.1   Classification Based on Sparse Representation

Sparse representation, which indicates that account for most information with a linear combination of a small number of elementary signals, has proven to be an extremely powerful tool for pattern classification. Finding a sparse representation can be solved as the following optimizing problem:

$$\hat{x}_0 = \arg\min \|x\|_0 \quad \text{s. t.} \quad Dx = y \qquad (4)$$

Where $\|\cdot\|_0$ denotes the l^0-norm, which counts the number of nonzero entries in a vector. Seeking the sparsest solution to $Dx = y$ is a NP-hard problem. The theory of sparse representation and compressed sensing reveals that if the solution x_0 sought is sparse enough, the solution of the l^0-minimization problem is equal to the solution to the l^1-minimization problem[16].

Given sufficient training palmprint samples of the i-th object hand class, $D_i = [d_{i,1}, d_{i,2}, ..., d_{i,n_i}] \in \mathbb{R}^{m \times n_i}$, a test palmprint sample $y \in \mathbb{R}^m$ from the same hand will approximately lie in the linear span of the training palmprint samples associated with object i $y = D_i x_i$ for some coefficient vector $x_i \in \mathbb{R}^{n_i}$. Usually, the small nonzero entries in the estimation associated with the columns of D from a single object class I, and can easily assign the test palmprint feature y to that class. Based on the prior sparse representation of palmprint images, one can treat the test feature can be treated as a linear combination of all training features of each object. And, one can identify the right class from multiple possible classes.

Therefore, given a new test palmprint template y from one of the classes in the training template set, we first compute its sparse representation via basis pursuit. And then, we compute the coding residual. It can be computed as follows: For each class i, let $\lambda_i : \mathbb{R}^n \rightarrow \mathbb{R}^n$ be the characteristic function which selects the coefficients associated with the i-th class, one can obtain the approximate representation $\hat{y}_i = D \lambda_i (\hat{x}_1)$ for the given test sample y. We then classify y based on the approximations by assigning it to the object class that minimizes the residual between y and $\hat{y}_i$: $\min r_i(y) = \|y - D \lambda_i (\hat{x}_1)\|$ [12].

3.2 Palmprint Recognition Using Directional Representation and Compresses Sensing

From discussion above, as illustrated in Fig.5, the proposed framework can be briefly summarized as follows:

Step 1: For reliable feature measurements, the gaps between the fingers as reference points to determine a coordinate system is used to extract the region part of a palmprint image. Fig.4 is an example.

Step 2: The ROI parts of palmprint image are further processed to obtain the directional representations. To be specific, the directional representations using multiple anisotropic filters will be obtained on the ROIs via winner-take-all rule.

Step 3: The PCA is employed to reduce dimension to reduce redundancy efficiently and extract the palmprint feature of the directional presentations for next match stage.

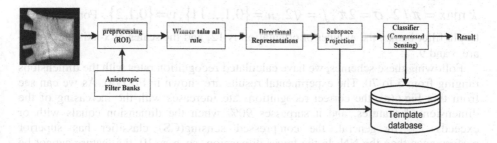

Fig. 5. The flowchart of the proposed palmprint recognition system

Step 4: By computing the coding residual, the minimum residual is identified as for recognition. Then recognition rates for the compressed sensing in a palmprint database are used to evaluate the performance.

4   Experimental Results and Analysis

In this section, we present experiments on PolyU publicly available databases for palmprint recognition, which demonstrate the efficacy of the proposed directional representations and validate the claims of the proposed sections. We will first examine the role of feature extraction within our framework, comparing performance across various feature spaces and feature dimensions. We will demonstrate the robustness of the proposed algorithm to noise corruption. Also, we test the speed of the proposed algorithm and compare with the traditional algorithms.

4.1   Experimental Settings

In PolyU Palmprint Database, there are 600 gray scale images captured from 100 different palms by a CCD-based device(http://www.comp.polyu.edu.hk/ biometrics). Six samples from each palm are collected in two sessions: the first three samples were captured in the first session and the other three were captured in the second session. The average time interval between these two sessions was two months. The size of all the images in the database was 384 × 284 with a resolution of 75dpi. In our experiments, a central part (128×128) of each image is extracted for further processing. The results have been generated on a PC with an Intel Pentium 2 processor (2.66GHz) and 3GB RAM configured with Microsoft Windows 7 professional operating system and Matlab 7.10.0(R2010a). In all of our experiments, we employed a highly efficient algorithm suitable for large scale applications, known as the Spectral Projected Gradient (SPGL1) algorithm, to solve the BP problems[17].

4.2   Recognition Performance

The first three samples of each palm are selected for training, and the remaining three samples are used for testing. The feature vector of the input palmprint is matched against all the stored templates and the most similar one is obtained as the matching result. In the implementation of Gabor filters, the parameters are set as

$k \max = \pi / 2$, $\sigma = 2\pi$, $f = \sqrt{2}$, $u = \{0,1,...11\}$, $v = \{0,1,2\}$. For the anisotropic filter banks, the number of orientation is 12, and the scales of x-axis and y-axis are 3 and 9.

Following these schemes, we have calculated recognition rates with the dimensions ranging from 5 to 70. The experimental results are shown in Fig.6(a). As we can see from this Fig.6(a), the correct recognition rate increases with the increasing of the dimension of features, and it surpasses 90% when the dimension equals with or exceeds 25. In general, the compressed sensing(CS) classifier has superior performance than the NN. In the lower dimension, such as 10, the feature cannot be

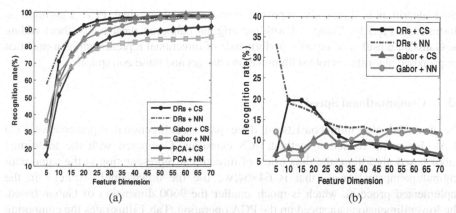

Fig. 6. Performance Original palmprint database (a) Recognition performance of different approaches with varying feature dimension (b)Performance comparison between directional representations, Gabor-based and the results obtained by PCA as a reference baseline

represent the intrinsic the palmprint image, therefore, the NN seems better. The Fig.6(a) also suggests that the recognition rate of Ours has better performance than all the other approaches under the same condition. For the feature dimension is lower than 45, the directional representations based approaches has better performance than Gabor methods. When the feature dimension is larger than 45, the performance of directional representation and Gabor are nearly the same. This is also been illustrated in Fig.6(b), in which PCA as a reference baseline(the same classification algorithm).

To test the performance under the noisy circumstances, we add the Gaussian white noise to each train samples and test samples. In the following experiments, the noise level is 25. As shown in Fig.7, we have calculated recognition rates with the dimensions ranging from 5 to 70. The Fig.7 also suggests that the recognition rate of our proposed method (Ours) has better performance than all the other approaches under the same condition. From the Fig.7(a), the CS classification is better than NN for the same features. If PCA is acted as a reference baseline, we will find we find the directional representation and NN

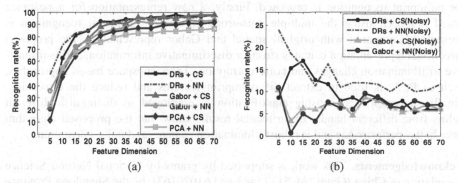

Fig. 7. Performance Original palmprint database under noise corruption (a) Recognition performance of different approaches with varying feature dimension (b) Performance comparison between directional representations, Gabor-based and the results obtained by PCA as a reference baseline

380   H. Li, L. Wang, and Z. Zhang

based algorithm is drastically improved compared with original PCA algorithm, as illustrated in Fig.7(b). Compared with the original palmprint database (without adding the Gaussian noise), The superior performance of directional representations means that our proposed algorithm is robust illumination changes and noise corruption.

4.3 Computational Speed

The proposed algorithm consists of three parts: (1)directional representations (2) PCA- dimension reduction method(3)CS classifier. Compared with the traditional Gabor-based algorithm, the dimension of directional representation is the same with original palmprint image, that is 64×64(we use the downsample strategy in the implemented process), which is much smaller the 9600 dimension of Gabor–based. The lower dimension can speed up the PCA operation. Tab.1 illustrates the computing time of the proposed approach and other approaches. Form the Tab.1, the computational running time of the proposed approach for feature extraction and classification is shorter than the Gabor based approaches. The performance of approaches based on directional representations is a little better than the Gabor-based. However, the running time of Gabor based palmprint recognition algorithms is 1.5 times of that Directional Representations based algorithms.

Table 1. Running time with different approaches(feature dimensions: 40)

Algorithms	PCA+NN	PCA+CS	Gabor+NN	Gabor+CS	DRs+NN	DRs+CS
Recognition rate	82.33%	87.33%	94.33%	96.67%	95.67%	97.00%
Time (s)	6.52	14.08	54.63	61.78	33.49	43.27

5   Conclusions

In this paper, a novel image representation approach, named directional presentations for palmprint recognition is proposed. Firstly, a new representation for appearance based approach using the multiple anisotropy filters for palmprint recognition is presented. Compared with original spatial and Gabor representation, the proposed directional representation contains stronger discriminative information, and is insensitive to illumination changes and noise corruption. Then, subspace based approaches, such as PCA, is used to extract the palmprint features and reduce the dimension. Finally, a compressed sensing classification is employed to distinguish different palms from different hands. Experimental results show that the proposed algorithm have better performance and is robust illumination changes and noise corruption.

Acknowledgements. This work is supported by grants by National Natural Science Foundation of China (Grant No. 51175443 and 61070163), by the Shandong Province Outstanding Research Award Fund for Young Scientists of China (Grant No. BS2011DX034) and by the Shandong Natural Science Foundation (Grant No. ZR2011FQ030 and ZR2011FM023).

References

1. Zhang, D., Kong, A., You, J., Wong, M.: Online palm print identification. IEEE Transactions on Pattern Analysis and Machine Intelligence 25, 1041–1050 (2003)
2. Kong, A., Zhang, D., Kamel, M.: A survey of palmprint recognition. Pattern Recognition 42, 1408–1418 (2009)
3. Lu, G., Zhang, D., Wang, K.: Palmprint recognition using eigenpalms features. Pattern Recognition Letters 24, 1463–1467 (2003)
4. Wu, X.Q., Zhang, D., Wang, K.Q.: Fisherpalms based palmprint recognition. Pattern Recognition Letter 24, 2829–2838 (2003)
5. Ekinci, M., Aykut, M.: Palmprint recognition by applying wavelet subband representation and kernel PCA. Journal of Computer Science and Technology 23, 851–861 (2008)
6. Ekinci, M., Aykut, M.: Gabor based kernel PCA for palmprint recognition. Electronics Letters 43, 1077–1079 (2007)
7. Kong, A.W., Zhang, D.: Competitive Coding Scheme for Palmprint Verification. In: Proceedings of the 17th International Conference on Pattern Recognition, pp. 520–523. IEEE Computer Society, Washington, DC (2004)
8. Jia, W., Huang, D.S., Zhang, D.: Palmprint verification based on robust line orientation code. Pattern Recognition 41, 1504–1513 (2008)
9. Li, H., Zahng, J., Zhang, Z.: Generating cancelable palmprint templates via coupled nonlinear dynamic filters and multiple orientation palmcodes. Information Sciences 180, 3876–3893 (2010)
10. Shen, L.L., Bai, L.: A review on Gabor wavelets for face recognition. Pattern Anal. Appl. 9, 273–292 (2006)
11. Hu, Q., Yu, D., Xie, Z.: Neighborhood classifiers. Expert Systems with Applications 34, 866–876 (2008)
12. Wright, J., Yang, J., Ganesh, A.Y., Sastry, A., Yi Ma, S.S.: Robust Face Recognition via Sparse Representation. IEEE Transactions on Pattern Analysis and Machine Intelligence 31, 210–227 (2009)
13. Eleyan, A., Ozkaramanli, H., Demirel, H.: Complex Wavelet Transform-Based Face Recognition. EURASIP Journal on Advances in Signal Processing, Article ID 185281, 13 pages (2008)
14. Li, H.J., Wang, L.H.: Chaos-based cancelable palmprint authentication system. Procedia Eng. 29, 1239–1245 (2012)
15. Yang, J., Zhang, D.: Two-Dimensional PCA: A New Approach to Appearance-Based Face Representation and Recognition. IEEE Transactions on Pattern Analysis and Machine Intelligence 26, 131–137 (2004)
16. Chen, S., Donoho, D., Saunders, M.: Atomic decomposition by basis pursuit. SIAM Review 43, 129–159 (2001)
17. van den Berg, E., Friedlander, M.P.: Probing the Pareto frontier for basis pursuit solutions. SIAM J. Sci. Comp. 31, 890–912 (2008)

FPGA-Based Image Acquisition System Designed
for Wireless

Haohao Yuan[1,2], Jianhe Zhou[1,2], and Suqiao Li[1]

[1] College of Computer Guangxi University of Technology, Guangxi, China
yuanhao_1027@163.com
[2] College of Information Engineering Wuhan University of Technology, Wuhan, China

Abstract. Introduced micro-wireless transmission system image acquisition the overall structure, each part of the overall design ideas and works theory. Designed SBBC camera interface and data cache control module, analysis of the TCP / IP communication protocol standard structure and characteristics, and on this basis, the design of a non-standard short-range wireless communication protocol. System which controlled by the input and output module is based on FPGA, the system collected data alternately to the outer SDRAM, this realized the image data collection; Ethernet interface controller also designed in FPGA to achieve the image data transmit through wireless. The results show that: under normal conditions, the wireless communication module can be complete, accurate, stable wireless data transceiver functions.

Keywords: Digital image sensor, SBBC interface, FPGA, wireless transmission.

1 Introduction

Along with the fast developing of computer, network transmission technology and communication technology, wireless remote video transmission has become a trend, this technology has gradually be applied in remote monitoring, aerial photo taken, conference call, etc. Video monitoring now generally use the following solutions [1,2]: use high imaging quality of image sensor camera, through the S-VIDEO terminals transmission data in real-time, this plan need a camera and acquisition device connectivity, and the monitoring center also need large storage space to store image and video clips, this scheme is high real-time, but big energy consumption and high cost, so it is suitable for public security and monitoring. While the wireless image acquisition system based on FPGA with its low power consumption and reconfigurable characteristics being perfect choice. In the condition of the wireless application protocol has been developed, the center of gravity of the research is to image sensor used by the hardware platform and streams of data processing, at present the main schemes are: ① DSP+FPGA, use FPGA to complete data storage and preprocessing work, using DSP chip high-speed processing power to compress and process the image, this plan suitable for need data parallel processing and more numerical JPEG data flow. ② FPGA+ video codec chips, use the parallel processing ability of FPGA transmit and processing of images and video data simultaneously, because FPGA is reconfigurable in hardware, this plan suitable for the experimental stage. ③CPU + FPGA, use as the central control and interface CPU control role, use the FPGA to complete image processing tasks. MicroBlaze32 a soft processor core of Xilinx company is the industry's fastest soft handling solutions, running in the 150 MHz clock, it can provide 125 D-MIPS , very suitable for design in view of the network, data communication, embedded of complex system, so that the plan three with the CPU inside FPGA to design the wireless image acquisition system [3].

2 Design Thought

Wireless image acquisition system Based on FPGA is mainly collection site image signal, and the collected image data will be transmit to upper computer in real-time through Ethernet, at the same time can receive instruction data from upper computer, the whole system diagram shown as in figure 1. System mainly consists of four modules: image sensor module, image acquisition and control module, image storage and processing module, network data transmission module. In the process of operation, the FPGA will storage the CMOS sensor data to the expansion SRAM [4], every full of a frame processor read data from SRAM and FPGA kernel processor by MicroBlaze will sent the data to the upper machine through the Ethernet chips.

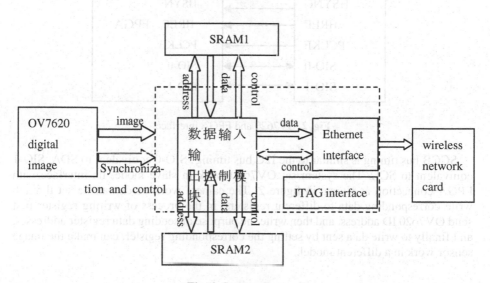

Fig. 1. System diagram

3 Control and Processing Module Design Use FPGA

The system use FPGA made by xilinx company the chip models is XC3S1500, the input and output of the camera data control module was designed, SRAM reading and writing control module and based on the soft nuclear Ethernet interface were designed, below will make a detailed description for each part .

3.1   Image Acquisition and Interface Design

In the system the Omni vision company OV7620 type CMOS image sensor [5] was adopted as image sensor, it support continuous and interlaces the two scanning mode, with 10 double channel A/D converter, output eight image data; the highest pixels is 664 x 492, frame rate is 30 FPS; data formats including YUV, YUV, RGB three types, can satisfy the requirement of general image acquisition system, can programming configuration internal registers through SCCB bus. Meet the power supply and working in the crystals can output digital streaming of video at the same time also provide pixel clock PCLK, line reference signal----HREF□vertical synchronized signal----VSYNC, can use the clock signal above to determine each frame output image starting bit and end bit, meet the requirements of the external circuit read image.

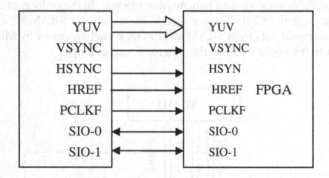

Fig. 2. OV7620 and FPGA interface

SCCB bus timing is similar to the I2C bus timing, SIO-0 equivalent to SDA, SIO-1 equivalent to SCL. The system set OV7620 work in slave mode, its interface with FPGA connection as shown in figure 2. The camera work mode can be set through write corresponding data to different registers, in the process of writing register first send OV7620 ID address, and then write the purpose of sending data register addresses, and finally to write data sent by setting the corresponding register, can make the image sensor work in a different model.

3.2   Data Storage Module

The data storage use company of ISSI IS61LV25616AL type chip, it is the static RAM, system of each frame resolution for the 640*480 grayscale image need 2457600 bit, less than its storage capacity 4194304 bit, can meet the storage requirements. To meet the real time requirement, data storage using high-speed SRAM switch mode, called "table tennis" mode. The specific process as follows: the data of image data flow distribution to two SRAM through FPGA output and output control module. In the first period, the input data cache to SRAM1, in the second period through the input data flow choice unit storage the date to SRAM2, at the same time, to sent the first cycle of data cached in SRAM1 to subsequent module through the output data choice unit, the third cycle the input data stream select units will storage data to SRAM1, at the same time, to sent data in SRAM2 last cycle through the output data choice unit to the fol-low-up module. So again, to realize the seamless data buffer, ensure the integrity of the image data.

3.3   Transmission

In the network transmission module, the chip BCM5221 made by BroadCom company was used in the physical, use the soft MicroBlaze32 nuclear as controller in the EDK development environment to generate the basic framework and related documents by generation wizard , the architecture of the soft nuclear as shown in figure 3.

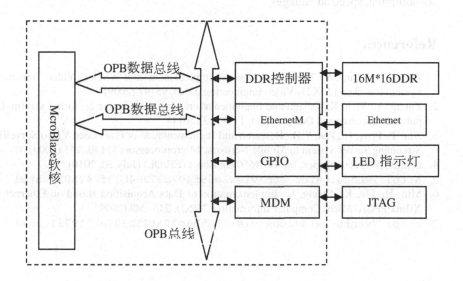

Fig. 3. Soft nuclear basic framework

In the system, the DDR (double-data-rate) use the OPB DDR controller connect with OPB high-speed bus, and the Ethernet physical chip----BCM5221 connect with OPB bus through the Ethernet MAC, they are all OPB bus high-speed equipment; GPIO (General Purpose I/O) is the low speed terminal controller also can be connected in OPB bus, links such as LED and some terminals not high demand for the speed. MicroBlaze Debug modules MDM (MicroBlaze Debug Module) is responsible for download and test software.

The system adopts the TCP/IP protocol, divided into fluctuation two parts to realize [6, 7]. The physical layer realized by the hardware BCM5221 chip, the MAC layer realized by xilinx company Ethernet MAC IP nuclear ; The top of the transport layer by the software code realization, the LwIP agreement module, this module is one set for the embedded system is an open source TCP/IP protocol, the main purpose of LwIP is to keep the TCP/IP protocol in the main functions of the case to reduce the use of RAM, at present LwIP is widely used in all kinds of top-grade of embedded system. LwIP also support server mode and client mode, in the upper part of the agreement we adopt LwIP agreement module.

4   Conclusion

After research of the traditional image acquisition module deeply, based on disadvantages of transmission to the traditional image acquisition module, designed the image acquisition and wireless transmission module for portable embedded system. The system realized the original data collection and image storage, ensure the image data of wireless transmission continuously and completely, with small size, low power consumption, speed advantages.

References

1. Zhang, Z., Zhang, Y., Zhu, X.: Design and Implementation of Wireless Video Transmission System Based on i.MX27. Video Engineering 33(2), 95–97 (2009)
2. Zheng, Y., Tian, X.: Design and implementation of wireless image capturing system. Computer Engineering and Design 32(1), 111–113 (2011)
3. Xia, T., Peng, H., Fan, Y.B.: Research and Implementation of Embedded Video Surveillance Streaming Server System in Mobile Network. Microprocessors 1, 140–143 (2009)
4. Omnivsion Technologies, Inc.OV9650 datasheet (EB/OL) (July 20, 2010)
5. http://wenku.baidu.com/view/e65f407d27284b73f2425073.html
6. Min, H., Liu, K., Huang, L.: Implementation of Data Acquisition Based on Ethernet With Xilinx FPGA. Micro Computer Information 25(62), 246–248 (2009)
7. http://wenku.baidu.com/view/7d5b6c20af45b307e8719731.html

A Context-Aware Multi-Agent Systems Architecture for Adaptation of Autonomic Systems

Kaiyu Wan[1] and Vasu Alagar[2]

[1] Xi'an Jiaotong-Liverpool University, Suzhou, PRC
kaiy.wan@xtjlu.edu.cn
[2] Concordia University, Montreal, Canada
alagar@cse.concordia.ca

Abstract. An important requirement of autonomic systems is that they self-adapt, both with respect to internal self-healing and with respect to external environmental changes. In order to fulfill this requirement autonomic systems must have awareness abilities, context gathering mechanisms, context-dependent adaptation policies, and the ability to react with respect to each adaptation requirement. In this paper we provide an insight into these issues and propose a multi-agent system that interact faithfully among themselves in order to self adapt with safety and security.

1 Introduction

The autonomic computing system (ACS) concept was initiated at IBM [6], as a solution to reduce the cost of resource management and system administration of large evolving systems. Since its proposal in 2001, several prototype examples of ACS are reported to have been built. These include IBM Tivoli Management Suite [10], SUN Microsystems [11], Hewlett-Packard's Adaptive Enterprise [7], and Microsoft's Dynamic Systems Initiative [9]. We believe that the self-adaptation issue for ACS is still an open and challenging problem. Motivated by this belief we provide a multi-agent system architecture in which context awareness and context-dependence are central to adaptation. The important characteristics of ACS [2,3] are as follows:

Self-configuring: It automatically adapts to dynamically changing environments.
Self-healing: It detects, diagnoses, and recovers from any damage that occurs during its life cycle.
Self-optimizing: It monitors and tunes its resources optimally.
Self-protecting: It detects and guards itself against damage from accidents and outside attacks.

Among these characteristics, the self-adaptability of ACS is the most effective approach necessary to tackle its immense software management complexity. It is called for during self-configuration, self-healing, and self-protecting scenarios. The system can self-configure when (1) due to its self-monitoring it has *sensed* that some internal updates have become available, or (2) it *perceives* changes to its environment or (3) when *stimulus events* are received from the environment. In the first two cases *anticipatory* adaptation is called for and in the third case a *reactive* adaptation is called for. To enact

Z. Shi, D. Leake, and S. Vadera (Eds.): IIP 2012, IFIP AICT 385, pp. 387–397, 2012.

an anticipatory adaptation, the ACS has to be *context-aware*, and to trigger a reactive adaptation the system must act in a *context-dependent* manner. Anticipatory adaptation is called for during self-healing as well. The system should monitor its internal contexts which include resources, users, and programs, and include externally perceived contexts for adaptive action when a failure happens. The main goal here is to restore system consistency without disrupting services. Although the adaptive actions can be triggered only after failures, it is important that the system is able to see the sign of failure before it happens. That is the reason we classify this adaptation type as anticipatory. Self-protection requires anticipatory adaptive actions. The system should protect its critical resources from unauthorized use and detect external attacks before damage is caused. So, we conclude that a broad classification of adaptation types can be 'anticipatory' (*before*) and 'reactive' (*reactive*), and both adaptation types involve *context* and *awareness*. Anticipatory adaptation requires *context-awareness* and reactive adaptation requires *context-dependence*.

1.1 Our Contribution

Basic concepts on context and context awareness are discussed in Section 2. Sections 3 and 4 describe the main contribution of this paper. An adaptation logic in a formal setting is given in Section 3.1. In Section 4 we discuss a MAS architecture in which context-aware and context-dependent adaptive actions are possible, and give a detailed agent interaction protocol. The agent collaborations in MAS offer a more effective solution for any large scale autonomic system adaptation.

Traditionally, self-adaptation is realized in a system by adding adaptation logic to the managed system. In our work we embed adaptation in a logic of context. In simple terms, adaptation may be viewed as a sequence of activities *"monitor (M), analyze (A), plan (P), and execute (E)"*. Monitoring includes self-monitoring (internal monitoring) and external monitoring. The analysis stage explores the available options for responding to the perceived situations. Often business policies and security policies will constrain the available options. The planning stage puts the options in a certain order of execution, may be in sequential or in parallel, and gives the plan to the execution stage. This sequence $\langle M - A - P - E \rangle$ of activities may have to be performed cyclically, where one cycle may include many smaller inner cycles. In [12] a mechanism to manage the intra-cycle and the inter-cycle coordination for the $\langle M - A - P - E \rangle$ paradigm has been proposed, and is illustrated for a traffic management system. In contrast, our approach is to lift $\langle M - A - P - E \rangle$ cycles to agent collaborations within a multi-agent system (MAS). System adaptation is achieved via agent collaboration. Consequently, the agents are made context-aware and their reactions become context-dependent.

2 Basic Concepts

Adaptation logic will include specific rules as well as meta inference rules. Both types of rules are usually applicable in many different contexts. Context information is both heterogeneous and multi-dimensional. A formal representation of context is essential in order to make agents detect and react to contextual changes. We use the syntax of context and the logic of context introduced by Wan [13]. In Section 2.1 we give a quick review of context definition, representation, and calculation of context expressions. Awareness for computing systems should mean perception, collection, and a rational use of the information towards a specified goal. The system should not only understand why some information is collected but also be intelligent to use it with a justifiable reason. Information related to internals such as computing resources, computational states, and policies for state change, and information related to its externals such as physical devices, other systems, and humans with which it interacts provide the awareness envelope. In Section 2.2 we discuss the internal monitoring, called *self-awareness* of the system. In Section 2.3 we discuss the external monitoring, called *context-awareness*. Adaptation, regardless of the specific rules binding it, should be based on contextual conditions.

2.1 Context and Situation

Context types are defined in [13] by defining a finite set of dimensions and a tag type for each dimension. Examples of dimensions are $LOCATION$, $TIME$, and WHO (or $NAME$ or $ROLE$). The tag type for $LOCATION$ can be the set of cities or geographical zones, and these are chosen to suit the system need. In general, given the set of dimensions $DIM = \{D_1, D_2, \ldots, D_n\}$, and X_i is the tag type for D_i a context c over these dimensions has the representation $[D_1 : x_1, \ldots, D_n : x_n]$, $x_i \in X_i$. As an example, $[HC : H123456, LOC : NewYork, DATE : 2003/09/30]$ defines a context with dimensions HC (health center), LOC (location), and $DATE$ (calendar date). Many events or situations that might happen in this context. A calculus of context and a toolkit based on it have been extensively used in many context-aware system applications [8,1]. In the rest of our discussion we will be using the context representation defined here.

A context in itself is not very useful unless it is associated with *events* or *situations* of interest in a system. We formulate a situation as a logical formula involving dimension names and system variables. In general, if α is a logical formula denoting a situation, we write $vc(c, \alpha)$ to mean that the situation is valid in context c. As an example, the situation $VERYWARM = TEMP > 40 \land HUMID > 80$ is true in context $[TEMP : 41, HUMID : 81]$, false in context $[TEMP : 38, HUMID : 85]$, and cannot be resolved in context $[HUMID : 78, WINDSPEED : 35]$. In general, (1) there may exist zero or more contexts in which a situation might be true, and (2) many situations may be valid in a context.

For self adaptation, the system must be able to evaluate situations automatically at different contexts. To enable this automatic evaluation we insist (1) a situation involves only dimension names and system variables, and (2) the situation is formulated as a Conjunctive Normal Form (CNF). The validation is then automated with the following steps:

1. Write the situation as $\alpha = p_1 \land p_2 \ldots, \land p_k$, where each p_i is a disjunctive clause involving the $DIMENSION$ names and system variables.
2. To evaluate α in a context c, evaluate the p_is in context c until either all of them evaluate to true or a p_i evaluates to false, or a p_i cannot be evaluated. In the first

case, the situation α is true in context c. In the second case the situation α is false in context c. In the third case the context or system information is inadequate to evaluate the situation α. The evaluation of each p_i is done as follows:

- if D is a $DIMENSION$ name in context c, and D occurs in the predicate p_i then the D in p_i is replaced by the tag value associated with D in c,
- substitute the variables in p_i by their respective values, as provided by the system in the environment of evaluation,
- if p_i still has $DIMENSION$ names or variables, then p_i cannot be evaluated in context c; otherwise p_i is now a proposition which evaluates to either true or false.

2.2 Self-awareness

We restrict to the self-awareness behavior related to protecting the critical resources and states of the system, and regulating services. It is necessary to include the following information for modeling self-awareness contexts:

1. *Roles* are defined on the subjects. Each subject must have a role in order to be active in the system.
2. *Resources* are the objects (files such as patient files, policy documents) that are under use. Organizational policies will constrain the access to resources. With each resource is associated a list of roles, indicating for each role the level of access to the resource.
3. *Permission* for each object in the system and for each role is specified. An access control list exists for each resource that specifies explicitly the actions that a role is authorized to perform on the resource, as well as the actions not allowed to perform.
4. *Obligation* is the set of mandatory actions to be performed by the subjects at certain stages.

From the above modeling elements, the context type for self-awareness can now be constructed. We call this set of contexts as *internal contexts (IC)*. Assume that a set of roles is $RC = \{RC_1, \ldots, RC_m\}$. Let $DC = \{DC_1, \ldots, DC_k\}$ be the set of resource categories which are to be protected. We regard RC_i's and DC_js as dimensions. As an example, the context $[RC_1 : u_1, DC_j : \texttt{Salary\_File}]$ might be constructed to indicate that user u_1 plays the role RC_1 while in possession of the resource $\texttt{Salary\_file}$. Let PC denote the purpose dimension. Assume that the tag type for PC is the enumerated set $\{\texttt{Technical, Auditing, Research}\}$. An example of a self-awareness context with these three dimensions is $[RC_1 : u_1, DC_2 : \texttt{Salary\_File}, PC : \texttt{Auditing}]$. The context describes the snapshot in the system when the user u_1 is requesting access to the resource $\texttt{Salary\_File}$ for auditing purposes. The organizational policy should be evaluated at this context before allowing or denying the request.

2.3 Context-Awareness

Contexts that characterize external awareness are constructed automatically from the information gathered by the sensors and user stimulus to the system. When a stimulus is received by the system, the sensed dimensions might include LOC, $TIME$,

WHO, WHY which correspond respectively to the location from where the system is accessed, the date/time at which the access happens, the user (role) of the subject accessing the system, and the reason for requesting the service. Context, as raw information gathered by sensors and other system stimuli, are aggregated into context representations by the context toolkit [13]. We call such contexts as *external contexts (EC)*. The twin issues related to the set EC are (1) the contexts are constructed in anticipation of future behavior, and (2) the system reactivity is according the adaptation policies set by the system. As an example, consider external system attacks against which the system should adapt. For such anticipatory adaptation the system should construct *security* contexts from information available in its environment. This might include classes of people, their locations, sets of remote systems, and types of communication links. Based on this information the system constructs dimensions, and tag types for each dimension. Reactive adaptation is triggered when a specific situation arises.

The sets IC and EC should have a *bridge* in order to help construct internal contexts from external contexts. We assume that an ontology exists or the dimension names and their tag types are the same for constructing the IC and EC context types.

3 Adaptation

The system administration is driven by policies, which include rules relevant to accessing the system from its environment and rules for accessing its resources. Adaptation policies are rules. An adaptation rule r can be formalized as a pair (α, A), where α is the situation and A is an action. The meaning is that the rule r must be applied in a context c if $\mathbf{vc}(c, \alpha)$ is true, and the application of the rule requires the fulfillment of the action A. As explained earlier the situation α governing a rule can be evaluated at different internal as well as external contexts constructed dynamically during system evolution. Authorizing access, performing obligatory actions, and providing services are examples of actions to be performed in context c upon the validity of the situation α in context c. Thus, our semantics of adaptation satisfies both anticipatory and reactive adaptations. Given this semantics we suggest a few rules of inference for reasoning with adaptation policies.

3.1 Inference Rules for Adaptation

In an imperative programming style we can write an adaptation policy as do A if α is true in context c. In logic programming style we can write $A \Leftarrow \mathbf{vc}(c, \alpha)$. This later style is more conducive to express a set of inference rules for adaptation. Note that in our discussion the symbol $\rightarrow$ denotes 'logical implication', and the symbol $\Leftarrow$ denotes 'the imperative command do'.

Rule 1. If α is true in a context c then it is true in every context that includes c. A context c' that includes c is an *outer* context of c. This rule is necessary to ensure that an adaptation that is valid in a context remains valid in its enclosure. Without this rule, contradictions might arise in executing incremental adaptations. We write this formally as follows

$$\frac{A \Leftarrow \mathbf{vc}(c, \alpha)}{\exists c' \bullet A \Leftarrow \mathbf{vc}(c', \mathbf{vc}(c, \alpha))} \tag{1}$$

Rule 2. We use the Law of Excluded Middle to give two adaptation rules. There exists an adaptation rule that states that action A can be initiated in context c whenever α is true in context c. If it is true that in context c, $\alpha \rightarrow \beta$ is also true then we can apply action A in context c whenever β is true. This follows from the predicate logic modes ponen rule applied to context c.

$$\frac{(A \Leftarrow \mathbf{vc}(c, \alpha)), \mathbf{vc}(c, \alpha \rightarrow \beta)}{A \Leftarrow \mathbf{vc}(c, \beta)} \tag{2}$$

In the second case, there exists a rule that S can be initiated in context c whenever $\alpha \rightarrow \beta$

$$\frac{(A \Leftarrow \mathbf{vc}(c, \alpha \rightarrow \beta)), \mathbf{vc}(c, \alpha)}{A \Leftarrow \mathbf{vc}(c, \beta)} \tag{3}$$

Rule 3. Adaptations corresponding to conjunction and disjunction of situations are easy to construct. An example rule to deal with situation conjunctions is

$$\frac{(A_1 \Leftarrow \mathbf{vc}(c, \alpha)), (A_2 \Leftarrow \mathbf{vc}(c, \beta))}{(A_1 \circ A_2) \Leftarrow \mathbf{vc}(c, \alpha \wedge \beta)} \tag{4}$$

where $\circ$ means that both actions A_1 and A_2 should be performed, however they may be performed in any order. We give two adaptation rules enabling to infer adaptation actions across different contexts.

Rule 4. Let there be an adaptation rule for initiating action A in context c, provided α is true in context c. Assume that it is also true that whenever α is true in c, α' is true in another context c'. The rule below states that action A can be initiated in context c'.

$$\frac{(A \Leftarrow \mathbf{vc}(c, \alpha)), (\mathbf{vc}(c', \alpha') \rightarrow \mathbf{vc}(c, \alpha))}{A \Leftarrow \mathbf{vc}(c', \alpha')} \tag{5}$$

The above rule can be generalized to multiple adaptations.

$$\frac{(A_1 \Leftarrow \mathbf{vc}(c, \alpha)), (A_2 \Leftarrow \mathbf{vc}(c', \alpha')), \mathbf{vc}(c', \alpha') \rightarrow \mathbf{vc}(c, \alpha)}{(A_1 \circ A_2) \Leftarrow \mathbf{vc}(c', \alpha)} \tag{6}$$

The ability to automatically detect changes, construct contexts, and evaluate situations, combined with the ability to trigger policies at different contexts make our approach very powerful. The adaptation inference rules can be applied diligently to reduce the overload involved in choosing the right adaptations for different contexts.

4 Multi-Agent System - Structure and Analysis

In this section we discuss an agent architecture and explain how they collectively cooperate in the system to improve awareness and facilitate adaptation.

4.1 Agent Architecture

We define agent types, instead of agents. An agent type consists of agents which have identical role and behavior. All agents are black-box agents. Many agents can be generated from each agent type. Agents belonging to two different types have distinct behavior. We use the notation $a : T$ to mean a is an agent of type T. The agent types and the relationships among agents of different types are enumerated below.

1. *CA Type:* Agents of this type will construct and manage contexts in a specific domain, according to the context calculus implemented in the context toolkit [13]. The type CA includes expert agents who can convert context types. The functionality of an expert agent is somewhat similar to the services provided by *Jena 2* ontology system [7].
2. *DA Type:* The system might have several database(s), each storing policies in a specific domain. Security policies, adaptation policies, and work flow policies belong to different domains. Type DA denotes the set of database agents, such that one agent is in charge of policies in one domain.
3. *MA Type* Autonomous agents who continuously monitor the internals of the system are of this type. An MA agent is a memory bank, tracking history of actions initiated by and performed by different agents in different contexts. They will feed valuable information for other agents to adapt.
4. *IA Type:* Each sensory device is an agent of type IA. Agents of this type are autonomous agents that focus on external information gathering. They aggregate raw data, and communicate with agents of type CA.
5. *RA Type:* All reactive agents are of this type. They receive stimulus from the environment, context information from CA agent types, and interact with the environment to provide the services that are allowed by the adaptation rules. When it receives a stimulus the context associated with the stimulus is sent to a context toolkit agent. It gets back the context constructed for that stimulus.
6. *TA Type:* Agents of type TA have the knowledge of adaptation rules and the ability to reason with them at different contexts. A TA type agent will get the adaptation policies from DA type agents, and dispatch the adaptation action to agents of type RA.
7. *PA Type:* Agents of this type guard the resources against unauthorized access and regulate the granting of resources to subjects in different contexts according to the satisfaction of system policies. Each agent may be in charge of guarding resources of any one type.
8. *BA Type:* Agents of this type guard the boundary of the system, such as firewall. Agents of this type act as bridges between the *outside* and *inside* of the system. They enforce the fire wall policies for all requests and services crossing the boundary.

The set EC of external contexts are constructed by agents of type CA at the request of the agent types $\{IA, RA\}$. The set IC of internal contexts are constructed by agents of type CA with the assistance from agent types $\{BA, MA, PA\}$. Agents of type BA, in collaboration with agents of type CA will be able to transit between EC and IC. An agent of type DA is a slave for the authorized agents of types $\{TA, PA, MA\}$ for providing access to the policies stored in it. Agents of type TA determine adaptation activity for a context by looking up with databases MA and by applying reasoning rules relevant to a current situation. Agents of type MA will monitor all activities that happen in the system, and feed the current system context to TA agents. Thus, collectively the agent types enumerated above will act in covering the adaptation requirements of the system.

4.2 Adaptation Protocol

Assume without loss of generality that there are many agents of each type in the system, distributed if necessary. We do not get into the specifics of agent communication, instead we discuss a high level interaction scenario in which the agents collaborate to adapt and react to specific requests from the system environment.

Step 1: Collect External Context Information- An agent $ia : IA$ corresponds to one sensory type of information. An agent $ra : RA$ interacts with clients external to the system and collects information corresponding to the identity or role (WHO), the service ($WHAT$), the place where it is to be delivered ($WHERE$), at what time/date the service is to be delivered ($WHEN$) etc;. Until the ra is able to collect such information, the information automatically sensed for that session by ia will be repeatedly sensed. Both ia and ra forward the collected information to $ca : CA$. It is the responsibility of ra to react to the specific request S (stimulus) received by it.

Step 2: External Context Construction- Context information received may be fuzzy, especially for sensory data. Consequently, this step may require many iterations between ca and agent ia, until agent ca is able to determine the dimensions and the types associated for the context. Eventually, agent ca constructs a context c_e for stimulus, and a context d_e where response is expected, and sends to ra the tuple (d_e, c_e, S).

Step 3: Boundary Analysis- Agent ra forwards the pair (c_e, α) to agent $ba : BA$. The boundary agent ba forwards to agent $ca : CA$ the context c_e and the dimensions of interest to the system boundary. Interesting dimensions at the boundary might include role of client, client category, and purpose of use. Agent ca constructs from c_e a 'boundary context' c_b and an internal c_i and returns them to agent ba. This is possible because of our assumption that cas have semantic expertise and also have ontologies. We skip the actual details involved for such a construction. The boundary agent ba applies the fire wall security policy β for context c_b. That is, it evaluates $\mathbf{vc}(c_b, \beta)$. If the authentication fails, agent ba returns context c_e to agent ra. In turn ra will inform the authentication failure to the client from whom the stimulus was received. If the validation is successful, agent ba saves c_e and forwards service stimulus S with contexts d_e and c_i to $ma : MA$.

Step 4: Resource Requirements- The agent $ma : MA$ is autonomous and has a supervisory role in the system. In that capacity it can track data and communication traffic, monitor resource protection agents, and different agent interaction scenarios. In particular agent ma can get from $da : DA$ the history of requests serviced in every context. So, ma acts as follows:

– It sends the current execution environment as context situation and c_i to ca and receives c_i', where c_i' is the smallest enclosure [13] containing c_i.

- It requests da and receives from it the set of policies $R = \{r :: (A \Leftarrow \eta) \mid \mathbf{vc}(c'_i, \eta)\}$ that are relevant for context c'_i. It computes the set R', $R' \subseteq R$, such that all actions specified in the rules of R' are necessary to fulfill the stimulus request S. This step is essentially a query processing steps, whose details are not important for us now.
- If $R' = \emptyset$, then the ma sends the message 'service cannot be fulfilled now' to agent ra, which in turn will inform its client outside the system.
- If $R' \neq \emptyset$ then for each action A in $r \in R'$ it determines the set of resources Q_A required by it. Then it determines the set of resource agents $Q_{pa} = \{pa_q \mid pa_q : PA, pa_q \text{ protects } q \in Q_A\}$.
- It requests the resource agents in the set Q_{pa} to evaluate the access rights for the role specified as part of context c'_i.
- If a resource is denied, the resource agent in charge of it communicates the decision $grant$ otherwise communicates the decision $deny$ to agent ma.

Step 5: Analysis by ma Once decisions are received from all *pas*, agent ma reasons about their decisions. If the service was successfully authenticated at the fire wall, but access to resources is denied at the context c'_i then ma can conclude that the role of subject or CLASS (a dimension in BSC context) to which the subject was classified at the fire wall might have changed. Agent ma records this information, and informs agent ba of the denial of resource access for servicing the request S. If all the resources are granted, ma forwards contexts c_i and d_e and the set of rules relevant for servicing α to an adaptation agent $ta : TA$ requesting it to process the service adaptations.

Step 6: Informing Client of Service Denial- Agent ba forwards the contexts c_i and c_b to the reactive agent ra and informs the reason for service denial. In turn, the ra agent informs its client that the requested service cannot be provided because access to resources are denied.

Step 7: Adaptation- In order to adapt, agent ta has to select the adaptation rules from the set R'. There are three ways in which agent ta can resolve an adaptation.

1. *Specific Reactions* For every rule $r :: (A \Leftarrow \eta) \in R'$, it evaluates $\mathbf{vc}(d_e, \eta)$. If it is true then it selects A for adaptation. Once this is done for all rulers in R' there may be zero or more such adaptive actions. These are communicated to ra for execution.
2. *Reasoned Actions* When there is no direct match with an adaptation rules, an inference rule (from Section 3.1) must be invoked. The following are possibilities.
   - It is known that $\mathbf{vc}(d_e, \alpha)$ is true for some α and it can be proved that $\eta \to \alpha$ and there exists a rule $r \in R'$ such that $r :: (A \Leftarrow \eta)$. By Rule 2 in Section 3.1 adaptation action A can be initiated.
   - There exists rules $r \in R'$ such that $r :: (A \Leftarrow \eta)$, and $r' \in R'$ such that $r' :: (A' \Leftarrow \eta')$, and $\mathbf{vc}(c_e, \alpha)$ is true where $\eta \wedge \eta' = \alpha$. By Rule 4 in Section 3.1 adaptation actions A and A' can both be initiated.
   - It is known that $\mathbf{vc}(d_e, \alpha)$ is true for some α, the rule $(A \Leftarrow \mathbf{vc}(c', \eta))$ exists in R, and it can proved that $\mathbf{vc}(c_e, \alpha) \to \mathbf{vc}(c'_i, \eta)$ then initiate adaptation A. This is justified by Rule 6 in Section 3.1.

An adaptation need not be successful. When the effort to adapt fails, agent ta informs ma of the failure, and the reason for the failure. In turn ma informs ra through ba. In case an adaptation is successful, both ra and ba are given the set of adaptation actions.

5   Conclusion

Self-adaptation is a must for autonomic computing systems. Most of the current work have only limited capabilities. An example is the Personal View Agent (PVA) whose design is based upon some empirical calculations tracking user profiles. As brought out in [5] there is a great challenge ahead in developing self-adaptive systems based on sound software engineering principles. To the best of our knowledge the contributions in this paper are new. In particular, the proposed logic of contextual adaptation has the potential to immensely increase the adaptation capabilities of any context-aware software system. By importing it to a multi-agent system architecture we only increase its enforcement power, because we can add the intelligence inherent for agents with the ability to reason using the adaptation logic. Within the adaptation logic the traditional $\langle M - A - P - E \rangle$ activities can be embedded as rules. Thus, the adaptation logic is both an abstraction and a generalization of the traditional adaptation activities. The additional merits of our approach are the following:

1. By designating roles for agents, we have separated design from implementation issues. Context implementation, for example, is independent from how it will be used. Defining situations formally we are able to evaluate situations in different contexts, which in turn make adaptations truly dynamic.
2. Adaptation logic is simple for the present. Experts may be able to add more adaptation rules. Actual domain specific rules, such as security and business work flow policies, may be added/deleted independent from the rules of inference for adaptation.
3. As opposed to the looping activities of $\langle M - A - P - E \rangle$ system [12], many adaptation strategies such as parallel or sequential, allowed by the policies and the logic are possible.

Some of the challenging issues for further research include the application of the proposed architecture to a large case study, developing an implementation framework for orchestrating different sequences of multiple adaptations, and assessing the effectiveness of the proposed architecture.

Acknowledgments. This research is supported by Research Grants from National Natural Science Foundation of China (Project Number 61103029), Natural Science Foundation of Jiangsu Province, China (Project Number BK2011351), Research Development Funding of Xi'an Jiaotong-Liverpool University (Project Number 2010/13), and Natural Sciences and Engineering Research Council, Canada.

References

1. Hnaide, S.A.: A Framework for developing Context-aware Systems. Master of Computer Science Thesis. Concordia University, Montreal, Canada (April 2011)
2. IBM White Paper. Autonomic Computing Concepts,
   http://www-03.ibm.com/autonomic/pdfs/AC_Concepts.pdf
3. Bantz, D.F., et al.: Autonomic personal computing. IBM Systems Journal 42(1), 165–176 (2003)
4. Chen, C.C., et al.: PVA: A Self-Adaptive Personal View Agent System. In: ACM SIGKDD 2001, San Francisco, CA, USA (2001)
5. Cheng, B., et al.: Software Engineering for Self-Adaptive Systems: A Research Roadmap. In: Cheng, B.H.C., de Lemos, R., Giese, H., Inverardi, P., Magee, J. (eds.) Self-Adaptive Systems. LNCS, vol. 5525, pp. 1–26. Springer, Heidelberg (2009)
6. Horn, P.: Automatic Computing: IBM's Perspective on the State if Information technology. IBM Corporation, October 15 (2001)
7. Packard, H.: HP's Darwin Reference Architecture Helps Create tighter Linkage Between Business and IT, San Jose California, May 6 (2003),
   http://www.hp.com/hpinfo/newsroom/press/2003/030506b.html
8. Ibrahim, N., Alagar, V., Mohammad, M.: Managing and Delivering Trustworthy Context-dependent Services. In: The 7th International Workshop on Service-oriented Applications, Integration, and Collaboration (part of the 8th IEEE International Conference on e-business Engineering), Beijing, PRC, October 19-21
9. Microsoft Corporation. Microsoft Dynamic Systems Initiative, White Paper (October 2003),
   http://download.microsoft.com/download
10. Murch, R.: Autonomic Computing. Prentice Hall Professional Technical Reference, pp. 235–245. IBM Press (2004)
11. Sun Microsystems, ARCO, N1 Grid Engine 6 Accounting and Reporting Console, White Paper (May 2005),
    http://www.sun.com/software/gridware/ARCO_whitepaper.pdf
12. Vromont, P., Weyns, D., Malek, S., Anderson, J.: On Interacting Control Loops in Self-Adaptive Systems. In: Proceedings of SEAMS 2011, Waikiki, Honolulu, HI, USA (May 2011)
13. Wan, K.: A Brief History of Context. International Journal of Computer Science Issues 6(2) (November 2009)

Eyes Closeness Detection
Using Appearance Based Methods

Xue Liu, Xiaoyang Tan, and Songcan Chen

College of Computer Science & Technology, Nanjing University of Aeronautics &
Astronautics, P.R. China
{liuxue,x.tan,s.chen}@nuaa.edu.cn

Abstract. Human eye closeness detection has gained wide applications
in human computer interface designation, facial expression recognition,
driver fatigue detection, and so on. In this work, we present an extensive comparison on several state of art appearance-based eye closeness
detection methods, with emphasize on the role played by each crucial
component, including geometric normalization, feature extraction, and
classification. Three conclusions are highlighted through our experimental results: 1) fusing multiple cues significantly improves the performance
of the detection system; 2) the AdaBoost classifier with difference of intensity of pixels is a good candidate scheme in practice due to its high
efficiency and good performance; 3) eye alignment is important and influences the detection accuracy greatly. These provide useful lessons for
the future investigations on this interesting topic.

Keywords: Eye closeness detection, Eye state measurement.

1 Introduction

As one of the most salient facial features, eyes, which reflect the individual's
affective states and focus attention, have become one of the most important
information sources for face analysis. Efficiently and accurately understanding
the states of eyes in a given face image is therefore essential to a wide range
of face-related research efforts, including human computer interface designation,
facial expression analysis, driver fatigue detection [1][2], and so on.

However, this task is challenging due to the fact that the appearance of eye
regions can be easily influenced by various variations such as lighting, expression,
pose, and human identity. To meet these challenges, numerous eye closeness detection methods have been proposed during past few decades [3][4][5][6][7][8][9].
The ideas of these methods can be roughly categorized into two types, i.e., detecting the closeness of eyes directly through various pattern recognition methods
or doing this indirectly by checking whether the eyes are actually open. Since
an open eye in general exhibits more appearance evidence (e.g., visible iris and
elliptical shape of eyelids), most methods (e.g., [6][7][8]) understand the state of
the eyes using this later philosophy, i.e., if the local evidence collected from the
image supports the conclusion that the eyes are open, then they must be not

Z. Shi, D. Leake, and S. Vadera (Eds.): IIP 2012, IFIP AICT 385, pp. 398–408, 2012.

closed. However, methods in this line may have their own problems. For example, for some subjects, their irises may be largely occluded by their eyelids. As a result, it would be very hard to reliably detect irises under this situation [5].

In this paper, we adopt the first philosophy, i.e., detecting whether the eyes are closed directly using their appearance evidence. The major advantage of this lies in its efficiency and robustness [10][11]. Current development in computer vision has allowed for robust middle-level feature description for eye patches despite of various changes in appearance, and the remaining variations can be addressed with powerful machine-learning-based classifiers. However, this appearance-based strategy can be implemented in numerous ways and involves many practical considerations. Therefore, an evaluation of popular methods for eye closeness detection is needed.

The major contribution of this paper is to make an extensive comparative study on this approach from the engineering point of view. In particular, we investigated in-depth the influence of several crucial components of an eye closeness detection system, including eye patch alignment, feature extraction and classifiers. Four types of representative feature sets including gray-values, Gabor wavelets, Local Binary Patterns (LBP,[12]), Histograms of Oriented Gradient (HOG,[13]) are compared with respect to three types of classifiers, i.e., Nearest Neighbor (NN), Support Vector Machine (SVM) and AdaBoost on a large benchmark dataset. Both LBP and Gabor feature sets have been used for eye representation before [14][15][16] but HOG not. However, since these features represent different characteristics (local texture, global shape and local shape, respectively) of eyes, it is beneficial to use all of them. Indeed, our experimental results show that fusing various feature descriptors significantly improves the performance of eye closeness detection. In addition, we also show that properly eye patch alignment is important for the performance. These provide useful lessons for the follow-up investigations on this interesting topic.

2   Overall Architecture of Our System

The overall pipeline of our eye closeness detection system is given in Fig. 1. For a given test image, we first detect and crop the face portion using the Viola and Jones face detector [17], then the eyes are localized using the method introduced in [18]. After this, we crop the eye region and align it with those in the training set with method of [19]. On the aligned eye patch various feature sets are extracted and input into the classifier for final decision.

One of the key components of our system lies in the inclusion of eye patch alignment module. This is based on the observation that eyes in a face image may undergo various in-plane/out-of-plane pose changes, or scale changes. Although some feature descriptors are not sensitive to these, other feature sets (e.g., LBP, Gabor wavelets, and HOG) don't have built-in mechanism to handle such variations. Therefore, performing geometric eye normalization is necessary for these descriptors. However, one difficulty for this is that it is hard to find anchor points for eye patches and hence traditional anchor-points-based alignment

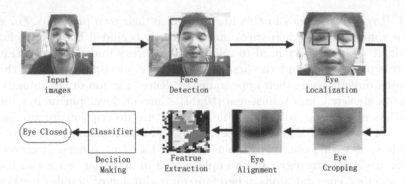

Fig. 1. The overall architecture of our eye closeness detection system

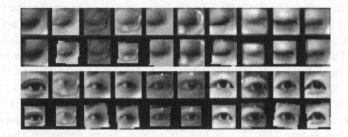

Fig. 2. Illustration of eye patches normalized with the congealing method, where patches in the top two rows are original images of closed eyes and their corresponding normalized versions respectively, and patches in the bottom two rows are original images of open eyes and their corresponding normalized images, respectively

method can not be applied. Here we adopt an information-theory geometric normalization method originally proposed for medical image registration, i.e., the congealing method [19]. This is an unsupervised image normalization method which learns a particular affine transformation for each image such that the entropy of a group of eye images is minimized. Fig. 2 gives some illustrations of eye patches normalized using this method, from which we can see that the locations of eyes are centered and their sizes are scaled.

3   Feature Sets

Four types of features are used in the experiments, they are gray-value features, HOG features, Gabor features and LBP feature respectively. The gray feature of an $M \times N$ image patch is simply a $MN \times 1$ column vector. In what follows we briefly describe the LBP feature, Gabor wavelets and the HOG feature.

LBP Feature. Local Binary Pattern (LBP) proposed by Ojala [12] is widely used feature descriptor for local image texture. The LBP descriptor has achieved considerable success in various applications such as face recognition and texture recognition, due to its capability to efficiently encode local statistics and geometric characteristics (e.g., spot, flat area, edge and corner) among neighborhood pixels and its robustness against noise by picking up only 'uniform patterns' for feature description. In this paper, we partition each 24×24 eye patch into 6×6 blocks and represent each block as a 59-dimensional histogram. Therefore for each eye patch we have a 944-dimensional LBP vector (16×59).

Gabor Wavelets. Gabor wavelets were originally developed to model the receptive fields of simple cells in the visual cortex and in practice they capture a number of salient visual properties including spatial localization, orientation selectivity and spatial frequency selectivity quite well. They have been widely used in face recognition. Computationally, they are the result of convolving the image with a bank of Gabor filters of different scales and orientations and taking the 'energy image' (pixelwise complex modulus) of each resulting output image. The most commonly used filters in face recognition have the form,

$$\Psi_{\mu,\nu}(z) = \frac{\|k_{\mu,\nu}\|^2}{\sigma^2} * exp(-\frac{\|k_{\mu,\nu}\|^2 * \|z\|^2}{2\sigma^2}) * [exp(ik_{\mu,\nu}, * z - exp(-\frac{\sigma^2}{2}))] \quad (1)$$

where μ and ν define the orientation and scale of the Gabor kernels, $z = (x, y)$, $\| \|$ denotes the norm operator, and the wave vector $k_{\mu,\nu}$ is defined: $k_{\mu,\nu} = k_{\nu}e^{i\varphi_\mu}$, where $k_\nu = k_{max}/f^\nu$ and $\varphi_\mu = \mu\pi/8$. k_{max} is the maximum frequency, and f is the spacing factor between kernels in the frequency domain [20]. We use 40 filters with eight orientations and five scales on 24×24 eye patch, then down-sample the resulting vector by 16 to a 1440-dimensional vector.

HOG Feature. The aim of Histogram of Oriented Gradients (HOG) feature proposed in [13] is to describe local object appearance and shape within an image by the distribution of intensity gradients or edge directions. To implement these descriptors, we first divide the eye patches into small connected cells of 4×4 in size, and for each cell a histogram (with 9 bins) of gradient directions is calculated, which is then undergone a contrast-normalization within each block, leading to better invariance to changes in illumination or shadowing. Through these steps, we have a 900-dimensional histogram for each patch.

The three aforementioned feature representations of an open eye and a close eye are illustrated in Fig. 3, respectively. One may notice that the LBP feature is good at characterizing detailed texture information of the image, and Gabor wavelets highlight the differences of eye images with different states with respect to global spatial frequency, while the local shape information is best described by the HOG feature.

4 The Classifiers

In this work, we use the Nearest Neighbor, SVM and AdaBoost as our classifiers. The nearest neighbor method is a simple and effective non-parametric classification method and used in this paper as our baseline.

Fig. 3. Illustration of original eye images, their LBP, Gabor wavelet and HOG feature representation, respectively (top row - open eye, bottom row - closed eye). Note that this is only for illustration purpose and the sizes of the eye images are different from those used in the experiments.

Support Vector Machines. Support Vector Machines (SVMs) is the state-of-the-art large margin classifier which has gained popularity within visual pattern recognition. One problem we should handle is imbalance problem. That is, the number of images of closed eyes and open eyes is different [1], which tends to increase the bias of trained SVM classifier to the class with more samples. To overcome this, before training, we set the penalty respective coefficients for the positive and negative samples to be $\omega_1 = \frac{N^+ + N^-}{2N^+}, \omega_2 = \frac{N^+ + N^-}{2N^-}$, where N^+ is the number of positive samples and N^- is the number of negative samples. We used the LIBSVM package [21] with RBF kernel for the SVM-related experiments.

AdaBoost with Pixel-Comparisons. As a final classifier compared, we use the AdaBoost, it provides a simple yet effective approach for stagewise learning of a nonlinear classification function. In this work, we use the "difference of intensities of pixels" proposed in [22] as our features. More specifically, we used five types of pixel comparison operators (and their inverses) [23]:

1) $pixel_i > pixel_j$;
2) $pixel_i$ within 5 units (out of 255) of $pixel_j$;
3) $pixel_i$ within 10 units (out of 255) of $pixel_j$;
4) $pixel_i$ within 25 units (out of 255) of $pixel_j$;
5) $pixel_i$ within 50 units (out of 255) of $pixel_j$;

The binary result of each comparison, which is represented numerically as 1 or 0, is used as features. Thus, for an image of 24×24 pixels, there are $2 * 5 * (24 * 24)(24 * (24 - 1))$ or 3312000 pixel-comparison features.

To handle the large number of features, we use Adaboost for feature selection while learning a strong classifier. This is done by mapping each feature to a weak classifier and then selecting the most discriminative weak classifier increasingly for an additive strong classifier at the same time. For more details, see [22]. In our experiments, 2000 weak classifiers are trained and some of them are randomly selected for evaluation in each iteration. We examined the performance achieved from evaluating 1%, 10% and 100% of all possible weak classifiers per iteration.

---

[1] In practice, it is much easier to collect images of open eyes than those of closed eyes.

5   Experiments

5.1   Data and Settings

The data for our experiments are collected from Zhejiang University blinking video database [24]. There are a total of 80 video clips in the blinking video database from 20 individuals, four clips for each individual: one clip for frontal view without glasses, one clip with frontal view and wearing thin rim glasses, one clip for frontal view and black frame glasses, and the last clip with upward view without glasses. We manually select images in each blinking process, including eye images of open, half open, closed, half open, open. In addition, images of the left and the right eyes are collected separately. Some samples of the dataset are shown in Fig. 4. We can see that these images are blurred, with low resolution and may be occluded by glasses.

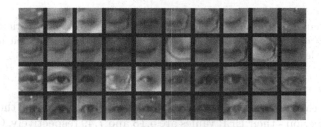

Fig. 4. Illustration of some positive (top two rows) and negative (bottom two rows) samples used for training

The collected eye images are then divided into two separate sets for training and test purpose. The training set consists of images from the first 16 individuals. The test set consists of the images from the remaining 4 subjects. Note that there is no overlapping in images of subjects between the training set and test set. To further increase the diversity of training samples, various transformations such as rotation, blurring, contrast modification and addition of Gaussian white noise are applied to the initial set of training images, yielding about 6,600 new images in total. Finally, the training set contains 7360 eye images in all, with 1590 closed eye images and 5770 open eye images respectively. The test set is constructed with 410 closed eyes and 1230 open eyes. All these images are geometrically normalized into images of 24 × 24 pixels.

5.2   Experimental Results

Fig. 5 (left) gives the overall performance of SVM and AdaBoost on our datasets in terms of Receiver Operating Characteristic (ROC) curves. From the results, we can see that the best performer is the SVM using HOG features. In particular, the AUC value of "HOG+SVM" achieves 97.7%, compared to 97.0% for AdaBoost with 2000 weak classifiers, while the next best performer is the "LBP+SVM". We

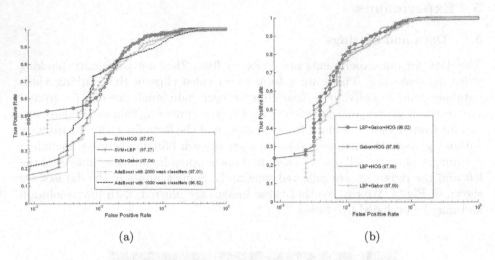

(a)   (b)

Fig. 5. ROC curves of various feature and classifier (left), and of various feature fusion strategies using the SVM classifier (right). AUC values are given in the end of corresponding legend texts.

also see that the LBP features are better than Gabor features for the task of eye closeness detection - their EER values are 5.13 and 7.49 respectively. One possible reason is that the LBP features give a detailed account on the appearance of eye regions while being insensitive to the lighting changes. Actually, when one screws up his eyes, it is really very difficult to make a decision on whether his eyes are closed or not. In these cases, the global shape feature represented by Gabor features is less discriminative than local texture/shape information characterized by the LBP/HOG features.

Table 1 gives the overall comparative performance of various methods and feature sets (the recognition rate is obtained by fixing a threshold learned from the training set). The various nearest neighbor classifier-based schemes serves as the baseline. Again we see that "HOG+SVM" performs best in AUC value but the "LBP+SVM" wins in recognition rate. However, in terms of testing efficiency, the AdaBoost classifier performs faster by at least four times over SVM but with slightly worse ROC performance. This suggests that Adaboost with difference of pixel features is a very attractive candidate in practice due to the high efficiency it provides in testing.

Table 2 and Fig. 5 (right) give the comparative performance using various feature combination schemes. Since LBP, Gabor and HOG feature characterize different aspects of eye patches (i.e, local texture, global shape, and local shape, respectively), it would be useful to fuse the information from three of them. Feature combination could be performed either at the feature level or at the score level and our previous experience shows that score level fusion is simpler

Table 1. Comparative performance of various features and classifiers

Approach		Recognition Rate(%)	Time(m sec)	AUC(%)	EER(%)
NN	Gray	73.50	220.25	-	-
	LBP	91.21	195.97	-	-
	Gabor	85.65	297.31	-	-
	HOG	91.88	268.62	-	-
SVM	Gray	75.03	12.97	50.00	99.6
	LBP	**95.42**	14.65	97.27	5.13
	Gabor	91.45	12.77	97.04	7.49
	HOG	91.88	19.24	**97.67**	**5.05**
AdaBoost(1000)	1%	92.06	**2.890**	96.48	8.30
	10%	92.91	2.912	96.52	8.71
	100%	92.31	2.980	96.50	8.45
AdaBoost(2000)	1%	92.37	2.943	96.67	7.65
	10%	92.67	3.059	96.73	7.48
	100%	93.47	3.132	97.01	7.08

Table 2. Comparative performance of various feature combination schemes (with the SVM classifier)

Approach	Recognition Rate(%)	AUC(%)	EER(%)
LBP	95.42	97.27	5.13
Gabor	91.45	97.04	7.49
HOG	91.88	96.67	5.05
LBP+Gabor	94.69	97.69	5.94
LBP+HOG	95.05	97.89	**4.80**
Gabor+HOG	94.81	97.98	5.05
LBP+Gabor+HOG	**95.42**	**98.02**	5.05

to implement (no need to handle the high dimensional problem due to feature concatenation) and usually leads to better results [25]. Hence we adopt the strategy of score level fusion with z-score normalization in this work. We can see from the table that combining all of the three feature sets gives the best performance both in terms of ROC curve and in terms of recognition rate. It is worthy noting that although the performance of the widely used HOG feature is inferior to the LBP feature in terms of recognition rate, its EER score is higher than LBP, which indicates that the local shape features play an important role in the detection of eye closeness alone. Furthermore, combining the local shape features with local texture descriptor (LBP) significantly improves its performance.

Fig. 6(a) illustrates some of images which are correctly identified as closed eyes using the "HOG/LBP/Gabor+SVM" scheme. Notice that there is large amount of diversity exhibited in the appearance of these closed-eye images. Fig. 6(b) shows some false negative images (upper two rows) and some false positive images (bottom two rows). By carefully examining these images, we can see that they look even confusing to human beings when deciding whether these eyes are

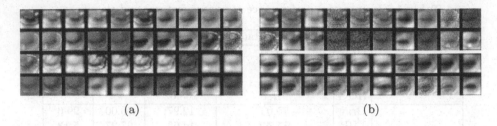

(a)   (b)

Fig. 6. (a)Illustration of images which are correctly identified as closed eyes (true positive), and (b)(upper two rows) images of closed eyes failed to be recognized (false negative) and (bottom two rows) images of open eyes incorrectly recognized as closed eyes (false positive). All the results are with "HOG + LBP + Gabor and SVM" scheme.

Table 3. Comparison of recognition rate w/o eye alignment with SVM

Processing	LBP	Gabor	HOG
Without alignment	92.7	89.2	89.5
With alignment	**95.4**	**91.5**	**91.9**

open or not. This helps us understand the challenges of eye closeness detection in the real world.

Finally, we investigate the influence of eye patch alignment on the performance of the system. As Table 3 shows, adding the module of alignment improves the performance consistently over all the feature sets tested. Although the LBP feature is known to be rotation-invariant and the HOG feature is robust against slight perturbation in the image, they are not robust to general affine transformations. Indeed, Table 3 shows that it is beneficial to do geometric normalization for eye patches before extracting features from them.

6   Conclusions

In this paper, we systematically evaluate several feature sets and classifiers for the task of eye closeness detection. Our experimental results indicate that fusing various feature descriptors significantly improves the performance of the detection system while the AdaBoost classifier with pixel comparisons is a good candidate scheme in practice due to its high efficiency and satisfying performance. In addition, our results show that eye alignment is important and influences the detection accuracy greatly.

Acknowledgements. The work was financed by the (key) National Science Foundation of China (61073112, 61035003).

References

1. Noor, H., Ibrahim, R.: A framework for measurement of humans fatigue level using 2 factors. In: International Conference on Computer and Communication Engineering, pp. 414–418 (2008)
2. Eriksson, M., Papanikotopoulos, N.: Eye-tracking for detection of driver fatigue. In: IEEE Conference on Intelligent Transportation System (TSC), pp. 314–319 (1997)
3. Mitelman, R., Joshua, M., Adler, A., Bergman, H.: A noninvasive, fast and inexpensive tool for the detection of eye open/closed state in primates. Journal of Neuroscience Methods 178, 350–356 (2009)
4. Sun, R., Ma, Z.: Robust and efficient eye location and its state detection. Advances in Computation and Intelligence, pp. 318–326 (2009)
5. Valenti, R., Gevers, T.: Accurate eye center location and tracking using isophote curvature. In: Computer Vision and Pattern Recognition (CVPR), pp. 1–8 (2008)
6. Wang, H., Zhou, L., Ying, Y.: A novel approach for real time eye state detection in fatigue awareness system. In: Robotics Automation and Mechatronics (RAM), pp. 528–532 (2010)
7. Jiao, F., He, G.: Real-time eye detection and tracking under various light conditions. Data Science Journal 6, 636–640 (2007)
8. Orozco, J., Roca, F., Gonzàlez, J.: Real-time gaze tracking with appearance-based models. Machine Vision and Applications 20, 353–364 (2009)
9. Li, S., Chu, R., Liao, S., Zhang, L.: Illumination invariant face recognition using near-infrared images. IEEE Transactions on Pattern Analysis and Machine Intelligence 29, 627–639 (2007)
10. Liu, Z., Ai, H.: Automatic eye state recognition and closed-eye photo correction. Pattern Recognition, 1–4 (2008)
11. Dehnavi, M., Eshghi, M.: Design and implementation of a real time and train less eye state recognition system. EURASIP Journal on Advances in Signal Processing 30 (2012)
12. Ojala, T., Pietikainen, M., Maenpaa, T.: Multiresolution gray-scale and rotation invariant texture classification with local binary patterns. IEEE Transactions on Pattern Analysis and Machine Intelligence 24(7), 971–987 (2002)
13. Dalal, N., Triggs, B.: Histograms of oriented gradients for human detection. In: Computer Vision and Pattern Recognition (CVPR), pp. 886–893 (2005)
14. Cheng, E., Kong, B., Hu, R., Zheng, F.: Eye state detection in facial image based on linear prediction error of wavelet coefficients. In: IEEE International Conference on Robotics and Biomimetics (ROBIO), pp. 1388–1392 (2009)
15. Zhou, L., Wang, H.: Open/closed eye recognition by local binary increasing intensity patterns. In: Robotics Automation and Mechatronics (RAM), pp. 7–11 (2011)
16. Wang, Q., Yang, J.: Eye location and eye state detection in facial images with unconstrained background. J. Info. & Comp. Science 1, 284–289 (2006)
17. Viola, P., Jones, M.: Robust real-time face detection. International Journal of Computer Vision 57, 137–154 (2007)
18. Tan, X., Song, F., Zhou, Z., Chen, S.: Enhanced pictorial structures for precise eye localization under incontrolled conditions. In: Computer Vision and Pattern Recognition (CVPR), pp. 1621–1628 (2009)
19. Huang, G., Jain, V., Learned-Miller, E.: Unsupervised joint alignment of complex images. In: International Conference on Computer Vision (ICCV), pp. 1–8 (2007)
20. Lades, M., Vorbruggen, J., Buhmann, J., Lange, J., von der Malsburg, C., Wurtz, R., Konen, W.: Distortion invariant object recognition in the dynamic link architecture. IEEE Transactions on Computers 42(3), 300–311 (1993)

21. Chang, C., Lin, C.: Libsvm: a library for support vector machines. ACM Transactions on Intelligent Systems and Technology 2, 27 (2011)
22. Baluja, S., Sahami, M., Rowley, H.: Efficient face orientation discrimination. In: International Conference on Image Processing (ICIP), pp. 589–592 (2004)
23. Baluja, S., Rowley, H.: Boosting sex identification performance. International Journal of Computer Vision 71, 111–119 (2007)
24. Database, Z.E., http://www.cs.zju.edu.cn/gpan
25. Tan, X., Triggs, B.: Fusing Gabor and LBP Feature Sets for Kernel-Based Face Recognition. In: Zhou, S.K., Zhao, W., Tang, X., Gong, S. (eds.) AMFG 2007. LNCS, vol. 4778, pp. 235–249. Springer, Heidelberg (2007)

Author Index